Pietro Carlo Cacciabue

Guide to Applying Human Factors Methods

Human Error and Accident Management in Safety-Critical Systems

 Springer

Pietro Carlo Cacciabue, PhD
EC, Joint Research Centre
Institute for the Protection and Security of the Citizen
Varese, 1-21020
Italy

British Library Cataloguing in Publication Data
Cacciabue, Pietro C.
    Guide to applying human factors methods:human error and accident management
    in safety-critical systems
    1. Human engineering
    I. Title
    620.8'2

Library of Congress Cataloging-in-Publication Data
Cacciabue, Pietro C.
    Guide to applying human factors methods:Pietro Carlo Cacciabue.
        p. cm.
    Includes bibliographical references and index.

    1. Human engineering.   I. Title.
    TA166.C33 2004
    620.8'2 – dc22

                                                                        2004044972
    ISBN 978-1-84996-898-0   ISBN 978-1-4471-3812-9 (eBook)
    DOI 10.1007/978-1-4471-3812-9

springeronline.com

Typesetting: SNP Best-set Typesetter Ltd., Hong Kong

34/3830-543210 Printed on acid-free paper SPIN 10896211

To Cathy

Contents

Acknowledgments

This book contains work carried out in collaboration with a group of outstanding people over a period of 10 years.

What makes the Human Factors Group special is that it was a set of constantly changing temporary researchers, scientific visitors, and PhD or graduating students, who remained for a number of years to work in the group or collaborated from their universities and labs before going on either to start a family or to become researchers, managers, university professors, engineers, or psychologists.

All these people shared one common characteristic: They fell in love with the study of human factors. The professor who guided me in my graduation thesis in nuclear engineering was very careful in warning me that neutron physics is a science that traps people and makes them want to do nothing else in their professional life. This is what happened to me. However, all that changed when I discovered human factors in the early 1980s. Nowadays, I say to all the students I work with: "Be careful, because you may fall in love with human factors."

This common interest and locus of attention is what makes this group of people remarkable – probably not unique, but certainly exceptional. The results of the work that we performed together are condensed in this book.

It is impossible to name all the members of this ever-changing group without risking that someone may be left out. It is much better to thank the Human Factors Group as a whole.

I am in debt to all of them for their support, help, contribution, enthusiasm, and encouragement, especially when the group has had to be temporarily dismantled at various stages. But the group has always been rebuilt with new faces, brains, and hearts because the work may have sometimes been interrupted, but the spirit never died.

Thank you, Human Factors Group – this book belongs to us all.

Pietro Carlo Cacciabue
Varese, Italy

1

Introduction

It could be argued that Human Factors (HF) is not a pure science, as it is transversal to other well-established sciences, such as physics, mathematics, psychology, and sociology.

Strictly speaking, this is correct as Human Factors is not self-referential, as all pure sciences are, and does not base its theories on unique and exclusive conservation principles and postulates.

Like all modern sciences, such as computing, economics, communication, and management, Human Factors spans several existing theoretical backgrounds and requires strong connections with practical areas of application. This ability represents an essential attribute for ensuring efficiency and effectiveness in real working environments. In particular, Human Factors extend over four essential domains, namely, engineering, psychology, sociology, and computer science. Each of these domains requires a profound theoretical knowledge and is equally linked to practical applications in technology and everyday life.

At the same time, it is limiting to identify Human Factors as a discipline, as it goes far beyond the mere application of engineering, psychology, sociology, and computer science principles, and requires blending existing theoretical methods in all these four fields, generating new and specific theoretical formulations and paradigms. In this way, it becomes possible to represent working contexts and sociotechnical aspects in a theoretical form, which then needs further simplification and elaboration in order to develop practical applications and quantifications for implementation in real systems and for the assessment of real working contexts.

Human factors, together with computer science, is the most relevant science that has been developed over the past 50 years, as a consequence of the technological development and the role assigned to human beings in the management of processes that have progressively become more and more complex and demanding in terms of control ability and skill. This latter aspect has increased attention on humans as the "weak point" in accident chains and in process control, and has generated the need to study more accurately the interaction of humans with machines.

In addition to these aspects, Human Factors bears a fundamental connection with the "mother" of all sciences, i.e., philosophy. This connection is quite obvious as the core subjects of Human Factors are human beings and their behaviour, which depends on their thinking and beliefs. This is indeed the subject matter of philosophy and consequently, correlates with the major philosophers and thinkers of the past and more recent times.

1.1 Aim of This Book

This book is dedicated to readers that have already acquired basic knowledge in the domain of human–machine interaction and cognitive ergonomics. It does not offer an introductory overview and basic understanding in Human Factors, as the literature is rich with very valuable textbooks that offer precisely this basic knowledge (Sheridan and Ferrell, 1974; Sanders and McCormick, 1976; Rouse, 1980; Kantowitz and Sorkin, 1983; Rasmussen, 1986).

The aim is to guide readers how to perform Human Factors analysis and how to select and implement existing methods and techniques for solving practical problems.

The process of selection and implementation is usually associated with a methodology that offers a roadmap for practitioners and requires formal and accurate application in order to ensure that models, methods, and data are adequately selected and fruitfully exploited. The results of applying such a methodology are consistently associated with implementations, recommendations, and assessments aimed at improving and augmenting systems safety.

This is why the book rotates around a reference methodology, called Human Error Risk Management in Engineering Systems (HERMES) that is described in detail in the first part and guides the practical applications to real case studies in the second part.

Human factors methods are applied in four main areas of application, namely, design, safety assessment, training, and accident investigation. The main domains of application of Human Factors are energy production (nuclear and conventional), transportation systems (aviation, railway, automotive, maritime), medicine, economic system, chemical and petrochemical environments, manufacturing, and economical systems.

These domains are usually based on large complicated plants and organisations. Consequently, the scenarios of practical applications of a Human Factors analysis are very complex and demand relatively extensive processes and lengthy implementations of methods and techniques. A crucial requirement that derives from such complexity is that the methodology guiding the Human Factors application must be strictly and formally put into practice, in order to avoid trivial implementations leading to unrealistic or even misleading results.

1.2 How to Read This Book

This book is divided in two parts, dedicated respectively to the theoretical description of the methodology and associated methods and techniques, and to the development of a number of relevant real cases in various areas of applications and in different domains.

In particular, in the first part of the book, i.e., Chapters 2, 3, and 4, the theoretical and methodological bases of Human Factors are discussed with reference to design, safety assessment, training, and accident investigation. The basic hypothesis and theoretical stands that sustain the HERMES methodological framework are presented. The most common methods and models for describing and studying human–machine systems are briefly discussed and compared. The procedures for implementing HERMES in different areas of application are described.

In the second part, i.e., Chapters 5, 6, 7, and 8, four complex cases in real domains are described in detail, covering all steps of implementation and showing how the HERMES methodology has been applied.

The two parts of the book can in principle be read separately. The expert Human Factors reader, interested primarily in practical implementation, can selectively read chapters of the second part and may refer back to specific sections in the first part that describe the associated theoretical and methodological background.

The less expert readers may initially pay attention to the first part of the book, where theories, paradigms, and basic principles of Human Factors are presented. In this case, the detailed application and complexity of the case studies are of minor interest, even if reading these chapters offers a feeling of the amount of work that practical applications require.

Hopefully, the basic inference that can be drawn from reading the whole book, or even from considering either parts separately, is that the application of Human Factors to practical problems demands an accurate work of analysis and the implementation of formal methods and techniques, which are well established and consolidated. Therefore, Human Factors problems cannot be solved by simply applying common sense and practical experience.

Moreover, the intrinsic dynamic and evolutionary nature of organisations, with the turnover of personnel and management, and the changing of technical and control systems as well as production, requires that the Human Factors issues are to be dealt with not only once in the life of a system, but demand constant awareness and recurrent application of audits. This grants that adequate safety levels are reached and maintained at all times.

1.3 Brief Summary of Each Chapter

Chapter 2

Chapter 2 contains the basic definitions and all theoretical foundations of the methodology.

In particular, the basic definitions of Human Factors (HF), Human–Machine Systems (HMS), and Human Errors (HE) are worked out in relation to the vast literature available that has been developed over many years. A number of related conceptual subjects are elaborated from these definitions. In particular, the sociotechnical elements that exist in any working context are defined.

The chapter then concentrates on human error and human error management and these concepts are expanded to embrace the vital issue of accident management. This is the critical safety concern of all modern technological systems.

The concept of Human Error and Accident Management (HEAM) is developed. In particular, five "standpoints" are defined in association with the HEAM definition. These standpoints represent the premises and issues that have to be resolved and established clearly in any HF analysis, in order to restrict the scope of the outcome to well-defined objectives and goals, and to support the HF analysts both during the selection of methods and models and during their implementation for specific studies.

The HEAM concept and the associated standpoints lie at the base of the HERMES methodology, which is the core subject of this book and is described as a direct implementation of the HEAM measures and requirements for safety studies in real working environment and organisations. The relation of the HERMES methodology with the modem concept of ecology and cognitive ergonomics is also discussed.

Chapter 3

Chapter 3 contains a short review of theories and methods for studying Human–Machine Interaction (HMI).

Firstly, models and taxonomies that represent human behaviour are described. These are the reference paradigms that can be found in literature and permit representation of the vast majority of working contexts and conditions. However, many other models exist, and a quite extensive literature about them is reported in the references of the book.

In addition to models and taxonomies of human behaviour, the other basic methods supporting HEAM analysis and the HERMES methodology are considered in this chapter, namely, ethnographic studies, cognitive task analysis, root cause analysis, quantitative risk assessment, and recurrent safety audit. The Ethnographic Studies (ES) are the set of theoretical and empirical methods for evaluating real working contexts. ES are discussed with reference to retrospective and prospective types of analysis.

The theoretical evaluation of modern working procedures, based on the so-called Cognitive Task Analysis (CTA), is then tackled by presenting a number of techniques and procedures for CTA.

Ecology-based methods are then presented. In essence, ecological interface design and cognitive work analysis are discussed.

The basic approach for performing accident investigation is Root Cause Analysis (RCA). Different techniques for performing RCA are considered and compared.

The basic methods for performing safety studies, and, in particular, Quantitative Risk Assessment (QRA) and Recurrent Safety Audit (RSA) are discussed. A variety of approaches are presented and their correlation to the field studies based on ES is discussed in order to highlight the crucial issue of data.

Chapter 4

Chapter 4 concludes the first part of this guide, which contains the description of theoretical and methodological approaches.

In particular, this chapter merges the methodological architecture of HERMES, described in Chapter 2, and the methods and models discussed in Chapter 3, and offers a roadmap for practical implementation in different areas of application.

There are four areas of application of HF methods, namely, design, training, safety assessment, and accident investigation. The design of human–machine systems covers control, emergency, and protection systems, as well as their interfaces and standard and emergency operating procedures. Training focuses on nontechnical skills and covers classroom and simulator training in Human Factors. Safety assessment implies the evaluation of the human contribution to design basis accident, quantitative risk assessment, and recurrent safety audits. Accident investigation considers the human contribution to the etiology of an accident and the identification of root causes.

For each of these four areas, a detailed stepwise procedure is developed that accounts for the methodological framework of HERMES and defines how the methods and models selected for the Human Factors analysis are correlated to one another and sequentially or iteratively applied in order to reach the objectives of the study.

Chapter 5

Chapter 5 contains the case study of the application of the HERMES methodology for the design of the Human–machine Interface (HI) of an anticollision warning system in the automotive environment.

The methodology is applied in three phases and closely follows the user-centred design principles and the procedure for designing HEAM measures described in the first part of this guide. The study of this design process will not enter into details of the technological development, but will focus on the contribution of Human Factors analyses for the design and implementation of the system.

In particular, Phase 1 focuses on the characteristics of the tool and warning strategies from a safety perspective, on the identification and definition of user needs, and on the selection of the model of driver (the user) and cognitive functions.

Phase 2 discusses the selection of characteristics of the Road–Collision–Warning System (RCWS) and the definition of the experiments and scenarios for testing and evaluation in relation to the model and taxonomy selected in Phase 1.

Phase 3 discusses the development of a safety matrix to be utilised during experiments in vehicles as well as on the performance of road tests and on the feedback provided to the final design iteration.

Chapter 6

Chapter 6 presents a real extended application of the HERMES methodology in the area of Human Factors training. In particular, the development of a Human Factors course, called Crew Resource Management (CRM), for the pilots of a large airline is described in detail.

The five standpoints that sustain any Human Factors application are respected and the procedure for developing training, discussed in the first part of this guide, is followed in accordance with the consideration of the HERMES architecture.

In particular, the logical interplay between retrospective and prospective approaches is granted by ensuring that information and data from field assessments and from studies of past events are taken into consideration in specifying the detail content of the CRM and in identifying markers and indicators of areas of concern for safety improvement.

The methodology is applied in three phases. Phase 1 focuses on the evaluation of the sociotechnical context existing within the airline and on the data collection of particular flying practices and habits. These ethnographic studies are framed by a theoretical model of pilot behaviour, considered adequate for guiding the development of the CRM and classifying behaviours and situations.

Phase 2 describes the detailed evaluation of all data collected by field studies and retrieved from past events. This leads to the detailed design of the CRM course.

Phase 3 describes the development and preliminary implementation of the training material for the CRM and the generation of a specific software for an additional Computer-Based Training (CBT) course in support to the more formal CRM. Testing and evaluation of the quality and effectiveness of the CRM course is also discussed.

Chapter 7

Chapter 7 presents a practical application of the HERMES methodology in the performance of the safety audit of a large railway company.

In this case, the five standpoints developed in the first part of the book are respected. In this case, the goals and objectives of an audit will be considered mostly from the perspective of HEAM. The importance of the selection of an adequate model for describing and classifying human behaviour in accordance to the characteristics of the organisation under study is discussed.

Moreover, the logical interplay between retrospective and prospective approaches is applied by ensuring that the definition of the factors influencing performance (Performance-Influencing Factors, PIFs) is correlated to information derived from visits to depots, cabin rides, interviews, and questionnaires, and from studies of past events.

The final step of the safety audit will consider the development of safety recommendations relative to the state of the company and the definition of the Indicators of Safety (IoS) and IoS-matrices typical of safety audits.

The methodology is applied in three phases. Phase 1 of the work initially focuses on the evaluation of the sociotechnical context existing within the railway company. These ethnographic studies are complemented by the development of a questionnaire for the evaluation of train drivers attitudes and safety climate. A model representing the interactions within the company is devised.

Phase 2 focuses on the collection and analysis of data and the review of reports of past events. The process leads to the definition of the PIFs and the development of scenarios for potential accidents related to Human Factors.

Phase 3 considers the actual safety audit based on indicators of safety and IoS-matrices and on the development of safety recommendations.

Chapter 8

Chapter 8 presents a practical application of the HERMES methodology in the area of accident investigation for the analysis of a real accident that occurred in a complex plant for the combined production of electrical power and thermal energy.

Accident investigation is a typical retrospective type analysis, which aims at identifying the root causes of a specific occurrence, and defines possible modifications that can contribute to improve the safety level of the whole system. Therefore, also in the case of accident investigations the recommendations represent the most important outcome of the detailed application of the methodology.

In order to reach these objectives, the assessment of the working context and the evaluation of the social climate existing within the organisation will be performed.

A cognitive model and associated taxonomy for classifying "human errors" and "system failures" will be selected and applied throughout the investigation.

The methodology is applied in two phases. Phase 1 of work focuses on the evaluation of the sociotechnical context of work and the familiarisation with the practices, habits, and tasks existing within the company (ethnographic studies). This leads to the definition of a model and formal approach for structuring the accident sequence and classifying human behaviour.

Phase 2 focuses on the detailed evaluation of the causes of the accident, mainly from the human factors and accident management viewpoints, and on the development of recommendations for improving the safety and for setting the basis of future audits and assessments of the plant and organisation as a whole.

2
Elements of Human–Machine Systems

2.1 Human Factors Role in Modern Technology

In the last 30 years, technology has expanded enormously especially in scope and efficiency of the operations that can be performed by machines alone, exploiting their imbedded autonomous "decision-making" rules and mechanisms.

Similarly, roles and duties of human operators have undergone tremendous changes, and nowadays, operators are mainly supervisors and monitors of procedures carried out automatically, once these have been set up by the operators themselves (Sheridan, 1992, 1999). In such configurations of Human–Machine Systems (HMS), the design of automatic systems and the control of the interaction with human operators have become much more complex. In particular, the consequences of a "human error" or of a "misunderstanding" between human and automation can be unrecoverable and catastrophic (Nagel, 1988).

Two main factors have contributed to generating relevant concern and attention on the human factor role in safety: the *enormously improved reliability of hardware*, and the *extensive use of automation*. Advances in hardware technology have vastly reduced mechanical faults and enabled the management of plants, even in the presence of severe system faults and malfunctions. In this way, the contribution of human factors to safety analysis has been enhanced and "human error" has become the primary "cause" of most accidents in all technologically developed domains. It is nowadays very common to hear "human error" quoted as a possible or likely cause of accidents, by media and even by safety authorities, immediately after the occurrence of an accident. Unfortunately, this is only a shallow, and in many cases inappropriate, explanation of the real causes of an accident and is frequently utilised just for assigning responsibility and blame, rather than finding explanation and remedies. In some unfortunate cases, human errors of front line operators are abusively indicated as primary causes by certain managers in order to protect themselves and cover their own faults and responsibilities.

Considering the role of automation, it is widely accepted that in normal conditions modern plants are easier to operate than their predecessors. Human-centred design principles are utilised by manufacturers to varying degrees of accuracy when designing control systems and interfaces. These principles aim at maintain-

9

ing a central role for the operator in the management (supervisory) control loop, and require that operators are constantly "ahead" of a plant's performance, controlling and supervising the automatic system in all its functions and procedures (Norman and Draper, 1986; Billings, 1997). However, designers do not always respect this essential requirement. Systems behave and respond via the automation and follow the rules and principles provided by their designers. These are not always totally known or familiar to front line operators. Moreover, in abnormal or emergency conditions, the dynamic characteristics of the sequence of events add to the inherent complexity of the situation and further complicate the decision-making process. If the expected response does not occur, a mismatch arises between the operator's understanding of the dynamic evolution (situation awareness) and the automatic system. Thus, working environments are much more demanding in terms of cognitive and reasoning abilities than simple sensory-motor skills (Rankin and Krichbaum, 1998; Hollnagel, 1993; Ralli, 1993).

In such scenarios, and aiming to offer valuable and consolidated ways to improve safety and control of HMSs, the role and interplay of humans and automation is vital and needs careful consideration. This discussion leads to two main considerations. On one side, while automation is necessary, as it supports human tasks performance and can successfully replace human activity, it should also be developed with consideration for the consequences of inappropriate reasoning or misunderstandings on the part of the operators. This is particularly important when an operator's knowledge and beliefs are deeply rooted in the social, organisational and technical context, usually called sociotechnical working context, in which they are born and developed. These inappropriate reasoning or misunderstandings are very difficult to trace and eliminate. On the other side, the occurrence of "human error" is an intrinsic characteristic of the management of any system, and it is impossible to conceive a plant that is totally "human error" free.

Consequently, the improvement of safety of a system cannot be achieved by tackling any actual inappropriate performance that has occurred or may have happened during an accident, but rather by understanding:

1. "why" operators took certain steps, and "what" are the root causes that may have generated, or may trigger in the future, inappropriate human behaviours;
2. "what" forms of inappropriate behaviour was produced, or could result, from such socio-technical root causes; and
3. "how" can systems be developed and humans be trained in order to:
   (a) anticipate and prevent accidents and incidents initiators;
   (b) manage accidents that still occur, and possibly recover normality; and
   (c) limit or protect other humans and environment from accident consequences, when prevention and recovery did not happen.

In addition, the possibility for systemic and component failures remains, and it would be unwise and unacceptable to consider technical systems fully failure free and focus only on human errors.

This is why, in order to ensure safety and efficiency of modern technological systems, a much wider process of evaluation and study of human–machine interaction has to be developed, rather than simply tackling human inappropriate performances. Such a type of approach may be called *Human Error* and *Accident Management* (HEAM), and it involves the thorough development of measures at human machine system level for (a) *prevention* of conditions that favour and lead to system failures and/or human errors; (b) *recovery* of a plants' normal or safe performance, once a failure/error could not be anticipated and avoided, and an accident/incident has started; and (c) *containment* of the consequences and protection of the human beings and environment, in the case that neither prevention nor recovery succeed.

These fundamental goals of HEAM can be achieved and maintained at different stages of the development and implementation of a technical system, namely at design level, as well as during its lifetime. In this way, it is possible to grant continuous improvement of normal operations and emergency management in complete safety at all times. In particular, the consideration and analysis of human error and accident management must influence the following four areas of development and application of HMSs:

- *design* of human machine interactions and interfaces;
- *training* of operators and main actors for managing nontechnical risky states;
- *safety assessment* of systems and organisations; and
- *accident/incident investigation.*

Improving the design of Human–Machine Systems implies ameliorating the design process in order to ascertain that the basic principles of human–centred automation are respected. Improving *training*, in nontechnical issues, intends to increase the ability of operators to capture, notice, and deal with those factors and indicators of the context that favour the occurrence of errors or mismatches between human situational awareness and automation performance. Implementing accurate *safety assessments* of systems and organisations, at the design stage as well as at periodic intervals during the lifetime of a plant/system, represents the most complete method for ascertaining and maintaining high levels of safety within an organisation. Safety assessments make it possible to identify and discover at an early stage the relaxation of certain expected and critical safety measures, as well as the appearance of new factors that may favour the occurrence of accidents. Finally, *accident/incident investigation* should focus nowadays on methods by which it is possible to trace, in addition to the human erroneous performances, primarily the root causes of accidents that are deeply imbedded in the sociotechnical contexts and specific working environments. Only with such a spectrum of approaches is it possible to achieve safe and efficient management of a complex human–machine system.

In this scenario, it is quite obvious that the design and the assessment of safe and effective systems and technological assets is no longer the sole responsibility of engineers, but implies the consideration of different perspectives and the contribution of a variety of specialists, especially from the human-related sciences. In particular, several disciplines must collaborate synergistically to reach such objectives, and this implies combining engineering know-how, psychology, and sociology principles, fundamentals of information technology, practical skill in normal and emergency operations, and acquaintance with real system behaviour.

This chapter is structured around two main correlated topics: (1) elements of HEAM; (2) and areas and types of application of HEAM. Firstly, the elements of complex technologies will be analysed, and a number of basic definitions for representing human–machine interaction will be developed. Particular attention will be dedicated to the concepts of human error and to the wider context of accident management. Then the two possible types of application of error and accident management studies will be considered. These are retrospective and prospective analyses. The need to preserve and ensure consistency between them will be discussed in detail.

The variety of areas of application of HEAM will then be considered. The correlation between *types of application* and *areas of application* will be dealt with in order to show which types of applications are necessary for different areas of application in order to develop consistent HMS analyses.

Finally, a methodology that can be applied in practice in different areas of application and merges the two basic types of application will be developed. This methodology is called Human Error Risk Management in Engineering Systems (HERMES) and represents a reference architecture that will be utilised throughout this book for discussing real applications of HEAM analyses.

The variety of methods, approaches, and techniques that can be applied for HEAM analyses will be described in the next chapter of the book.

2.2 Elements of Human–Machine Interaction

2.2.1 Definition of Human Factors

All complex technological systems, such as aircrafts, air traffic control rooms, chemical and energy production plants, and the like, operate in risky environments and share a number of characteristic elements, which affect their control processes (Maurino et al., 1995). In particular, the study of such systems from a human perspective implies the consideration for what is generally called the "human factor." Moreover, all HMS can be formally analysed by approaches similar to each other for what concerns the architecture and theoretical frame adopted to describe the Human–Machine Interaction (HMI).

As this chapter refers to the elements that govern human–machine interaction, it is important to define the concept of Human Factors (HF), which embraces all the subjects discussed in this book.

> Human factors may be defined as the technology concerned with the analysis and optimisation of the relationship between people and their activities, by the integration of human sciences and engineering in systematic applications, in consideration for cognitive aspects and socio-technical working contexts.

By this definition, Human Factors extends the concept of ergonomics, as the science of humans at work, beyond the workplace and behavioural performance to the cognitive and social aspects involved in human activity (Edwards, 1988).

Human Factors is conceived here as a "technology," emphasising its practical nature rather than its disciplinary character. In this sense, the difference between human factors and human sciences is the same that exists between engineering and physics. Physics and human sciences look at the basic principles and fundamental criteria that govern their locus of interest, while engineering and human factors concentrate on the implementation in the real world and working environment of these principles and criteria. This distinction is particularly important, as it is recursively called upon for distinguishing different subject matter, especially when one looks at HMI issues from a merely theoretical or a more applied perspective.

2.2.2 Human–Machine Systems

Human–machine interactions and processes always occur in realistic contexts. They are characterised by the plant or machine under control, in direct contact with the operator, and by the socio-technical working context, in which the interactions take place (Figure 2.1).

The plant interacts with the human operator through its *interfaces* and *controls*. They may be defined as follows:

- *Interfaces* are display panels, indicators, decision support tools. They transform the behaviour of the machine in visual, auditory and tactile information. These support the operator in perceiving and understanding the state and dynamic evolution of the system and in developing the strategies for its management and control.
- *Controls* are means by which it is possible to operate on the system and automation in order to implement the operator's intention and strategy. Interventions of controls are transformed in machine information by actuators.

The socio-technical working conditions, also called *context* and *environment*, comprise of the following:

- the actual environment in which operations take place, including noise, space, light, temperature, etc.;
- other operators, cooperating directly, or collaborating at a distance, with the decision maker; and
- the social context, represented by management policies, company rules, society, and cultural climate.

The plant interfaces and socio-technical working context are the main sources of *stimuli* for the operator. They affect the operator's allocation of resources and his or her knowledge base. They may modify the unfolding of the reasoning and cognitive processes as well as the performance of manual or control actions by, for example, causing errors or inappropriate behaviour. The loop of human machine

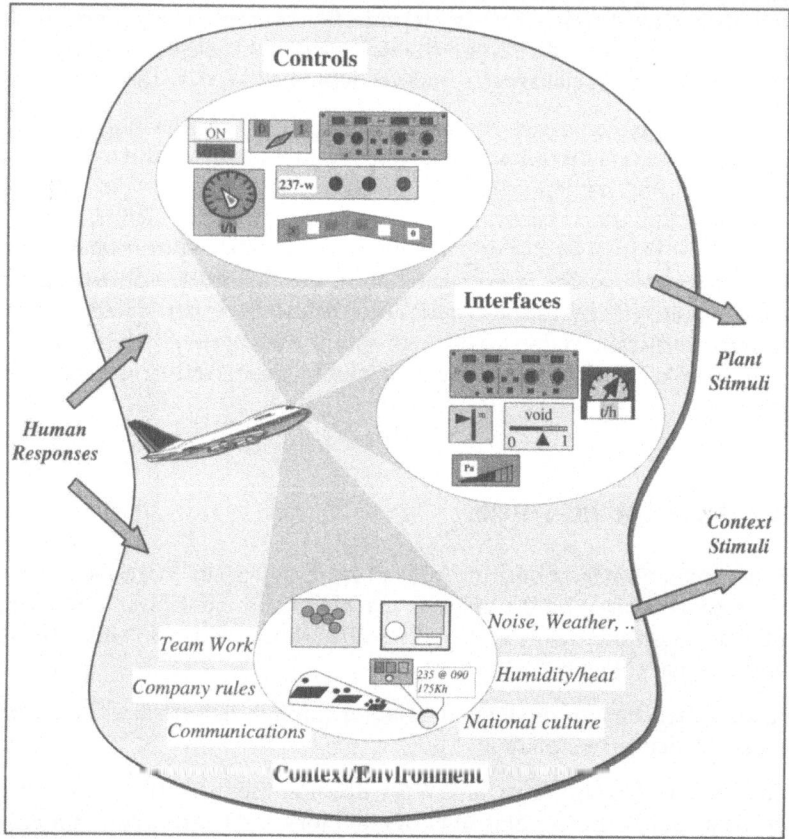

Figure 2.1 Structure of a human–machine interaction system.

interaction is then closed by the *human responses* that maintain the dynamic evolution of the interaction and generate new control actions, etc.

This structure captures and describes what may be defined as an HMS. A number of comprehensive definitions of a human–machine interaction system, or human machine system, have been proposed, as the notion of the combined human–machine element has changed over time and has become crucial for the development of all systematic safety theories. A quite complete definition of HMS can be found in the document MIL-STD-882B (DoD, 1984):

> A human–machine system (HMS) can be defined as a composite, at any level of complexity, of personnel, procedures, materials, tools, equipment, facilities, and software. The elements of this composite are used together in the intended operational or support environment to perform a given task or achieve a specific production, support, or mission requirement.

This definition is deemed very appropriate, as it embraces the effects of socio-technical environment discussed above. Moreover, it looks explicitly at crucial aspects derived from the use of modern computer technology for the control and

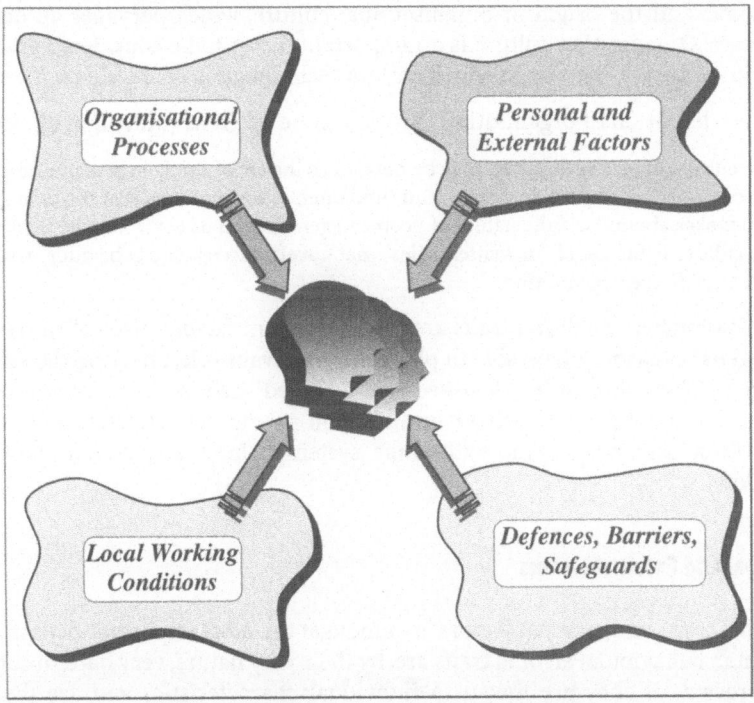

Figure 2.2 Socio-technical elements of a human–machine system.

management of plants and machines, in normal operations and emergency or transient conditions.

However, usually the terminology "system" is associated simply with the whole plant or with hardware and software components of a plant, and, thus, it represents a synonym of "machine." In the definition of HMS adopted above, the terminology "system" is used in a much wider sense, which includes also humans, context, and environment.

2.2.3 Sociotechnical Elements of a Human–Machine System

Four main socio-technical elements influence a human–machine system and intervene in all dynamic processes characterised by human–machine interactions. These four elements are: (*a*) *organisational processes*, (*b*) *personal and external factors*, (*c*) *local working conditions* and (*d*) *defences, barriers and safeguards* (Figure 2.2).

Organisational Processes

It is now well accepted that strategic decisions have deep consequences for the way in which a system is managed. Cultural traits are the root cause of corporate behaviour and the origin of organisational culture, which pervades all decision processes. Organisation culture is an important factor to be considered in assessing and designing the way in which certain technology is or should be managed.

A broad definition of organisation culture can be taken as (Mitroff et al., 1990):

> The culture of an organisation may be defined as the set of rarely articulated, largely unconscious beliefs, values, norms, and fundamental assumptions that the organisation makes about itself, the nature of people in general and its environment. In effect, the culture is the set of "unwritten rules" that govern "acceptable behaviour" within and outside the organisation.

Organisational and cultural traits are then important factors, also called "resident pathogens" (Reason, 1990), able to play a very relevant role, affecting the safety of a system. They have to be identified and detected early, so as to ensure correct understanding of people's behaviour and grant prompt and effective intervention when these factors combine with some system failures to generate dangerous situations.

Personal and External Factors

Personal traits and external factors are amongst the most important determinants of human behaviour. Personal traits are, by their very nature, very hard to consider and prevent, as they are linked to individual characteristics and are therefore impossible to generalise in stereotype constituents.

External factors are determined by contextual random events. Consequently, they are equally difficult to formalise and generalise, as they are directly related to human or systemic behaviours in specific environments.

However, these factors must be defined and considered in a methodological framework for the study of HMSs. They can only be approximately and imprecisely formalised by means of statistical algorithms able to capture their random nature.

The definitions of external and personal factors are proposed as follows:

> External factors can be considered as all random physical or system contingencies that may alter or impinge on local working conditions and safety measures, so as to foster inadequate system performance and erroneous human behaviour.

> Personal factors are individual, physical, or mental conditions that affect human behaviour, and are specific to each person. They can only be accounted for by a random variable affecting the generic behaviour of large classes or categories of people.

Personal and external factors should be considered as random variables in an overall framework of accident causal path, and their role can only be marginal in a structured analysis of an *organisation*. Their presence is anyhow considered and recognised by such random quantities.

In accident analysis, their contribution to event development and root causes is crucial in many cases, and the identification of their role needs adhoc assessment and evaluation.

Local Working Conditions

Local working conditions are "expressions" of physical and social contexts, including higher-level organisational and cultural traits. These are transmitted along various pathways of the *organisation*. They are probably the most relevant factors affecting the behaviour of front line operators, as well as people involved in decision making, as they are immediately and promptly related to the environment and dynamic evolution of the human–machine system.

The definition of local working conditions can be adapted from Maurino et al. (1995) as:

> Local working conditions are the specific factors that influence the efficiency and reliability of human performance in a particular work context.

Local working conditions affect the performance of tasks by influencing either the interface between operators and control systems and/or cognitive activities. Examples of local working conditions are: workplace design, interfaces with automation, tools and instruments, protective equipment, job planning, procedures, supervision processes, workload, training, and policies.

Defences, Barriers, and Safeguards

Defences, Barriers, and Safeguards (DBS) are all structures and components, either physical or social, that are designed, programmed, and inserted in the human–machine system with the aim of making more efficient and safe the management of a plant, in normal and emergency conditions.

In general, the following definition can be adopted:

> Defences, barriers, and safeguards are the measures developed by the organisation aimed at creating awareness, warning, protection, recovery, containment, and escape from hazards and accidents.

DBS are then a direct result of a high-level organisational process which includes planning, design of automation and emergency systems, definition of policies on training and procedures, and the like (Reason, 1997; Hollnagel, 1999; Polet et al., 2002). They fulfil a series of functions, namely:

- to create awareness and understanding of the hazards;
- to support the detection process of off-normal conditions;
- to support restoring normal operating conditions;
- to protect from injury;
- to contain the consequences of an event;
- to support escape in the case of loss of control or accident.

In order to offer here a formal distinction between different types and modes of DBS, the classification proposed by Hollnagel (1999) can be proposed as an example of guidelines for a safety analyst. According to this classification, DBS can be grouped into four main types:

1. *Material Barriers.* These types of DBS prevent the performance of dangerous actions or contain consequences of occurrences by physical constraints. Examples of *material* DBSs are doors, railings, fences, safety belts, filters, etc. These barriers aim at attaining their goals by simply being located in strategically relevant positions or by reacting to physical and environmental conditions.

2. *Functional Barriers.* These barriers require that a certain function occurs or that certain variables reach or are assigned predefined values to become active. In other words, a certain *function* has to be satisfied or fulfilled in order to make the barrier either effective or ineffective, depending on its purpose. Examples of *functional* DBS are air-locks, dead-man-buttons, passwords, safety codes, delays, etc.

3. *Symbolic Barriers.* These DBS are associated with a certain logic or conventional rule or habit that indicate the presence of a dangerous or safety relevant condition. In other words, *symbolic* DBS require knowledge of certain rules and regulations, or habits, and their interpretation in order to be effective. *Symbolic* DBS may not be respected or may be bypassed by users. Examples of *symbolic* DBSs are safety code sequences, instructions and procedures, signs, signals and warnings, clearances, joborders, etc.

4. *Immaterial Barriers.* These DBS are the most highly located barriers in a cognitive sense. They demand explicit interpretation by the user, as they are known but only in general form and are not present in any of the other DBS forms, i.e., symbolic, functional, or material. In general, these are the result of cultures, philosophies, or policies which develop within an organisation and are very difficult to modify or adapt to new situations and contexts. Examples of *immaterial* DBS are laws, general rules, standards, etc.

The differences that exist between these four categories are useful for supporting analysts or designers in developing DBS at different levels of depth. They are not totally independent from each other. However, the existence of certain overlapping amongst them does not affect the overall understanding and support that such classification may offer in the process of design and validation of a safety system.

2.3 Human Error and Error Management

Th previous sections have considered the building blocks of organisations from a socio-technical perspective. These factors represent the underlying conditions that foster the generation of human inappropriate behaviour and human errors. While many psychologists have discussed the fundamental nature of human error in detail (Norman, 1981; Reason 1986, 1987, 1990, 1997; Reason and Mycielska, 1982; Rouse and Rouse, 1983; Rasmussen, et al., 1987; Senders and Moray, 1991), the human factors perspective adopted in this book shifts the focus of attention on the effects of errors in a technological environment.

2.3.1 Human Error in an Organisational Perspective

Definition of Human Error

A definition and classification of Human Error (HE) which can be considered "classical" has been given by Reason (1990), and may be summarised as follows:

> Human error may be defined as the failure of planned actions to achieve their desired ends without the intervention of some unforeseeable event.

This failure can occur either when the plan is adequate but the actions deviate from the plan (*slips, lapses*), or when the actions conform to the plan but the plan is inadequate for achieving the desired ends (*mistakes*).

> Slips are associated with attentional or perceptual failures and result in observable inappropriate actions. Lapses are more cognitive events and usually involve memory failures. Mistakes are errors made at a high cognitive level, involving processes linked to the available information, planning, judging, and formulating intentions.

Another type of error is considered in Reason's classification: *violations*.

> Violations are deviations from safe operating practices, procedures, standards, or rules. Most violations are deliberate actions, even if sometimes they can be erroneous.

Errors defined in the Reason's theoretical framework, independently of their type, can take different modes according to the person that makes them and the role that this person occupies in the organisation. Errors made by front line operators and primary actors, in the control process of a system, emerge immediately and become very visible in the evolution of an event. These are called *active errors* and are the most obvious occurrences and the most rapidly identified human contributors in an accident.

Errors made at higher levels of the organisations, such as in the definition of policies or emergency procedures, or in remote and distant working systems such as at the maintenance level, are more complicated and difficult to spot at first sight. These errors lie inactive in the system and do not show their negative effects until specific conditions are encountered.

The higher the level of the organisation at which these errors are made, the more serious are the consequences at the front line operation. Indeed, errors of strategic nature, such as when defining company philosophises or policies, affect safety attitudes and the culture of operators and managers, creating working conditions that foster violations and inappropriate or careless performances.

These errors are defined as *latent errors* and are the most dangerous and serious errors to be tackled.

Types and Modes of Human Error

Types of Human Error

The definitions of human error discussed in the previous section concentrate on the types of inappropriate behaviour that can be identified primarily in an organisational perspective, as they identify generic manifestations of behaviour.

In this sense, *slips*, *lapses*, *mistakes*, and *violations* concern the individual behaviour, while *active* and *latent* errors are representative of the organisational perspectives in which individual behaviours are framed.

Other *types* of errors can be defined, for example, focusing on the specific performance of individual persons. In particular, a simple structuring of error types in errors of *omission* and *commission* allows the classification of a wide variety of inappropriate behaviours (Swain and Guttmann, 1983). Errors of omissions are, as the definition says, simple omission actions or steps in the performance of a procedure or a well-known process. Errors of commission are all the remaining possible manifestations of inappropriate behaviour that imply the actual performance of an inappropriate action.

Modes of Human Error

When studying human–machine systems, a concrete representation of actual inappropriate behaviours is necessary.

Error types can be complemented by the identification of the forms taken by inappropriate behaviours. These can be classified as error *modes* and are associated with error types in classifying and representing the behaviour of humans interacting with machines.

Examples of modes of errors are the actual amount of delays in performing certain actions, or the inadequate amount of force applied in performing a certain operation, etc.

In practice, when classifying human errors in an accident analysis, or when considering errors in safety studies, it is necessary to:

• frame inappropriate behaviours in the socio-technical environment in which they are made (*error types*); and
• define the actual forms that errors take when performing a certain action (*error modes*).

This representation of errors allows the complete consideration of errors in any type of study or analysis.

2.3.2 Human Error in Safety Practices

These concepts of human errors are nowadays accepted by designers and analysts of many organisations and technologically advanced systems. As a consequence, error management tools are implemented in practice.

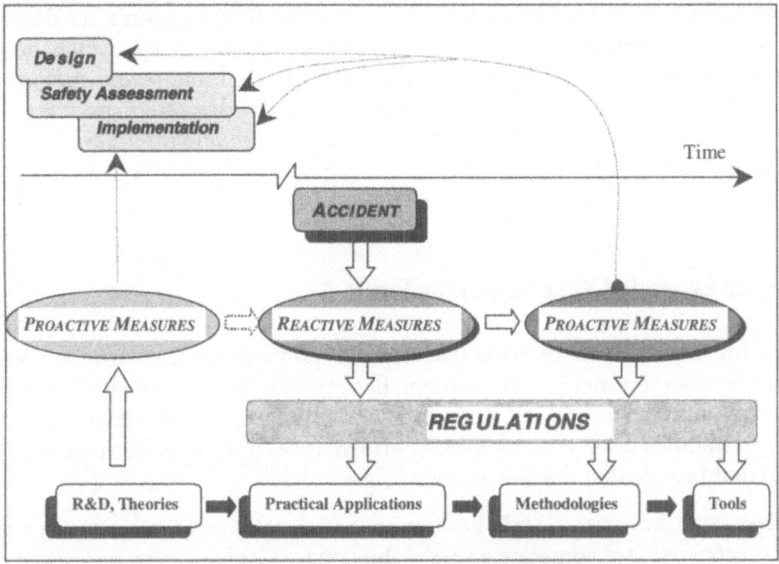

Figure 2.3 Proactive–reactive measures for safety management and accident prevention.

However, in many cases they still present crucial limitations. They are implemented as "piecemeal" rather than planned interventions; they are developed as "reactive rather than proactive" measures, i.e., they result from "event-driven" rather than "principle-driven" considerations (Reason, 1997, p. 126).

Consequently, certain types of safety and error management measures suffer major drawbacks for three main reasons:

- They consider errors as causes rather than consequences of accidents.
- They focus on people rather than situations.
- They rely on punishment and blame rather than improving safety culture and attitudes.

Human errors that are immediately visible in the case of accidents are simply manifestations of inappropriate behaviour. They require an explanation as much as the accident that is related to them. It is necessary to understand the context in which errors have occurred, the socio-technical and organisational environment and, sometimes, also the personal factors that have foster them. Finally, the results obtained by the "blame, exhortation, and disciplinary" sanctions are small when compared with the development of a (safety) culture through personal conviction.

In practice, the application of a safety method and the development of measures, including approaches for error reduction and containment of consequences, usually follow an evolutionary process (Figure 2.3). Safety methods are originally generated at the research and development (R&D) level, as proactive measures for improving safety. However, they have a limited impact on design, safety assessment, and implementation, and they are rarely immediately transformed into practically applicable tools. Only later, during system operation and lifetime, and following the occurrence of a severe accident, (reactive) measures are developed, as a reaction aimed at avoiding the repetition of the causes and circumstances of that specific accident. These are then further expanded into sound methods and are introduced as (proactive) mandatory measures by safety and regulatory authorities, with the precise aim of accident prevention and limitation (Cacciabue and Pedrali, 1997; Cacciabue, 1999).

2.3.3 An Expanded View on Human Errors

The definitions of human error discussed in previous sections, and the various types of errors that derived from them, fit very well the socio-technical elements of a human–machine system. They encompass the logical and strict connections between manifestations of erroneous performances and organisational and environmental factors that may be at their origin.

In many cases, however, certain behaviours, which are later identified as erroneous or inappropriate, are completely reasonable, unavoidable or even necessary, given the contextual conditions and the operator's appraisal of the situation ("situational awareness") at the time of their occurrence.

Therefore, studying safety and simply discussing in terms of human errors is not an effective way forward, while it is much more important to understand and analyse the overall human–machine conditions and the context in which accidents and human behaviour develop. Moreover, it is obvious that it is impossible to eliminate all errors or inappropriate behaviours that may occur during the management of a plant, especially in those cases where certain decisions and choices are made as a consequence of special contextual conditions. Therefore, it is equally important to accept that "errors" occur and to also consider and develop, in addition to preventive measures, adequate means for ensuring prompt errors recognition and recovery or even protection for humans and environment in case of accidents.

Following this line of thought, we will frame the human contribution to accident causal paths in a perspective that considers the "human error" not as the cause but as the consequence of other factors that reside at different levels of the organisation, as well as in the contextual and dynamic circumstances of the specific occurrence. These are the important causal elements, or root causes, that need to be identified and removed from the system, or at least minimised, in order to prevent their occurrence and their negative effects or to ensure their effective control and recovery or, eventually, protection in the case of an accident.

The safety of any system depends indeed on a combination of technical and social factors, which are deeply correlated and cannot be separated and dealt with independently from one another. They must be tackled by appropriate methodologies of human–machine interaction and human–machine system for the identification of "*safety critical factors*" and "*safety levels*" that enable evaluation of the safety state of a system and its "distance" from dangerous or unsafe conditions.

Any system, starting from its design stage and then, following its practical implementation and during its entire operating life, detains and attains different safety levels and develops a variety of new safety-critical factors. The safety of a system depends on these factors and levels. They must be determined and evaluated at the design stage, as well as during the life of the system, for ascertaining its operability and possibly discovering the need for improvements and/or new safety measures.

In practice, it is necessary to accept that human errors occur and cannot be totally prevented or eliminated. Consequently, the way to improve the safety levels of complex systems concentrates on three different means of intervention: (1) by *prevention* of risky or inappropriate circumstances; (2) by offering adequate ways of *recovery*, when prevention has not been possible; and (3) by *containment* and limitation of consequences, when neither prevention or recovery has been successful. To ensure maximum safety it is necessary to define or identify indicators or safety critical factors. These allow to "measure" the level of safety attained by a system at any time of its life and to "confront" these indicators with acceptable values, possibly for all three types of means of intervention.

This implies developing adequate human–machine interaction management, or human error and accident management, approaches within an organisation.

2.4 Human Error and Accident Management

Given the above discussion, the definition of Human Error and Accident Management (HEAM) that is adopted is as follows:

> Human error and accident management is the variety of methods and measures designed to reduce inappropriate and risky human–machine interactions at different stages of a system lifetime, by offering means and ways to recognise and prevent them, to control and recover from those that still occur, and to contain and escape their adverse consequences, when full recovery is not possible.

The clear understanding of the above definition and of the goals of HEAM is the *first fundamental standpoint* in the development of human–machine systems and effective safety measures.

The definition of HEAM is strictly coupled with the definition, developed in the previous section, of "human error" as inappropriate performance/behaviour, dependent on the context and dynamic contingencies and imbedded in a specific socio-technical environment. This definition of "human error" is integrated in a more general representation of HMI and modelling of human behaviour, which embraces all types of interactions, either adequate or inappropriate.

The understanding of the concept of human error in terms of a human–machine interaction process and the adoption of a model of HMS that considers humans and machines at the same level, in a sort of a "joint cognitive system," represents the *second fundamental standpoint* in the development of effective HMS and HEAM measures.

The objective of this book lies precisely in the structuring and in offering guidance to the application of methods for developing and implementing HEAM means and measures that can be proactively implemented within an organisation, to minimise the occurrence and to control human–machine related accidents.

2.4.1 Types of Analysis for HEAM

Prospective and Retrospective Analyses

Human error and accident management should be conceived as a set of proactive measures that improve the safety standards of an organisation. Proactive measures may be developed on the basis of creative thinking and safety-oriented attitudes of analysts who are able to imagine safety-critical scenarios and study appropriate ways and measures to prevent their occurrence, recovery normal system functions when they still happen, and protect humans and environment in case of accident. These analyses are performed with a prospective view of what may happen during an abnormal situation.

At the same time, proactive measures must be associated to the real socio-technical contexts in which they are applied. They require, therefore, a thorough assessment of the *local working conditions*, and *organisational processes*, including their dynamic and evolutionary aspects, and their history, in terms of past incidents and accidents that have involved failures of *defences, barriers, and safeguards* and *personal and external factors*. The studies of these socio-technical elements of HMS are basically retrospective analyses by which it is possible to learn the lesson from past experience and to understand the actual working conditions in which HMI takes place.

There are, therefore, two major types of analyses that support the development of HMS and HEAM measures, namely, *retrospective* and *prospective analyses*. They are complementary to each other, and contribute equally to design and safety assessment processes.

In a wider context of human–machine interaction studies, prospective and retrospective types of analyses can be defined as follows:

> Retrospective analyses consist of the assessment of events involving human-machine interaction, such as accidents, incidents, or "near-misses," with the objective of a detailed search for the fundamental reasons, facts, and causes ("root causes") that fostered them.

> Prospective analyses entail the prediction and evaluation of the consequences of human–machine interaction, given certain initiating events and boundary configurations of a system.

The clear understanding and consideration of the difference and synergy between prospective and retrospective analyses is the *third fundamental standpoint* in the development of effective HMS and HEAM measures. To make these concepts more clear a detailed discussion will now follow.

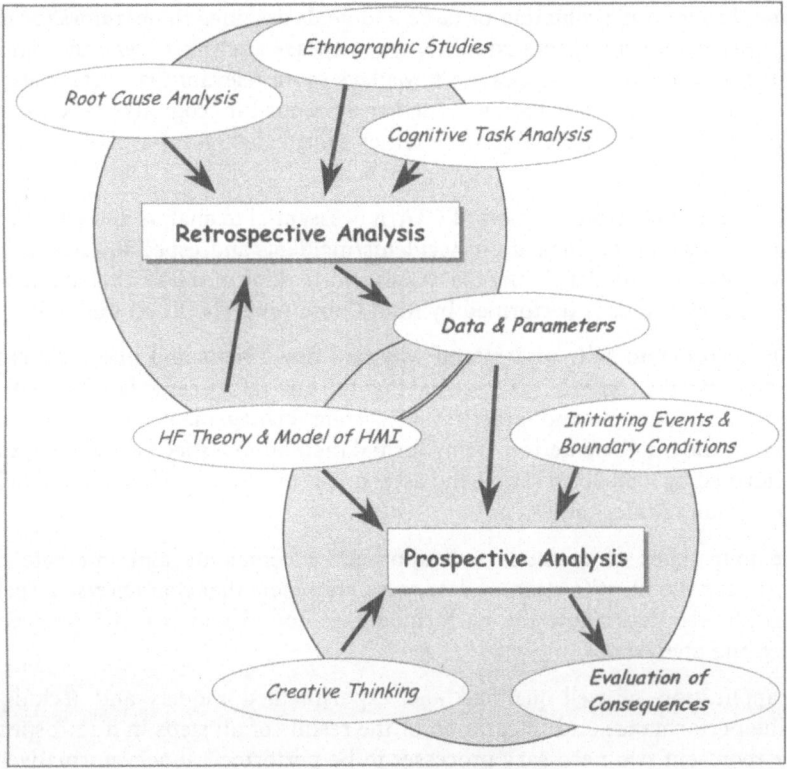

Figure 2.4 Prospective and retrospective analysis.

Retrospective Analyses

In practice, retrospective analyses are oriented to the identification of "data and parameters" associated with a specific occurrence and context.

They can be carried out by combining four types of methods and models that are extensively formalised and discussed in the literature, namely (Figure 2.4):

- root cause analysis;
- ethnographic studies;
- cognitive task analysis; and
- HF theories and models of HMI.

Human factors theories and models of HMI consider several reference paradigms that can be applied in studying a specific contextual environment. These are generic models and architectures of human–machine interaction that must be adapted to the specific context and system.

In order to select and apply the most appropriate HMI model, an analyst must primarily study objectives, formal plans, and procedures for operating processes. This is done through the evaluation of tasks and goals assigned to operators for managing normal and emergency conditions, i.e., by task analysis. Given that cognitive and decision-making processes are nowadays more relevant than actual actions, the standard task analysis has been further developed in "cognitive task analysis" (CTA) that focuses on the cognitive rather than behavioural aspects of human activity.

In addition to the formal process of CTA, it is essential to analyse and evaluate the outcome of past events in terms of accidents/incidents and, especially, near missies with the objective of identifying the causes and reasons of specific behaviours and HMI in general. This is performed by Root Cause Analysis (RCA) methods.

Finally, in order to fully understand why and how events and interactions take place in a specific context, it is essential that the analysts become familiar with the working environment and practices of system management. This is a crucial process in the implementation of any formal method to a specific real case, and it is performed by field observation and assessment of working practices and habits, i.e., by ethnographic studies.

These four types of methods and approaches contribute and integrate their outcome for the identification of data and parameters that characterise a specific context. These data allow the performance of sound and realistic prospective studies and analyses.

The application of well-qualified and experimented models and techniques, and the performance and integration of the results of all steps in a retrospective study, represent two necessary processes to be performed almost normatively in order to obtain a clear picture of the existing context and socio-technical working environment, as well as consistent sets of data and parameters for predictive studies.

Prospective Analyses

Prospective analyses aim at the "evaluation of consequences" of HMI scenarios, given a selected spectrum of (Figure 2.4):

- HF theories and models of HMI,
- data and parameters,
- initiating events and boundary conditions, and
- creative thinking.

The HF theories and models of HMI that sustain prospective analysis must be the same as those applied for retrospective analysis. The same conditions apply with respect to the generality of paradigms and specificity of domains of application. The models selected for application need to be transformed into simulations that can be practically implemented in (computer) programs and adapted to specific contexts in order to perform previsions and estimates of the likely consequences of accidents or prediction of special situations that develop from certain initiating/boundary conditions and HMI.

The data and parameters that sustain such models and simulations, as well as their validity and applicability, are the outcome of retrospective studies and strictly depend on the accuracy of the (retrospective) analyses that produce them.

The initiating events and boundary conditions give the spectrum of situations that are analysed by the prospective study and are developed by the analyst on the basis of his/her experience, expertise, knowledge and, specifically, creative thinking.

The latter is a fundamental component of any speculative and perspective type of analysis and plays a clear role in the selection of models, variables, data, and all other elements that combine in a prospective study. Creative thinking is the necessary condition for the development a "human-centred" prospective analysis, which is an essential contribution to any type of study where novelty and imagination are governing components, aiming at anticipating and predicting possible HMS behaviours and HMI.

As for the case of retrospective analyses, the application of a prospective study must be structured and formalised in order to develop a consistent methodology that may be recursively and effectively applied. The reliability of a methodology for prospective analysis can be observed through the evaluation of the consequences of HMI. These are the ultimate outcome of a prospective safety study, and represent the values and quantities that allow an analyst to draw conclusions and evaluations about the safety state and safety level of a system.

Differences and Commonalities Between Prospective and Retrospective Analyses

The procedures for developing prospective and retrospective analysis bear certain commonalties, but also contain important differences. The differences consist in the basic objectives of the two approaches. In prospective studies the analyst must look ahead and speculate in a creative way. In retrospective assessments the focus lies in understanding and extracting the lesson from past events and occurrences.

The commonalties between the two approaches lie in human factors theories and human–machine interaction models and sets of data and parameters. The same basic HF theory must be considered for prospective and retrospective studies and, consequently, coherent human behaviour and error models must be applied. In this way, data and parameters derived from retrospective studies of real events and evaluation of working environment can be consistently and coherently applied for prospective analyses. These common elements should be well identified, as they represent logical links between the two approaches.

In other words, to make prospective and retrospective analyses consistent with each other, it is essential that identical, or at least coherent, HF theories and HMI models should be utilised for both types of analysis. In this way, data and parameters derived from retrospective studies may be applied in prospective assessments without having to make inferences and judgement, which introduce further and unnecessary uncertainties on the evaluation of consequences.

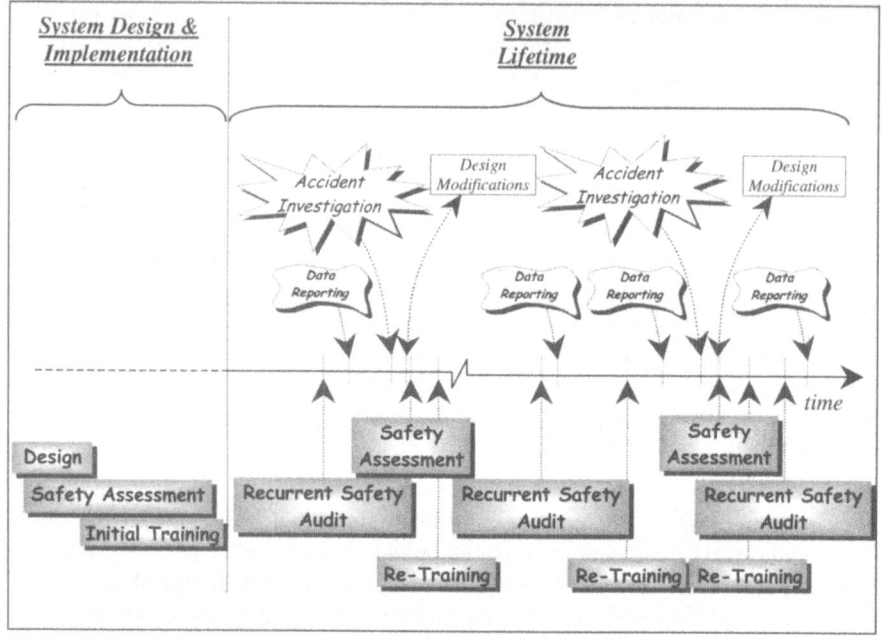

Figure 2.5 Safety evaluations during design, implementation, and lifetime of a system.

2.4.2 Areas of Application of HEAM

According to the definition of human error and accident management, the measures that ensure safety and prevent risky interactions within a human–machine system are considered and dealt with at different stages of the system lifetime (Figure 2.5).

This implies that before the actual implementation and installation of a system, i.e., during the *design* stage and *preliminary safety assessment*, the possible occurrence of incidents and human inappropriate behaviours have to be considered, and suitable defences, barriers, and safeguards must be devised in order to prevent them, control them, and minimize their consequences.

Similarly, the same conditions have to be considered for the development of *initial training* (before system installation) and *retraining* (during system lifetime), which should strengthen human perception and promote adequate reaction in the case of appearance of contextual conditions that may favour and generate inappropriate behaviours and possible incidents and accidents.

Moreover, during the operational life of a system, it becomes essential to learn the lesson that may be drawn from the occurrence of incidents, accidents, and near misses, as well as to be prepared to evaluate the dynamic evolution of the safety levels of the system. This implies that appropriate HEAM approaches have to be considered that favour *data reporting* and *collection*. These data support the *acci-*

Table 2.1 Application and basic requirements for HEAM

Area of application	Type of assessment
Design	• Design of control, emergency, and protection systems • Design of human–machine interfaces • Development of standard and emergency operating procedures
Training	• Classroom human factors training • Human factors training in simulators
Safety assessment	• Human contribution to design basis accident • Human contribution to quantitative risk assessment • Evaluation of safety levels of an organisation by recurrent safety audits
Accident investigation	• Human contribution to accident etiology and identification of root causes

dent investigation and *recurrent safety audits*, which consent to ascertain whether the conditions for safe and effective operations still exist, or some changes and improvements are needed in order to reestablish adequate levels of safety condition and operability.

In other words, HEAM approaches should be applied at four stages of the development and operation of a technological system, namely, *design, training, safety assessment*, and *accident investigation* (Table 2.1). The consideration of HEAM measures in these four areas of application requires recursive utilisation of appropriate methods for the evaluation of the safety level of an organisation in order to ensure the preservation of adequate safety levels throughout the lifetime of a plant or system.

Each of these four areas of application encompasses specific types of assessment. These will be briefly discussed in the following sections, while a complete analysis of specific tools and detailed procedures for application will be performed in the next chapter of this book.

The *fourth fundamental standpoint* in the development of effective HEAM measures lies in the appreciation that before and during the lifetime of a system a variety of tools and approaches must be applied for the continuous verification that adequate safety conditions exist and are maintained. These tools aim at sustaining and ensuring affective applications in the four areas of application of *design, training, safety assessment*, and *accident investigation*.

In performing safety evaluation of a system and in devising measures for prevention, recovery, and protection from accidents it is essential that adequate *indicators* and *markers* are identified that allow the measurement of the safety level of a system. Only generic indicators can be defined according to types of system. The definition of what *indicators* and *markers* are important and constitute valuable measures of safety for a specific system can be defined only by the individual organisations responsible for the management of the system, or family of systems. Therefore, a set of methods and approaches must be applied for the adaptation of generic indicators and for the definition of appropriate safety levels for each specific plant.

The appreciation of the importance and role of *safety indicators* or *markers*, and their specific values associated to each plant, working context, and organisation, represents the *fifth fundamental standpoint* in the development of effective HMS and HEAM measures.

Design

At the design level, HEAM can be tackled by developing *control, emergency, and protection systems* that are effective and useful. These systems are always coupled to appropriate *procedures* and *interfaces* (Stokes and Wickens, 1988; Degani and Wiener, 1994a). Applying models and numerical simulations of human–machine interaction, different procedures and interfaces can be designed, compared, and tested for a large variety of initial and transitory conditions generated by plant malfunctions, emergencies, and normal operations. The study of procedures and interfaces, for diverse human behaviours, is a typical application of prospective HMI methods.

Training

Training human factors insight has nowadays become common practice for highly specialised operators, such as nuclear power plant operators, pilots, air traffic controllers, etc. This type of training is performed in addition to, and is complementary to, the more classical training of technical skill and plant control performance. Therefore, it is usually called human factors, or "nontechnical," training, so as to distinguish it from the more classical formal training of ability to manage and control the plant from the physical and technical viewpoint.

In some cases, such as in civil and military aviation, regular and recurrent training procedures in human factors are already formalised by regulatory bodies and authorities and are integral part of the overall curriculum of expertise development. Two specific types of human factors training are considered: *classroom* and *simulator*. Classroom (human factors) training consists in introducing the concepts of human behaviour, human–human and human–machine interaction in very specialised discussions and lectures, as part of the standard and recurrent training (Wiener et al., 1993).

Simulator (human factors) training is carried out during practical, hands-on, sessions at a "full-scale replica" simulator. Operators are trained in these sessions with the objective to develop their "technical" skill in controlling and supervising the machine during abnormal conditions, but also to manage critical situations and exploit human competence and potentialities at their best, especially when working as a team.

In both these cases, i.e., classroom or simulator training, the instructor or facilitator must master different paradigms of human behaviour in order to be able to describe, review, and characterise different human performances.

Safety Assessment

A Safety Assessment study can be performed from three quite different perspectives: *Design Basis Accident (DBA)* analysis, *Quantitative Risk Assessment (QRA)*, also called Probabilistic Safety Assessment (PSA) or Probabilistic Risk Assessment (PRA), and *Recurrent Safety Audit (RSA)*.

Design Basis Accident

Design basis accident analyses consist in safety studies of specific accidents. The boundary and initial conditions are prespecified by the designer and are believed to represent the set of worse possible accidental scenarios, which encompass all other conceivable accident configurations.

Safety measures and protection devices are designed and dimensioned on the basis of the results of DBA studies, which imply the evaluation of all engineered safety devices, standards, procedures, and training, including human interactions and plant performances, in such worse possible conditions.

Quantitative Risk Assessment

Quantitative risk assessment methods evaluate the frequencies of occurrence, or probabilities, associated with certain accidents, in relation to predefined selections of initiating events. As basic methodologies for systematic safety analysis, they combine classes of erroneous behaviour and system reliability data, i.e., failure rates, in structured representations of events (Hollnagel, 1993; Parry, 1994; Cacciabue, 1997).

Quantitative risk assessment studies are essentially prospective types of analysis, as the analysts define hypothetical initiating events and failure/error probabilities and calculate the frequencies of a spectrum of consequences derived from different paths of accidents. In particular, the probabilities associated with human erroneous actions are needed to perform the evaluation of human interaction processes and to quantify success/failure of a performance of certain tasks or procedures.

The final objective of a QRA is the quantification of the risk associated with certain events and the evaluation of whether such risk falls within the limits set by regulations and standards. When this does not occur, new or more reliable safety measures must be considered in order to contain further the risk and improve safety.

Recurrent Safety Audits

The constitutive elements of complex technologies have been identified in the presence and interconnection of organisational and cultural traits; working conditions; defences, barriers; and safeguards; and personal and external factors. The assessment of the safety level throughout a system and an organisation requires that these factors be evaluated at periodic intervals, i.e., recursively, in order to examine the state and possible evolution of the system/organisation towards different levels of safety/risk conditions.

These types of evaluations focus on data, critical system functions, and specific human–machine characteristics that require particular attention and need to be evaluated in relation to each specific system/organisation.

Recurrent Safety Audits (RSA) of organisations attempt to evaluate the safety state (level) of an organisation with respect to a variety of safety indicators and markers associated with the current state of the constitutive elements of a system, i.e., organisational and cultural traits, working conditions, personal and external factors, and, above all, defences, barriers, and safeguards.

The recurrent assessment of the safety level of an organisation requires a methodological framework where different methods and approaches are combined and integrated for considering HMI. RSA of organisations are critical and essential key processes for preserving systems integrity and for preventing and protecting from accidents.

The absence or a poor practice in the application of RSA has been recognised as a major deficiency in organisations that have experienced serious accidents, which, in some cases, have led to lethal consequences for an entire technological domain (Cacciabue et al., 2000).

Accident Investigation

Accident investigations are oriented to the identification of the root causes of an accident, either related to human errors and mishaps and/or to system failures and malfunctions. From the human factors viewpoint, a method for accident analysis requires a methodological framework that comprises models of cognitive processes and organisations. These models lead to classification schemes, or taxonomies, that allow the categorisation of observed behaviours (ICAO, 1987, 1993; Hollnagel, 1991, 1998).

Accident studies and investigations are reactive types of study, as they usually point towards the definition of preventive measures against future events of the same type. Sometimes, however, accident investigations stop at the identification of root causes and correlations between causes–effects–consequences within the human–machine system. In this case, the reactive approach is limited at the level of retrospective analysis with no proposition for system improvements and feedback modifications.

It is important to distinguish between "accident investigation," or "accident analysis," and "root cause analysis (RCA) of events." The former represents a much wider type of study that embraces the study of previous events that have occurred within an organisation, and the assessment of all root causes of the accident under examination, both human and system related. On the other hand, "root cause analysis" implies a more focused technique on evaluation of causes and effects of a specific event that occurred. Consequently, "root cause analysis" is only one of the elements that constitute an "accident investigation."

In the case of human factors, the RCA of an event involves the evaluation of reasons and causes associated with a single inappropriate performance or error. On the other hand, the contribution of human performance during an accident demands the consideration of many events and interactions (positive as well as negative), which contribute to the dynamic evolution of the accident. This distinction is important and will be discussed in much more detail in later sections of this book.

2.5 Integration of Prospective and Retrospective Analyses: The HERMES Methodology

The need to integrate and ensure consistency between prospective and retrospective studies has been discussed earlier in this chapter and represents a fundamental standpoint for developing effective HMS and HEAM measures. Similarly, the approaches and methods for prospective and retrospective studies must be integrated during different stages of design, assessment, training, and accident investigation of a HMS.

The need to correlate prospective and retrospective studies in a logical analytical process that can support the consideration of sound HMI approaches in different areas of application, has led to the development of a methodology that respects all requirements and basic conditions for their integration and mutual correlation.

This methodology is called Human Error Risk Management for Engineering Systems (HERMES).

HERMES is structured in a number of steps that may be applied in order to follow and preserve the basic requirements of congruence and consistency between retrospective and prospective studies, as well as to underpin the correspondence between recurrent HMI analyses and system safety and integrity, which changes during the lifetime of a system (Figure 2.6).

2.5.1 Human Error Risk Management for Engineering Systems

As already discussed, both types of retrospective and prospective analyses rest on a common empirical and theoretical platform: the evaluation of the socio-technical context, and the theoretical stand with respect to modelling human–machine interaction.

The evaluation of socio-technical context represents an essential condition that leads to the definition of data and parameters for prospective studies, and supports the analyst in identifying the conditions that favour certain behaviours, which may foster accidents.

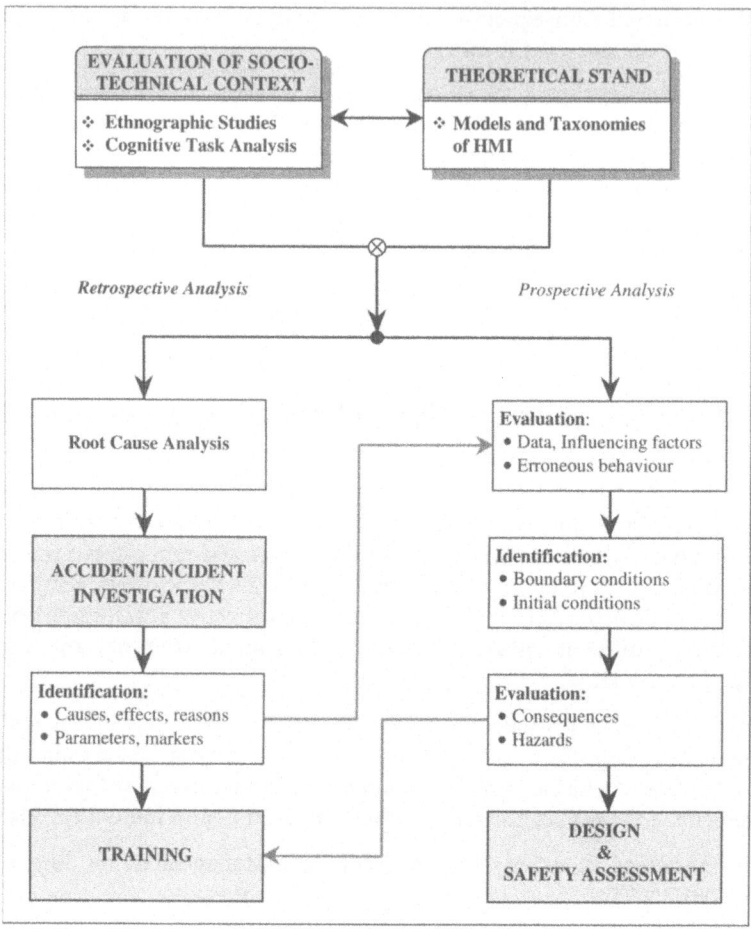

Figure 2.6 Methodology for human error risk management for engineering systems (HERMES).

The study of socio-technical contexts is performed by theoretical evaluation of work processes, i.e., "Cognitive Task Analysis" (CTA), and by field studies, such as interviews with operators, observations of real work processes, application of questionnaires, and analysis of simulator sessions. These represent a set of empirical methods that can be classified as "Ethnographic Studies" (ES).

The selection of a joint cognitive model of HMI and related taxonomy is equally important, as they define the correlation between humans and machines that are considered in order to structure formally the HMS and HMI in prospective studies. At the same time, in retrospective studies, taxonomies are essential in identifying items of HMI that affect incidental conditions. These two forms of assessment of HMS are correlated by the fact that all empirical studies and ethnographic evaluations can be implemented in prospective and retrospective applications only if they can feed their observations in a formal structure offered by the HMI model and taxonomy.

These initial common elements of the methodology, ensures coherency in performing either an accident investigation, a safety study, a design, or a training course for a certain system or organisation. Moreover, the performance of prospective and retrospective studies requires further correlation and exchange of data between the two types of analysis.

These commonalities and correlated steps can be discussed as follows:

- The investigation on past events and accidents requires that they are described according to the temporal sequence of events. With reference to each event, it is then necessary to identify human behaviours and/or organisational factors that may be considered inappropriate for the circumstances or systemic failures. Root Cause Analysis (RCA) governs this process.

- The combination of series of RCA leads to the identification of causes of accidents, as well as data and parameters that combine to generate the overall accident investigation or accident analysis.

- From the accident analysis it is possible to derive valuable information applicable for correlated prospective studies. In particular, it is possible to derive:

  —causes, effects and reasons of errors; and

  —parameters, indicators, and markers of erroneous behaviours.

- These data and parameters, derived from retrospective analysis, are the basis for the evaluation, in a prospective analysis, of:

  —data and factors influencing performance; and, in general,

  —possible forms of erroneous behaviour.

- These generic types and forms of behaviour are then further elaborated, in a perspective analysis by experience, expertise, and creativity of the analyst, in order to *identify* specific:

  —boundary and initiating conditions.

- Using these data, parameters, boundary and initial conditions in combination with the selected human behaviour model and error taxonomy, it is possible to apply risk methods to evaluate safety margins and outcomes of potential accidental scenarios. These represent the consequences and hazards associated to certain inappropriate human behaviours and systemic failures.

- The outcome of these perspectives analyses can then be further utilised for generating possible accidental scenarios useful for training purposes.

- In this way, coherence between retrospective and prospective analyses is preserved. Moreover, the synergism and correlation that exist between them may be adequately exploited for:

  —performing design evaluations and safety assessments, in order to develop, maintain, audit, and ensure safety standards of an organisation throughout its entire lifetime; and for

  —defining contents and objectives of nontechnical training of operators, permanently linked to the evolution of a system and organisation.

In the following section, we will show how the HERMES methodology enables, in a less formalised structure, to associate goals and methods for prospective and retrospective types of studies with the areas of application of human error and accident management measures.

2.5.2 Analyses and Areas of Application

This section considers the general connections existing between areas of application and types of analysis.

It has been argued that the distinction between prospective and retrospective analyses is only apparent, as they both are important and synergetic representations, in their own way, of a human–machine system. They are indeed the two sides of the same coin, as they must be coherently and consistently applied for obtaining sound results in terms of safety.

What matters is that the theoretical models applied for prospective and retrospective analysis are identical, or at least they are based on the same paradigm of joint cognitive HMS and HMI, so as to ensure complete correspondence and maximum feedback from one type of analysis to the other. Moreover, it is important that realistic and consolidated bodies of data and parameters can be drawn from retrospective analyses that can be consistently applied for prospective studies. This serves the purpose of granting reliability and coherence of results of prospective studies.

However, when different areas of application are considered, it turns out that different types of analysis are better suited than others to satisfy the requirements of each specific area (Figure 2.7). In particular, a number of considerations can be made in respect of each type of area.

Design

Design methods tackle all basic objectives of safety systems, namely: prevention of errors and accidents, recovery from malfunctions and safety critical conditions, and protection from hazards to humans and environment due to an accident.

Design methods are always applied in a prospective oriented view. Designs of control, emergency and protection systems, as well as interfaces and procedures that govern human interaction with systems, are always performed by estimating possible scenarios of application.

However, these are not developed in isolation and need reliable and consolidated data obtained from past experience and engineering knowledge.

This is the "normal" correlation that exists between prospective and retrospective analyses.

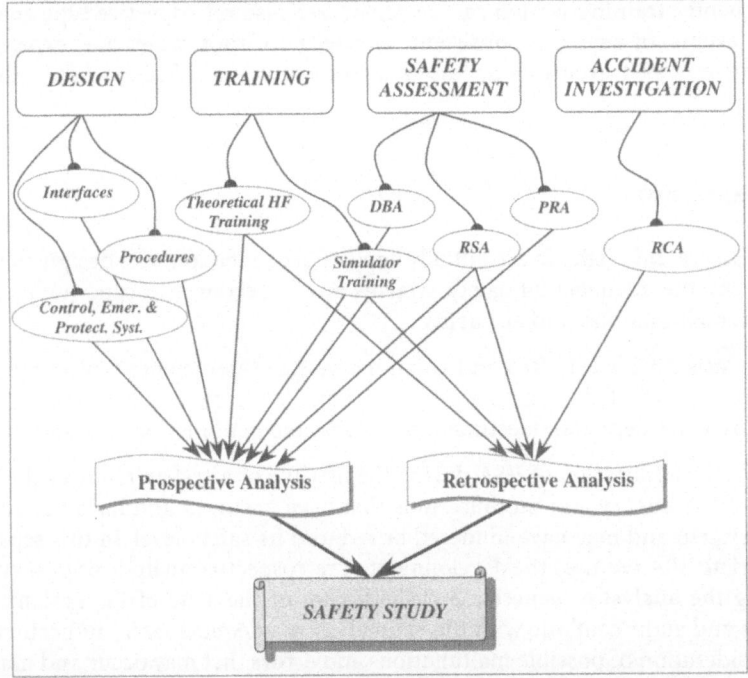

Figure 2.7 Types of applications, types of analysis, and types of simulation of human–machine systems.

Training

Training approaches aim at generating a safety culture and practices that are essential for accident prevention and management by improving and developing relevant technical and nontechnical skills in most technological environments.

This is particularly true in the aviation domain, where the very high reliability of components and the complexity of control tasks make the occurrence of malfunctions and faults very rare. This makes extremely difficult the prevention and management of emergency situations.

Moreover, the human factors contribution to accidents has been shown to be extremely relevant. Consequently, it becomes important to train operators to recognise and anticipate HF problems, and to deal with them as soon as they appear.

This is done in real simulators as well in classroom training sessions, making predictions and estimations of possible anomalous and emergency circumstances and occurrences. These are typical prospective estimates of scenarios.

At the same time, past accidents serve as a reference for the selection of possible training sessions. In such cases, it is possible to evaluate generic behaviours and attitudes related to distributed and shared organisational and national cultural factors in circumstances that have already been encountered in the past.

Consequently, training session can be considered also retrospective type analyses of past events. In any case, consistent knowledge of root causes and experience, derived from past events, is an essential contributor to effective non-technical training.

Safety Assessment

Safety assessment methods are initially applied in connection with design, in order to evaluate the adequacy of safety systems and procedures to cope with abnormalities, malfunctions, and accidents.

Design Basis Accident (DBA) and Quantitative Risk Assessment (QRA) methods are directly connected to the design process and, consequently, they are prospective approaches demanding estimations of possible malfunctions and errors.

However, the application of RSA during the lifetime of a system requires the evaluation of the history and modifications that have occurred and have been made on the system and may have hindered or reduced its safety level. In this sense, an appropriate RSA requires the development of retrospective analysis of past events, enabling the analyst to generate a clear picture of the state of the system. This analysis and audit combine with the estimation of adequate safety indicators and the consideration of possible malfunctions and errors that may occur and must be managed for the overall evaluation of the safety level of an organisation. This is why RSA can be considered a retrospective as well as prospective type of analysis.

In all cases, DBA, QRA, and RSA are effective only if supported by well-correlated prospective and retrospective analyses and by coherent data derived from studies of past event and occurrences. Once again, this is the essential requisite of coherent and sound safety studies.

Accident Investigation

This area of application can be considered as the prototype of the retrospective type of analysis.

Indeed, the objective of all accident investigation approaches is dedicated to the identification of root causes and reasons of accidents. In this sense, accident analysis methods offer a substantial contribution to defining the set of data and parameters that can be derived from past experience and field observation.

As already discussed, the performance of an accident investigation demands that the analyst acquires valuable and complete information of the socio-technical environment in which the accident developed. This offers the possibility to analyse human erroneous behaviours as consequences of other reasons deeply rooted in the organisation.

Moreover, even though accident investigations are substantially retrospective-type studies, their outcomes are used and exploited in prospective application for accident prevention and containment. Therefore, it is necessary that theoretical grounds and techniques, on which accident investigations are based, are coherent and consistent with the methods and techniques applied for prospective analyses. This is the "normal" correlation that exist between prospective and retrospective studies for ensuring consistency, seen this time from the side of retrospective analysis.

2.6 Ecology of Human–Machine Systems

The need to develop methodological frameworks for considering human–machine systems and human interactions in a consistent and coherent fashion with their working context has already been well recognised in the past.

The literature of the last decade is rich in approaches and methods that focus on this subject. An excellent comprehensive and short review of methodological approaches and theoretical construct has been given by Moray (1997), covering the last 30 years of research and development in the field of human factors applied to different domains and industrial settings.

In particular, the concept of ecology of human–machine systems has become of great relevance in human factors, as it embeds in a single expression a wide variety of contributors and influences on human behaviour, mostly related the role of situation and context.

In general, ecology concerns the dependence that exists between the different actors, human or animal, and the environmental constituents and peculiarities in which they live. This creates interdependence, primarily at natural level, that leads to establishing a dynamic equilibrium and explains the evolution of life and is essential for understanding, analysing, and designing artefacts that are to be utilised by human being to operate in the world.

Nowadays, looking more closely at the working contexts of modern technology and industrial systems, one can consider different *forms of ecology* (Rouse and Sage, 1999):

- *Information ecology*, which involves the context and impacts of "information technology" on people and organisation. This is primarily the role of computer and automation on everyday life.
- *Knowledge ecology*, which considers the way in which contingency and practical experience affect adaptation of information and normative systems to real-world application.
- *Industrial ecology*, which is the effective system engineering and management of industrial processes aimed at developing sustainable products and services. This requires adaptation of personal attitudes and cultures to higher demanding organisational goals and philosophies.
- *System ecology*, which affects planning and defining systems requirements and specifications in adequate considerations for human–machine interactions as a whole.

Designing or studying a human–machine systems is therefore recognised as an endeavour that covers many different domains and must be carried out in consideration of "ecological" aspects involving the interplay and interaction of individuals with other human beings and socio-technical working contexts.

2.6.1 Ecological Approaches to Cognitive Ergonomics

In cognitive ergonomics, a vast movement of research and development has been based on the ecological approach to psychology, derived from the original work of Brunswik (1956) and further developed by Gibson (1966, 1979).

The essence and global perspective of the ecology of HMS are contained in two milestone reference books for human factors analysts that describe the theoretical "global perspectives" (Flach et al., 1995) and show practical implementation of ecological approaches, in different industrial contexts and applications (Hancock et al., 1995).

In particular, the concept of *affordance* strongly affects these ecological approaches to human machine system (Gibson, 1979). *Affordance* implies that mutuality exist between the environment (object, substances, etc.) and the individual. *Affordances* are material properties of the environment that support and limit the potential activity and intentions of the individual. Thus they are measurable quantities, but can be considered only in relation to the individual.

In a modern socio-technical perspective, the concept of *affordances* needs to cover also more immaterial properties of the "work environment" that include cultural and social relations affecting human behaviour. These have to be associated with environmental properties that can be "measured" by identifying material indicators that give a quantifiable size of such immaterial dimensions, which are extremely relevant in bounding and characterising human activity (Zaff, 1995).

Another relevant concept and guiding principle pertaining to ecological approaches is the requirement that a "good psychological theory is an essential aid to design" (Kirlik, 1995), as it represent the reference notion and paradigm for human factors that are equivalent to the basic conservation principles for engineering design. An approach based on cognitive psychology that is capable of predicting environmentally situated behaviour is essential. This is particularly true in a "macroscopic" perspective that aims at enabling the representation of actual human behaviour in a working context, rather than focusing on the ("microscopic") description of the neural processes and personality aspects that develop in a human brain (Cacciabue and Hollnagel, 1995).

This perspective was already considered by Brunswik (1952) in the lens model (Figure 2.8), where the interplay of environmental structure and cognitive properties is critical for describing behaviour. In particular, in the lens model, there exist in our society and working contexts certain sets of *cues* (X_i) which bear specific relations and may take different values of ecological validity ($r_{e,i}$) with respect to the environment. The utilisation of such *cues* by the organism ($r_{s,i}$) depends then

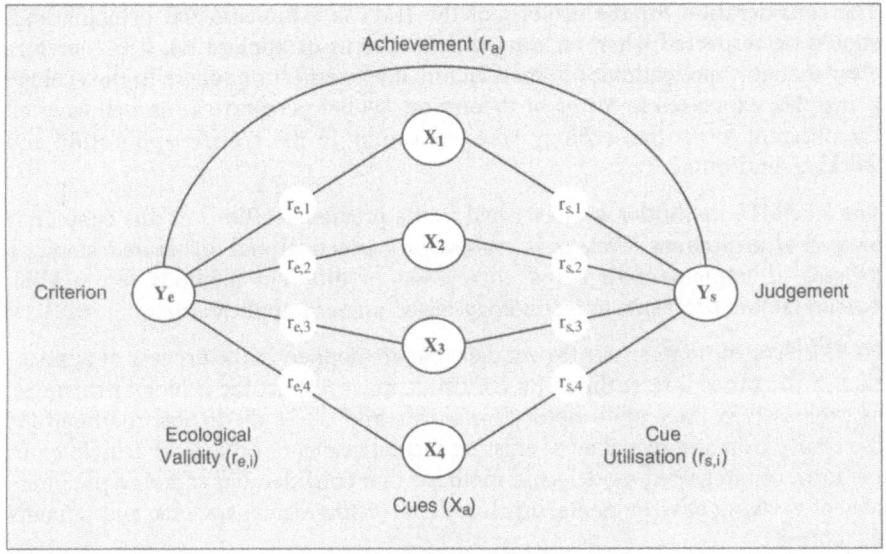

Figure 2.8 The lens model of Brunswik.

on the cognitive counterpart of the environment and leads to the actual implementation of performances in terms of achievements.

As in the case of "affordances," the symmetry identified by Brunswik between perception/judgement and action with respect to cognition and environment, expands, in modern research, to perception and learning of multiple cues of an environment affected by policies and organisational cultures, and by socio-technical aspects such as risk perception, collaborative or cooperative teamwork, communications, and interpersonal relations.

From a general perspective, all ecological approaches to cognitive ergonomics share the fundamental requirements associated with the need to integrate human activities, at the mental and physical level, with the socio-technical environment in which they are embedded. Consequently, the general label of "ecological approach" can be assigned to any HMS method that recognises such fundamental need in modern technology.

Several methods have been developed over the years in this frame, and some of them will be briefly revised in the next chapter that deals with specific methods and models for HMS.

2.6.2 Ecological Systems in HERMES Perspective

The large family of ecological approaches to human–machine systems represents a well-founded and conceptually sound formalism to approach to human factor issues in modern technological environments.

The consideration for the ecology of the HMS is a fundamental principle that should be respected when tackling different areas of application. It is therefore clear that any application of human factors must respect or adhere to the ecology principles, expressed in terms of theoretical "global perspective," as well as to all the different *forms* that ecology takes, according to the specific application and working environment.

The HERMES methodology, described in the previous sections of this chapter, is located at a different level, as it represents a practical and structured stepwise process of implementation and correlation of different methods for tackling human factors problems in technology-based areas of application.

HERMES, or more precisely the models that are applied in the process of application of the procedure outlined by HERMES, must respect the ecology principles, as expressed in their philosophical premises, and offers the logical roadmap for the connection and interplay of existing instantiations of ecological principles, in the form of integrated models and methods that consider and combine the interplay of working environments, organisations, technological systems, and humans in control.

The implementation of a design or safety analysis, as much as the study of an accident or the development of a training programme, for a human–machine system requires the combination of many different and specialised methods and approaches, with precise characteristics. The ecology of the human–machine system is one of them, and, although fundamental for its content of correlations between all socio-technical aspects of working contexts, it is not the only principle that human factors analysts must consider in the process of implementation and integration of methods.

A methodological framework like HERMES aims solely at clearing the way for the analyst in the process of stepwise and logical application of methods, models, and approaches for solving the problem at hand.

2.7 Summary of Chapter 2

In this chapter, a number of basic definitions and standpoints for performing Human Error and Accident Management (HEAM) have been considered. These have been developed starting from the consideration that any Human–Machine System (HMS) and Human–Machine Interaction (HMI) play a fundamental role in the process of design and assessment of any technological system, and that they involve the working context and socio-technical dynamic conditions, in addition to the direct interplay of the human operator with the plant under control and management.

The concepts discussed in this chapter rotate around *five standpoints* that represent the axioms and foundation on which any HMS and HEAM measure should be based (Figure 2.9).

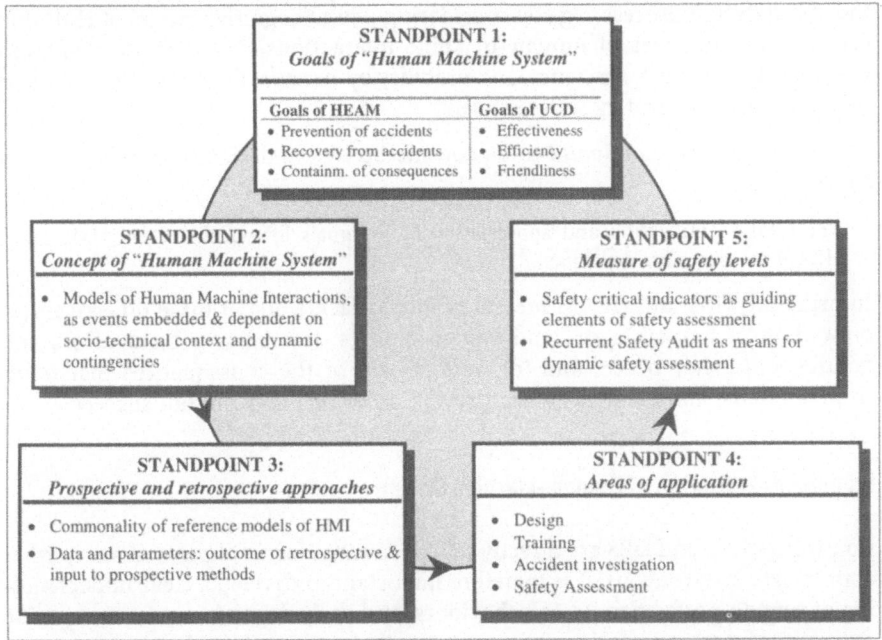

Figure 2.9 Standpoints for the development and assessment of defences, barriers, and safeguards for HEAM.

A methodology that guides the user and safety analyst in the application of Human Factors methods and respects these five standpoints has been developed. This methodology is named HERMES (Human Error Risk Management for Engineering Systems).

The five standpoints and HERMES methodology will now be briefly summarised.

Standpoint 1: Goals of Human–Machine System

The improvement of safety in any technological system demands that appropriate measures are developed aiming at creating awareness, warning, protection, recovery, containment, and escape from hazards and accidents.

These are usually defined as DBS and represent all structures and components, i.e., physical, human, or social that are designed, programmed, and inserted in the human–machine system with the objective of making the management of a plant more efficient and safe, in normal and emergency conditions. DBS must be developed in consideration of the overall design of a system.

Moreover, in the case of the design of HMS, three fundamental principles of modern technological systems must be accounted for, namely: *Supervisory Control, User-Centred Design (UCD)*, and *Systems' Usability*. These three principles must combine in such a way to enable the designer to include the user role already from the initial design process, e.g., by considering a joint cognitive model of HMI that accounts for all types of human–machine interactions, and then to verify and improve effectiveness, efficiency, and usability by iterative feedback processes and field tests with adequately selected users.

Therefore, the *first standpoint* to be considered by the designer or safety analyst consists in:

> The clear identification and appreciation of the goals for which certain HMS or HEAM measures are developed.

In principle, DBS and HEAM measures should tackle one of three objectives: (a) *prevention* of human errors and system failures; (b) *recovery* from errors and failures, once they occur; and (c) *containment* of the consequences that result from accidents and incidents when prevention or recovery did not succeed.

Standpoint 2: Concept of Human–Machine System

Control systems and DBS are directly related to some forms of performance, either appropriate or erroneous. It is therefore important to develop a clear understanding of human performances or behaviours, and their dependence on the specific dynamic context or contingencies and on the socio-technical environment in which they are imbedded.

Consequently, it is important to develop or consider a model, which allows the simultaneous, or joint, representation of the interactions of humans and machines. This includes also a general concept of what is meant by "human error."

From a designer viewpoint, a strict definition of "human error" is not necessary, and may be bounding or limiting the focus of DBS. What is essential instead is that DBS enable to tackle inappropriate performance/behaviour, in relation to the context and dynamic contingencies, and the specific socio-technical environment in which they are imbedded. In this perspective, the consideration for "human errors" expands the more classical definitions and embraces all behaviours that may engender dangerous configurations of a plant.

Consequently, the *second standpoint* in the development of effective HMS and HEAM measures demands that:

> Adequate models of human–machine systems and interactions must be applied for simulating dynamic interplays of humans and machines, as events embedded and dependent on the socio-technical contexts in which they are generated and evolve.

Standpoint 3: Prospective and Retrospective Approaches

The variety of models and methods that are necessary for the development of DBS and HEAM measures can be structured in an integrated framework that considers two types of analysis, i.e., retrospective and prospective studies. These are complementary to each other and equally contribute to the development and safety assessment of HMS and HEAM measures.

Retrospective analyses are oriented to the identification of "data and parameters," and are built on structured studies that combine RCA, observation and evaluation of working contexts (ES), CTA, and theories and models of HMI. Prospective analyses aim at the "evaluation of consequences" of HMI scenarios, given selected spectrum of: "initiating events and boundary conditions," appropriate "data and parameters," predictive models of HMI, and "creative thinking."

In practice, these analyses rest on a common empirical and theoretical platform: the evaluation of socio-technical context, and the model of HMI and related taxonomies.

The consideration and clear understanding of the differences and synergies between prospective and retrospective analyses is of fundamental importance in the development of any HMS and HEAM measure, in particular all defences, barriers, and safeguards that rest on human intervention and control.

Consequently, the *third standpoint* in the development of effective HMS and HEAM measures can be defined as:

> HMI models and theories, as well as data and parameters, derived from evaluation of real events and working environment (retrospective studies) must be consistently and effectively applied for predicting consequences and evaluating effectiveness of safety measures (prospective studies).

Standpoint 4: Areas of Application

Only applying specific methods at different stages of development and management of a system, it is possible to ensure efficiency, effectiveness, and user friendliness of HMS and DBS and preservation of adequate safety levels throughout the lifetime of a plant.

In particular, four areas of application must be considered, namely: *design, training, safety assessment*, and *accident investigation*. The *design* of human–machine interactions implies implementing basic principles of human-centred automation in the design process, as discussed in Standpoint 1. *Training*, and more specifically nontechnical training, intends to increase the ability of operators to manage safety critical situations, and to capture and notice those factors and indicators of the context that favour the occurrence of errors or mismatches between human situational awareness and system performance. *Safety assessment* of plants and organisations represents a basic requirement and the most complete method by which prevention and control of possible accidents can be performed. *Accident/incident investigation* aims at identifying systemic and socio-technical root causes that generate accidents.

Each of these four areas of application encompasses specific types of assessments and analyses.

The fourth standpoint for designers and analysts of HMS and HEAM measures is correlated to this issue and can be defined as follows:

The development of effective HMS and HEAM measures demands that a variety of tools and approaches are applied for the continuous verification that adequate safety conditions exist and are maintained before and during the lifetime of a system, at the stages of design, training, safety assessment, and accident investigation.

Standpoint 5: Measure of Safety Levels

In order to complete the process of appreciation and generation of measures to improving and safeguarding the safety of a system, a final standpoint is necessary. This is related to the definition of appropriate safety levels of a plant and ways to regularly assess them.

Indeed, in all types of analyses (retrospective and prospective), and for all areas of application (design, training, safety assessment, and accident investigation), it is essential that adequate *indicators*, *markers*, and parameters are identified that allow the estimation or measurement of the safety level of a system.

As each plant and organisation bears specific peculiarities and characteristics related to their context and socio-technical environment, appropriate methods and approaches must be applied for the definition of numerical, as well as qualitative, indicators, which are unique to the plant and organisation under scrutiny.

Moreover, the assessment of acceptable safety levels and standards cannot be limited to the design or plant implementation stage. It is essential that recurrent assessments are performed, in order to account for aging, technical updates and modification, improvements due to accidents or incidents, or simply ameliorations derived from implementing a different technology.

The continuous check and verification that a plant respects safety standards, by carrying out audits and evaluation of safety indicators, is of paramount importance in ensuring that hazards for plants, humans and environments are limited and contained within acceptable boundaries.

The *fifth standpoint* that sustain the activity of analysts and designers of HMS and HEAM measures refers to safety audits and is defined as follows:

> The continuous measurement and assessment of safety levels of HMSs is essential for ensuring minimum risk and effective performance of a plant. This process requires the identification of safety critical indicators, as guiding elements of safety assessment, and the performance of recurrent safety audits throughout the whole socio-technical system and organisation.

This last standpoint completes the generic framework of different topics that play a role in developing safety measures for a system. The analyst and designers should select within this framework the features and the most suitable methods and techniques that are of interest for the system under study.

In any case, all five standpoints discussed here need consideration, before developing, or implementing and assessing specific safety measures.

Human Error Risk Management for Engineering Systems

A methodology that offers a roadmap for safety analysts in respecting the five standpoints for applying methods and techniques for HF analyses has been discussed.

This methodology is called HERMES and demands that a series of field studies is performed in association with models and taxonomies of HMI in order to achieve the necessary knowledge of the system under study as well as to develop a consolidated database of information concerning the whole socio-technical working environment that can support predictive assessment of safety.

This methodology will be applied in all test cases and applications shown in the forthcoming Chapters 4–8.

3

Theories and Methods for Human–Machine Systems

3.1 Introduction

In this chapter the fundamental components of prospective and retrospective analysis are discussed. The objective is to offer a spectrum of possible approaches among which to select the most suitable for application to the case at hand.

Not all methods and techniques existing in literature are presented, as the objective of this chapter is not to be exhaustive, but rather to present well-known methods of reference and paradigms, as well as to discuss some innovative ideas that may be useful in following a logical and coherent structure, such as the one proposed in the HERMES methodology.

Moreover, given that prospective and retrospective studies must be based on the same models and theories of HMI, in order to ensure consistency of analysis, the chapter starts with reviewing and comparing the most common and well-known model and taxonomies of HMI that may be applied as foundation of any assessment and study of a Human–Machine System (HMS).

Then, the specific components of retrospective studies will be considered in sequence, namely Ethnographic Studies (ES), Cognitive Task Analysis (CTA), and Root Cause Analysis (RCA). For each type of method, specific approaches and techniques have been developed and some of these will be described and reviewed. A comparison will be performed on well-known and widely applied techniques and a specific section will be dedicated to the issue of ecological validity and ecology-based approaches.

Focusing on prospective analysis, there exist many standard and basic techniques, especially for design and risk assessment purposes, such as numerical approaches and algorithms, fault trees, and event trees. The analysis of these typical techniques is outside the scope of the present book, as the literature is full of well-developed and recognised monographs and textbooks on these subjects. The review of some modern techniques will be preformed instead, centred on the inclusion of human factors in quantitative risk assessment and recurrent safety audit.

In this way it is hoped that the reader will be able to develop the basic knowledge of existing theories and methods for studying HMSs.

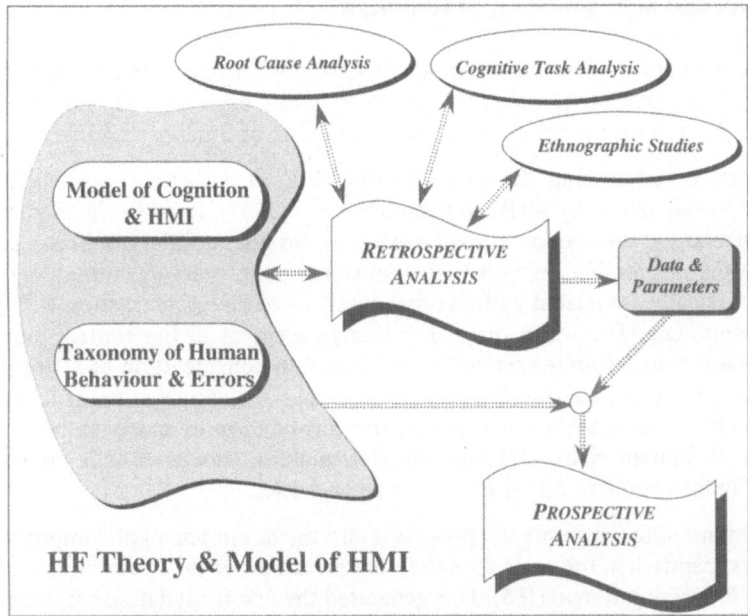

Figure 3.1 Models and taxonomies of human behaviour for retrospective and prospective analysis.

3.2 Models and Taxonomies of HMI

As discussed in the preceding chapter, the identification and selection of a model of human–machine interaction (HMI) and associated taxonomy, for classifying human behaviour and errors, are essential theoretical standpoints for structuring an HMS and developing consistent safety studies.

On the basis of taxonomy, the empirical studies and field observations, typical of ethnographic studies, as well as root cause analysis and task analysis, are developed and structured (Figure 3.1). According to the conceptual discussion relative to standpoint 3 (§2.4.1) for performing sound HF studies, the models and taxonomies are essential also for the performance of predictive (prospective) studies.

The ability and expertise of the human factor analyst resides also on the preliminary selection of the model and taxonomy to be applied as guides for the rest of the study. Making the wrong selection at the beginning may turn out to be very expensive and result in having to repeat the whole HF study. This would involve restarting and iterating several times between the selection of a model/taxonomy, the ESs and field assessment to ensure that the model adequately represents the working context and HMI.

In this section, a number of models and associated taxonomies of human behaviour will be reviewed so as to offer to the analyst a spectrum of possible alternatives that may be considered when tackling a specific HMI and HEAM problem in the framework of a methodology like HERMES.

3.2.1 Models and Paradigms of Cognition

In the context of modern HMSs, modelling human behaviour entails considering, primarily, cognitive processes and system dynamic evolution, i.e., mental activities, resulting from the time-dependent interaction of humans and machines.

The need of simulating the man-plus-the-controlled-element was enunciated already in the sixties by McRuer and colleagues (1965). Prior to those years, the most substantial theories on human behaviour did not consider predictive models outside the observable domain (behaviourism), as all relevant human activities were practically associated with manual control and direct interaction with physical phenomena. The inclusion of the human element in the control loop of a process was thus a simple exercise of mathematical consideration of action delays and parameters estimation. However, the progress of technology towards supervisory control and automation required the formulation of much more complex models of human reasoning and decision-making processes, able to account primarily for cognitive rather than manual activities.

The demand to focus on mental processes and the development of computer technologies inspired, in the early seventies, the metaphor of the operator as an *Information Processing System* (*IPS*). This generated the first formulations of theoretical models of cognition (Neisser, 1967; Newell and Simon, 1972), and, since then, a variety of paradigms of human behaviour have been developed (Rouse, 1980; Stassen et al., 1990; Sheridan, 1992). In the following, a number of such models that can be utilised for representing human behaviour will be described in some detail. They are only a small number of the vast variety of models and theoretical representations developed in the literature, but they can be considered stereotypes of such models as they cover different levels of complexity and depth in representing mental processes and cognitive functions as well as behavioural performances. These models are:

- *reference model of cognition* (*RMC*) (Wickens, 1984);
- *step ladder/skill, rule, knowledge* model (*SL/SRK*) (Rasmussen, 1986);
- *model of fallible machine* (Reason, 1990); and
- *contextual control model* (*COCOM*) (Hollnagel, 1993).

In addition to these models based on the IPS paradigm, another structure is widely applied and very useful to represent HMIs in socio-technical environments. It is called:

- *Software, Hardware, Environment, and Liveware* (SHEL) (Edwards, 1972).

SHEL is, in essence, a configuration of the whole socio-technical environment that characterises HMI in technological contexts, more that a model of human behaviour. However, it entails the human element as the focal point of the architecture, and, consequently, SHEL will also be reviewed and compared with the above-mentioned models.

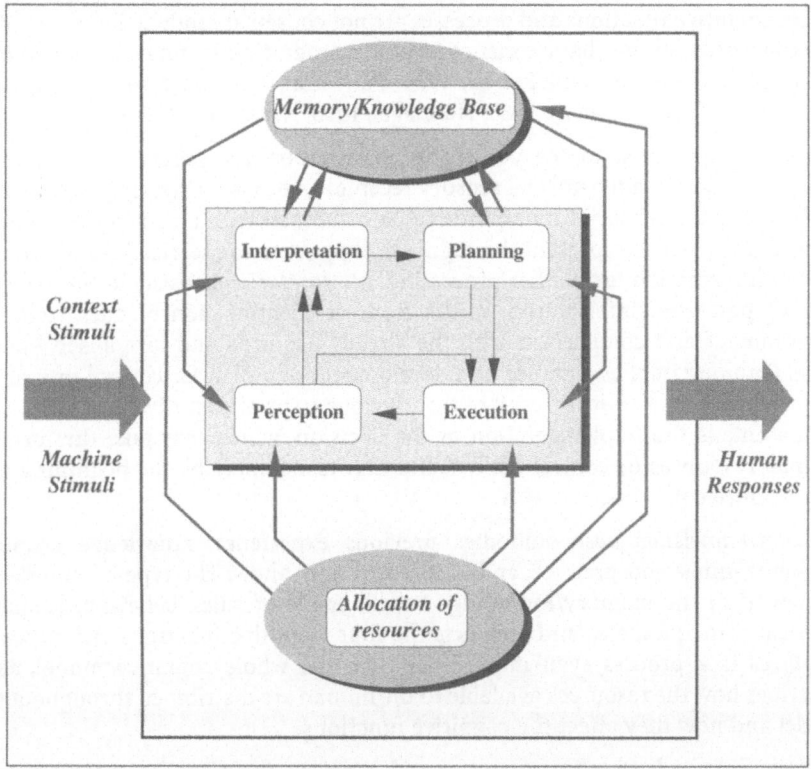

Figure 3.2 Reference model of cognition (RMC/PIPE).

Reference Model of Cognition

A *Reference Model of Cognition* (*RMC*) can be developed that contemplates all fundamental components of the paradigm human behaviour, based on the IPS metaphor (Figure 3.2).

In RMC, specific attention is devoted to stimuli and responses and to the mental operations that occur between them. *Stimuli* are generated by the plant under control or by the working environment and the context affecting human operators. *Responses* are the control or manual actions carried out in response to stimuli and/or following cognitive processes.

Four main *Cognitive Functions* are the basic elements that describe the mental and behavioural performances. These functions are: *Perception, Interpretation, Planning,* and *Execution* (PIPE). The relevance of the role of the four *Cognitive Functions* in the overall architecture has led this model to be identified also as RMC/PIPE (Cacciabue et al., 2003).

Two critical *Cognitive Processes* govern the four cognitive functions: *Memory/Knowledge Base* (KB) and *Allocation of Resources* (AoR) (Wickens, 1984; Wickens and Flach, 1988; Cacciabue, 1998).

These cognitive functions and processes are not chosen at random, but they reflect the consensus on the characteristics of human cognition that has developed over many years (Sheridan and Ferrell, 1974; Pew et al, 1977, Pew and Baron, 1983; Rasmussen, 1986; Sheridan, 1985; Hollnagel, 1993; Amalberti, 1996).

Perception is related to the content of information transferred to the human sensory activity, via the normal sensory receptors, i.e., vision, hearing, taste, touch, and smell. This sensorial function is to be combined with a contribution due to expectations, which represent the cognitive support to the search for and acquisition of information for further processing. *Interpretation* consists in the elaboration of perceived information, in the form of identification of cues from the environment and comparison with the already acquired and formalised knowledge. *Planning* implies a choice or a development of a plan for control or manual action to be carried out, as result of the previous steps of cognitive processes. *Execution* entails the implementation of the decision, which may take the form of manual responses or control actions, but may also simply be the beginning of a new cognitive process.

Memory/knowledge base embodies previous experience, knowledge acquired during training and practice, and theoretical know-how. The type of knowledge contained in the memory/knowledge base is made of rules, beliefs, procedures, physical principles, etc., and supports the four cognitive functions. *Allocation of resources* is a process even more crucial for the whole cognitive model, as it describes how the resources available to the human are distributed throughout the model and how they affect the cognitive functions.

The way in which *cognitive processes* and *cognitive functions* interact with each other and with the external world (machine and context) determines the dynamic evolution of the human model. Allocation of resources influences the way to perform any one of the other cognitive processes and functions, and generates or introduces shortcuts and shortcomings in the memory, knowledge base, and cognitive functions. Memory and knowledge base are only activated by the two cognitive functions of interpretation and planning, as these are considered "high-level" mental activities, while perception and execution are mainly related to sensorial and "low-level" mental activities. However, memory/knowledge base can feed data and information to all four cognitive functions in order to support their performances. The stimuli and responses appear as the sensory motor connection of the operator with the machine. The effects of the context and environment are considered in the cognitive processes and cognitive functions. In this way the socio-technical context can modify the whole cognitive process and can be linked to specific behaviour at all levels, from mental processes to actual performances.

The connections amongst the four cognitive functions of perception, interpretation, planning, and execution are established to maintain the cyclical nature of cognition, by which information perceived from external stimuli and knowledge are combined with reasoning over past events and anticipation of future plant responses, to produce further perceptions, reasoning, planning, etc.

The model RMC/PIPE has been utilised in the past as well as in more recent applications for simulating pilots and operators in the domain of aviation (Wickens and Flach, 1988; Cacciabue et al., 2003).

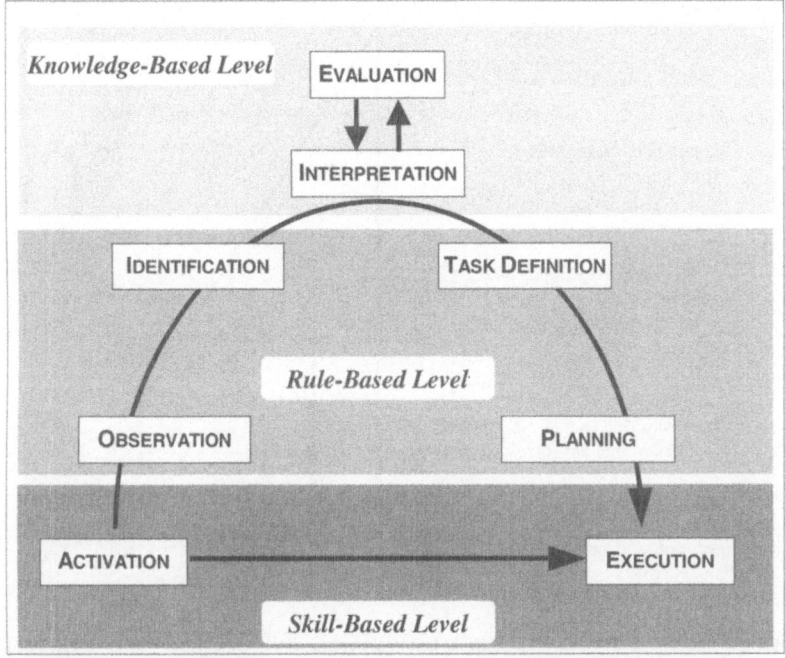

Figure 3.3 The step-ladder/skill, rule, knowledge (SL/SRK) model of decision making (from Rasmussen, 1986).

Step-Ladder, Skill–Rule–Knowledge Model

The most known and widely applied approach that implements the IPS metaphor was developed by Rasmussen (1986) in the seventies and eighties with the Step-Ladder (SL) model, which includes the notion of the three levels of cognitive behaviour, known as *Skill, Rule,* and *Knowledge Based* (SRK) (Figure 3.3).

The model indicates a set of states of knowledge in the decision-making process, and links these states of knowledge through a set of information-processing activities.

In a basic sequence, all information-processing activities of the operator defined in the model are exploited and the flow of information follows known patterns of rules and well-known procedures (*rule-based behaviour*). However, any of the activities may be abandoned by making shortcuts and/or bypassing intermediate stages, on the basis of heuristics or immediate activation of frames for action.

A typical situation of shortcut occurs at *skill-based behaviour* level, when the performance of an action is released and guided by the sensory perception of stimuli linked to well-experienced and established behavioural modes.

When an unfamiliar and unexpected situation is encountered, more sophisticated reasoning processes may be demanded, which go beyond the acquired rules and frames. These situations require the use of knowledge of systemic behaviour and causal functional reasoning (*knowledge-based behaviour*).

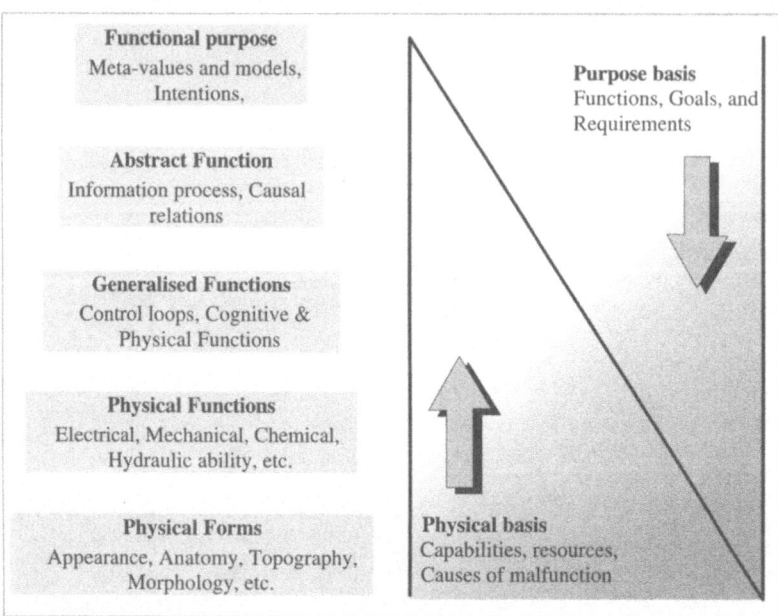

Figure 3.4 Means-ends/abstraction hierarchy for the functional properties of a technical system (from Rasmussen, 1986).

Another important feature of the model SL/SRK is the structured representation of the physical system, as it appears to the human supervisor and controller. The model considers a means-ends *abstraction hierarchy* for the identification of the functional properties of a technical system. In particular, five levels of abstraction are considered: physical form, physical functions, generalised functions, abstract functions, and functional purpose (Figure 3.4).

By means of the five levels of the abstraction hierarchy it is possible to define the type of reasoning that is carried out by the human operator when facing a decision-making process with respect to plant behaviour. In particular, according to the level of abstraction at which reasoning takes place, different plant aspects and components are accounted for in cognitive processes. In the model, moving from lower to higher levels of abstraction, general functions are added to the knowledge of the system, while physical appearances, process components, and standardised functions are progressively disregarded. The abstraction hierarchy is very important in the management of accidents, when different gaols and requirements may be conflicting, especially during early stages of the accidents.

Model of Fallible Machine

The paradigm of human behaviour called *model of a fallible machine* was developed by Reason (1990). The label reflects the two main features of this model, namely: the erroneous (fallible) behaviour of humans, which is the central theme

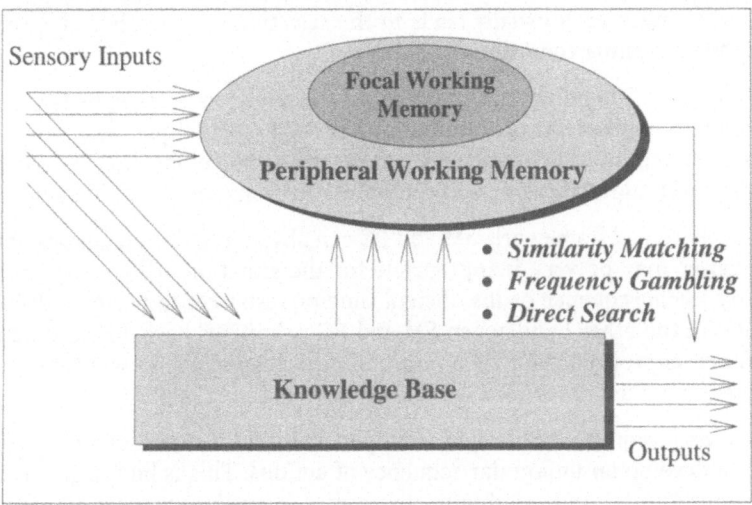

Figure 3.5 The model of a fallible machine of Reason (1990).

of Reason's research in cognitive psychology, and the idea of expressing the model in a potentially computable form, as is usual when dealing with "machines." In this sense, the theory is borrowing terminology and structures from engineering and computer sciences, such as artificial intelligence, and its application and computational implementation is expected to be more manageable.

The model of a fallible machine can be described by starting with the "machine" part, i.e., the structure and framework by which human cognition is described (Figure 3.5). The model has two principal constituents: *Knowledge Base* (*KB*) and *Working Memory* (*WM*). Working Memory is further subdivided into *Focal Working Memory* (*FWM*) and *Peripheral Working Memory* (*PWM*).

This subdivision preserves the distinction between two areas: an area of vast capacity, that is *PWM*, which receives information directly from the *KB* and the outside world and makes selection; and an area of limited capacity, that is *FWM*, which continuously receives filtered information, through the *PWM*. The filtering mechanisms which support the prioritisation of the information that reaches *FWM* are based on a number of sensory and cognitive principles, namely: "visual dominance," "change detection," "coherence principles," "activation principle," and "dedicated processors" (Reason, 1990, pp. 127–128).

The most original idea of this model lies in the retrieval mechanisms applied for comparing perceived information and KB content and for selecting between alternative solutions. These mechanisms are the "Primitives of Cognition" of *Similarity Matching* (*SM*) and *Frequency Gambling* (*FG*) and the process of *Direct Inference* or *Search* (*DI* or *DS*).

Similarity Matching is applied for searching within the KB for "knowledge frames" that better match the information perceived from the external world, or "cues" of the world. This search usually leads to the selection of a number of frames that satisfy the matching conditions of SM.

The *Frequency Gambling* primitive resolves the selection problem by favouring the candidate frame that is more familiar or has been most frequently encountered in the past. Sometimes, no matching process takes place and the FG mechanism becomes even more perverse and symptomatic of inappropriate behaviour.

In this theoretical framework, SM and FG can always provide an answer. However, this answer may be very inappropriate for the situation at hand, and might be induced as consequence of insufficient time or resources to perform deeper reasoning. On the other hand, when SM and FG primitives have difficulty to deliver a "definite" appropriate plan of action, and the situation allows accurate reasoning and planning, then the Direct Search mechanism can take place.

Direct Search requires frequent and complex iteration between KB and Memory in order to develop an unfamiliar sequence of actions. This is built utilising several knowledge "units" and combining them in a "novel" sequence, not previously exploited, and thus not yet imbedded into a well-structured "frame," that is ready to be applied in practice. In other words, these actions are already individually present in the KB in the form of knowledge "units," and they are put together to provide a new plan of action able to deal with the unfamiliar situation. This is the main difference between "frames" and "units," i.e., the latter are blocks of individual or simple sequences of actions which are combined to produce plans and complex sequences, while the former are already known plans of interventions, fully built into the KB and ready for application (Schank and Abelston, 1977).

In Reason's theory, the "Primitives of Cognition" of SM and FG, which guide human behaviour, are also the prime generators of human errors. The fallible machine theory expands in this direction, looking firstly at modes and types of human errors that can be directly observed during performance of activities (Marsden, 1987). These are the so-called "active" errors. The model then develops the issue of dynamic interactions within the organisations at different levels of decision making, with the aim of tracing and highlighting the hidden root causes of an accident. These are called "latent" errors. The focus of Reason's error theory has been towards the analysis of organisations and socio-technical contexts in addition to dynamic interactions of operators with the plants under control, during the unfolding of an accident (Reason, 1997).

In essence, the concepts and principles adopted in the fallible machine model can be applied either to study how latent errors creep through an organisation affecting human performance at the "sharp" end of the control chain (operators and controllers of plants and systems) and/or to handle situations in which changing contexts and working environments influence the behaviour of humans in direct contact of plants.

Contextual Control Model

The Contextual Control Model (COCOM) of cognition (Hollnagel, 1993) has been developed mainly as a theoretical framework for representing the behaviour of humans through two fundamental characteristics, namely: the cyclical nature of cognitive mechanisms and the inherent dependence of human actions on context and environment. The model considers human behaviour as characterised by two main elements: the *model of competence* and *the model of control*.

The *model of competence* is substantially characterised by the four basic *Functions of Cognition* and by the combined repository of knowledge base and memory retrieval process that have already been discussed for the RMC. This part of the model is considered by Hollnagel to be "very classical" and well described by literature and previous works. For this reason, the competence model has been formalised in a Very Simple Model of Cognition (VSMoC), which offers, in any case, a complete structure able to describe the most relevant functions of cognition, their interconnections, and their links with memory and knowledge base. The correspondence of the VSMoC model and the RMC is obvious and no further discussion is therefore needed. The important feature of this modelling framework is that the connections envisaged in VSMoC are able to handle a cyclic model of cognition, and, therefore, to simulate more complex and complete processes of human cognition and interaction with the machine and context.

The *model of control* operates in direct combination with the competence model and constitutes a sort of model of meta-cognition. The control model represents the most innovative feature of COCOM and contains the theoretical development and description of the mechanisms (and parameters) that govern the sequence and operability of the cognitive functions contained in the competence model.

Four different control modes are envisaged: *scrambled, opportunistic, tactical,* and *strategic*. They represent attitudes of the operator versus the use of his/her competence and determine ultimately the sequence of actions carried out. The environment strongly affects the control mode, by causing dynamic changes of control mode, and, affects performances through change over in the control mechanism (Figure 3.6). The theory (Hollnagel, 1993) enters in more detailed analysis of the parameters and factors that affect the control mode, by defining a number of main control parameters and additional dimensions of control.

In particular, two main parameters are identified, as a sort of primitives of meta-cognition or regulators of control mode: (a) the outcome of the previous task or action, and (b) the subjective estimation of available time.

Additional dimensions defined by Hollnagel as parameters of control mode are: number of simultaneous goals; availability of plans; event horizon; and mode of execution. These parameters represent the connection between control mode and context and allow to describe, dynamically, during the HMI process, the selection of a certain control mode as well as the transition from one control mode to another.

Although these main control parameters and additional dimensions may not be an exhaustive set of variables, as Hollnagel points out, they show how the theoretical background of COCOM has been very accurately developed, in consideration of all the different aspects for modelling cognition and human interactions.

The computation and implementation of COCOM has not been fully carried out. However, the author has quite extensively described all the necessary requirements

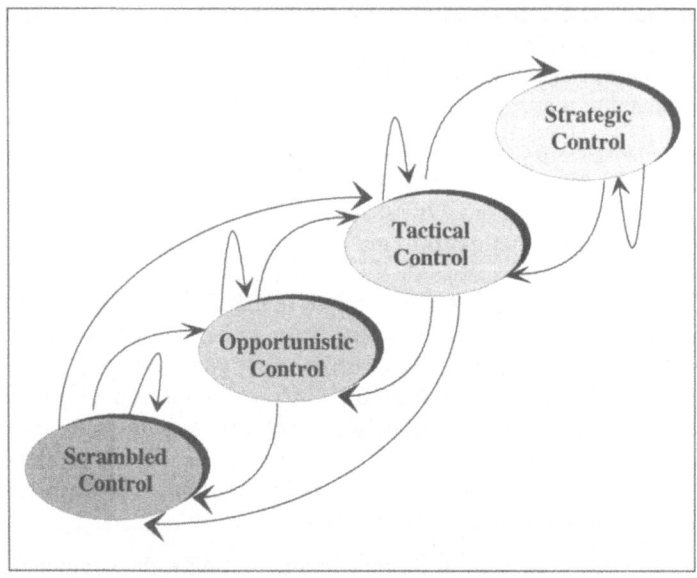

Figure 3.6 Control modes and transitions (from Hollnagel, 1993).

and also the possible specifications to follow in order to develop a computerised version of COCOM. A simplified version of COCOM, where the model of competence has been included and the contextual control mechanisms have only been considered superficially, has been developed in a computer simulation program (Hollnagel and Cacciabue, 1991).

This model has been applied to the analysis of transient conditions in civil aviation systems (Cacciabue et al., 1992) and in the nuclear environments (Yoshida et al., 1997).

The SHELL Family of Configurations

A family of configurations has been developed to describe the existing relationships between people, their working contexts, their organisations, and all sociotechnical activities in a human factors perspective.

Original SHEL Structure

The original SHEL structure considers mainly a framework to describe the connections existing between humans, called *Liveware* (L), and other elements of the working environments (Figure 3.7).

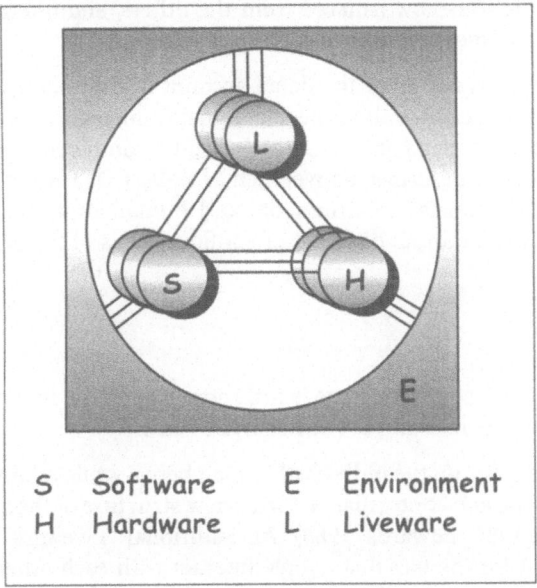

S   Software   E   Environment
H   Hardware   L   Liveware

Figure 3.7 The SHEL configuration (from Edwards, 1972).

These other elements consist of:

- Physical sources, such as equipment, materials, interfaces, and machines, which may be termed *Hardware* (H).
- Rules, regulations, laws, procedures, customs, practices, and habits governing the way in which a plant or a machine is operated, which may be called *Software* (S), so as to characterise the difference between the other class of elements, e.g., hardware made up of more visible physical characteristics.
- As *liveware, hardware,* and *software* operate in real contexts, a further element is considered to complete the basic "unit" of the SHEL structure: the *Environment* (E). The environment covers social, physical, and technical aspects of working contexts. Therefore, it considers all those factors over which a preliminary control can not be devised, but may only be evaluated and studied. In other words, the environment consists of physical, economic, political, and social factors that may indirectly influence human performance.

A basic SHEL "unit" consists therefore of three elements—a Liveware, a Hardware, and a Software, interacting with each other, and each of them interacting with the Environment. A system of humans and machines comprises several "units" operating in a certain environment. Thus, an additional dimension is considered, which accounts for the interaction within a group of people, and including this with machines generates a complete SHEL diagram.

An important characteristic of SHEL is that it can be applied for studying any interaction occurring in working contexts. All possible combinations between the four constituents of SHEL can be accounted for. However, it suffers the limitation that each "unit" is somehow isolated from the others, and interaction can occur only through a common environment.

Moreover, SHEL does not enter the detail account of the cognitive processes, i.e., does not contain a means to describe the human information processing mechanism that generates actions. In this sense it can be combined with the more cognitive models of HMI described above, such as RMC, COCOM, or SRK, in order to cover both socio-technical environments and human cognitive processes. This characteristics is valid for all different of configurations of the SHEL family.

The SHELL Structure

In the original SHEL structure there is little attention to the individual (liveware), which constitutes the human part of the SHEL "unit."

In order to focus more on the individual human being, while retaining correlations with organisational and contextual factors, a new structure of the model was developed (Hawkins, 1987; Edwards, 1988). An additional "Liveware" dimension was added to account for the fact that people interact with each other as well as with machines. In this way, the model becomes more focused on the characteristics and personal attitudes of a single human being and on the interactions that such a person develops with other human beings, while working with and within certain environments, hardwares, and softwares.

The original SHEL unit conceived by Edwards no longer exists, and has been replaced by a new unit with the human at the centre of the structure, and its interactions with surrounding socio-technical and human environments. The configuration in this form is called SHELL (Figure 3.8).

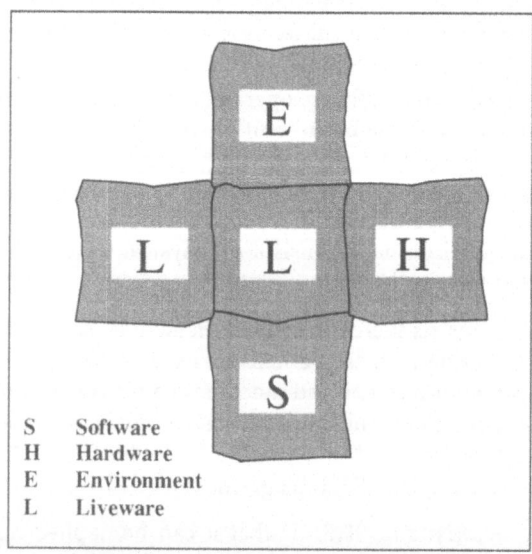

Figure 3.8 The SHELL configuration (from Hawkins, 1987).

This SHELL architecture is very flexible and becomes very useful for studying the links existing between persons, organisations, and society. These are of special interest to the analysis of modern complex systems such as aviation, energy production, and process systems.

The SHELL-T Structure

The complexity of working contexts and interplay of different actors is not easily considered in a SHELL unit. In practice, when different human beings collaborate and communicate from distant locations or cooperate in common working contexts to reach the same goal, it becomes very difficult to capture, by means of a single SHELL unit, the teamwork and correlations existing between different actors.

In order to consider in detail this specific characteristic, typical of teamwork, a modelling architecture that contemplates several interacting SHELL units can be developed. In this modelling architecture, called SHELL-T (for SHELL-Team), each unit shares certain elements with other units, according to the specific arrangement of the working environment and task requirements. In this way, each specific situation can be represented by an appropriate combination of the five basic elements of the unit, considering that the SHELL-T architecture puts the human being at the centre of the model.

As an example, the cockpit of an aeroplane with three crewmembers on board, namely pilot, co-pilot, and flight engineer, can be represented with a combination of three SHELL-T units, each one associated with a member of the team (Figure 3.9). In addition, a fourth SHELL-T unit can be considered in order to account for the air traffic controller, collaborating with the cockpit from a distant location, i.e., the control tower or air traffic control room.

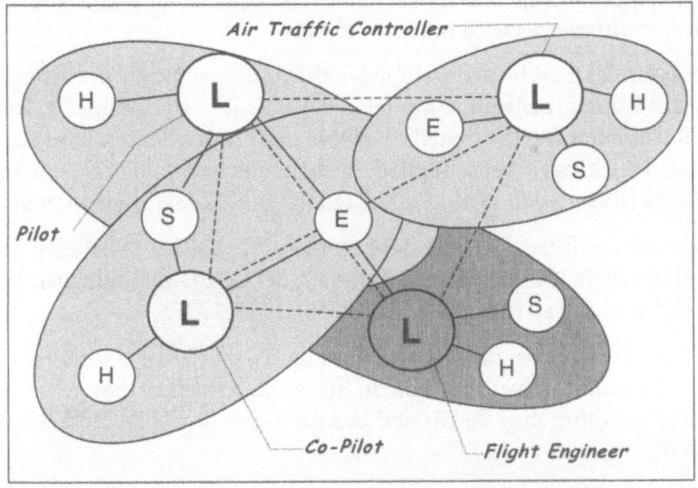

Figure 3.9 SHELL-T architecture for representing teamwork activities of three cockpit crewmembers and an air traffic controller.

In particular, in the case of cockpit crewmembers and air traffic controller the overall SHELL-T architecture considers that:

- Each crewmember has a dedicated interface (Hardware, H) to deal with.
- All three crewmembers share the same working Environment (E), in terms of socio-technical and organisational system: same company, same organisational culture, same policies, as well as same working context (the cockpit) and general physical variables (temperature, noise, etc.).
- The pilot and co-pilot share the same training and procedures, while the flight engineer is referring to a different set of norms (Software, S).
- All three crewmembers communicate with each other for the performance of their flying task, as well as with the air traffic controller.
- The air traffic controller communicates and collaborates with the crewmembers, but does not share any other elements of the SHELL-T unit, as he/she operates in a different working context (E), with different rules and regulations (S), and with a totally different control system and interfaces (H).

3.2.2 Taxonomies of Human Behaviour

In the analysis of human behaviour and errors in modern technology, the model of cognition acts as reference paradigm. Models of cognition are supported by corresponding taxonomies that enable to describe formally and to structure human (inappropriate) behaviour and performance.

A *taxonomy* is a classification, i.e., a set of categories and items that guide data collection. In safety assessment, a taxonomy aims at classifying information on an accident or incident in a systematic way. In so doing, a predefined scheme is followed that offers also a guide for the analysis of accidents.

The model of cognition lies at the core of a taxonomy and helps in understanding the cognitive mechanisms that characterise the sequence of events and shows the similarities or differences with other accidents.

Several taxonomies have been developed in the past, specifically in correspondence with relevant human behaviour models. In particular, the skill, rule, knowledge model of Rasmussen and the model of fallible machine of Reason have engendered taxonomies, which have been applied in different technological and industrial environments (Rasmussen et al., 1981; Bagnara et al., 1989; Hudson et al., 1994).

Two taxonomies will be analysed here, namely "Cognitive Reliability and Error Analysis Method" (CREAM), related to the model COCOM (Hollnagel, 1998), and ADREP, related to the structure SHELL (ICAO, 1987, 1997).

As in the case of cognitive models and paradigms, these two taxonomies are only two of the many alternatives that exist in literature. However, they cover the whole scope of analyses that may be carried out for safety purposes and human error management.

In particular, SHELL and its associated taxonomy focuses on a representation of the HMI at a sociotechnical level where human behaviour is framed in a broad context of interactions. On the other hand, COCOM and its relative taxonomy concentrates on the cognitive functions and mental processes that guide human behaviour and, thus, they focus on the "individual" level of HMI.

Both taxonomies describe the causal and logical links between individual behaviour and errors and the underlying factors and contextual circumstances that are the primary causes of unwanted events and accidents.

Taxonomy CREAM

The taxonomy related to the model COCOM is associated with the method CREAM. It has been developed for root cause analysis and risk assessment applications (Hollnagel, 1998).

In this taxonomy a logical sequence is preserved amongst the four *cognitive functions* of the model COCOM, namely perception, interpretation, planning, and execution, by carefully linking causes, effects and manifestations of behaviour in a circular dynamic interaction. This is the fundamental characteristic of the approach that allows the development of a formal procedure by which the model and taxonomy are applied for the logical construction of sequences and for the inferential consideration of decisional processes leading to manifestations of errors.

The causes of erroneous behaviour, also called "genotypes," are the roots of certain inappropriate behaviours. These can be further subdivided in internal causes, which depend on the individual characteristics (personality, attitudes, etc), and external causes, which depend on the human–machine system and context. The effects and manifestations of erroneous behaviour, also called "phenotypes," consist in the forms that the whole cognitive process takes in terms of human actions or explicit performances.

The distinction between causes and effects must be clearly maintained in the analysis. In the taxonomy associated with CREAM, this distinction is formally applied with respect to the four *cognitive functions* of the model of competence of COCOM, namely: perception, interpretation, planning, and execution. For each cognitive function the taxonomy defines (Figure 3.10):

- "Effects," i.e., manifestations of the specific cognitive function. These are the "phenotypes," in the case of the function *execution*.
- "Generic causes," which consist in a list of causes usually related to the cognitive function immediately linked to the one under analysis, in either direction of a cognitive process.
- "Specific causes," which are essentially made of environmental and contextual factors that may affect behaviour. These are a set of constant factors for all cognitive functions, but depend mainly on specific domains and working environments. Even if a general list of these factors can be developed, they have to be identified in relation to each domain of application.

The detailed process of exploitation of the taxonomy CREAM for identifying specific root causes (genotypes) of inappropriate behaviours (phenotypes) will

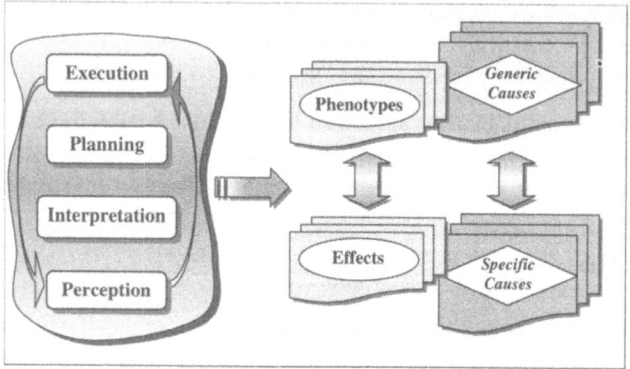

Figure 3.10 The taxonomy CREAM.

be described later on this chapter, in connection with the review of methods for RCA.

Taxonomy ADREP

The International Civil Aviation Organisation (ICAO) has adopted for a number of years a specific "Accident/Incident Data Reporting" (ADREP) system for the collection of information and data on aviation accidents.

A recent development of this classification has been specifically dedicated to improving the reporting of human behaviour and the new system has been called ADREP-2000 (ICAO, 1997). In the remaining of this chapter, all references made to ADREP will always refer to this new development and, in particular, to the human behaviour part of the new classification.

The taxonomy adopted in ADREP-2000 for structuring and classifying human factors is based on the SHELL architecture. In particular, the links existing between persons, organisations and society are of special interest to the analysis of aviation accidents and are well captured by the reference structure of SHELL. In addition to this, the model allows to consider latent and active errors, made at different levels within an organisation, and their effects on the active behaviour of front line actors in system management.

In the model considered by ADREP-2000, the human being (*Liveware, L*) plays a central role in a system made of other humans (*L*), environment (*E*), hardware (*H*), and software (*S*). The taxonomy based on this model considers therefore five major areas of analysis (Figure 3.11): *Human (L), Liveware–Liveware interface (L-L), Liveware–Hardware interface (L-H), Liveware–Environment interface (L-E)*, and *Liveware–Software interface (L-S)*.

These five areas of analysis can be described as follows:

- *Human (L)* covers all aspects associated with the individual characteristics and performances, including physical attributes, physiological, and psychological issues.

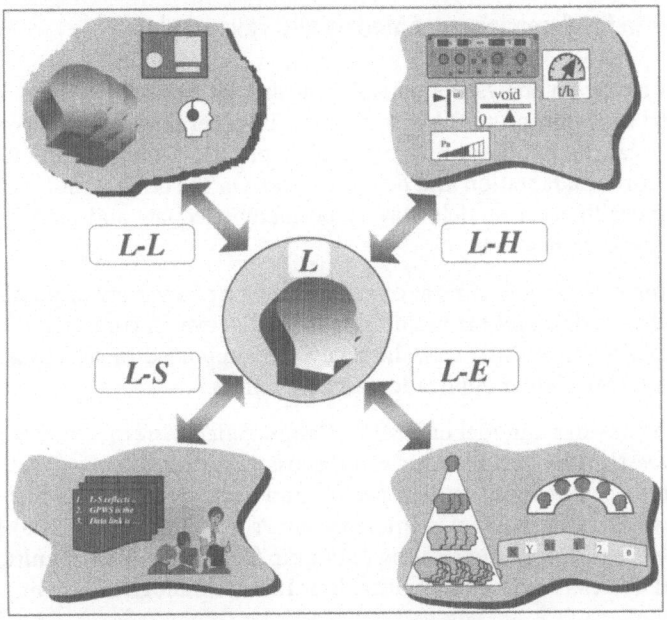

Figure 3.11 ADREP-2000 taxonomy of HMI.

- *Liveware–Liveware interface (L-L)* covers communications, supervision, and checks with other persons, in the immediate surroundings of the human being at the centre of the model.
- *Liveware–Hardware interface (L-H)* refers to the human interacting with the system. This includes the actual control room or working environment instrumentation, the equipment, and any supporting material which may be utilised to carry out a job or task. For example, in the case of aviation, the L-H interface considers the aircraft cockpit and control panels, as well as the equipment for emergency control, such as gas masks, standards, and emergency operating manuals, etc.
- *Liveware–Environment interface (L-E)* covers all aspects associated with the human being interacting with the environment. This includes the physical environment, the task environment, the social, and company/management issues which may affect work.
- *Liveware–Software interface (L-S)* accounts for indirect or nontangible issues affecting humans at work, such as training, procedures, etc. This part of the taxonomy is considered separately from the L-H interface, as these factors play an important role in HMI and need dedicated attention.

This taxonomy supports retrospective analyses and classification of human-related factors associated with an accident. In particular, the ADREP-2000 reporting system codes all factors according to the findings of an accident investigation (accident report). The accident investigation approach, based on SHELL and ADREP-2000 taxonomy proposed and supported by ICAO, will be described in more detail in a latter section of this chapter.

3.2.3 Criteria for Comparison of Models and Taxonomies

When performing an analysis of an HMS, the analyst must apply the most appropriate model of cognition to the case at hand. In general, while most comprehensive models can be applied to all types of application, their use requires quite extensive effort of adaptation and data selection. On the other hand, less sophisticated, but more focused models may be perfectly adequate and readily available for well-defined applications.

The following comparison aims at offering a review of a variety of possible applications of the models and taxonomies described above, in connection with their optimal exploitation. In this way, the analyst is supported in selecting the most suitable model for his/her application and objectives.

The need to consider a model of HMI in *Design, Safety Assessment, Training*, and *Accident Investigation* has already been discussed earlier. Moreover, two types of analysis have been defined, i.e., *Prospective* and *Retrospective*, for which a model is necessary. The metaphor of a Information Processing System (IPS) has been underpinned as the basis and starting paradigm for all models of cognition, which bear special correlation to human behaviour in a technological system.

In other words, it is assumed here that when an analyst tackles a HMI problem, then he/she is concerned with an integrated prospective and retrospective type of analysis, which is applied either for the design or safety assessment and safety audit of an organisation, or for training, or for accident investigation purposes. The IPS metaphor, as an appropriate paradigm of human behaviour, can take various modelling forms.

Moreover, the increased complexity of the human tasks in highly automated technological environments, asks for consideration of specific aspects of HMI, not contained in certain models.

The completeness of a study depends, in many cases, on the ability of the model to account for special features of the working context and cognitive processes.

Two major aspects characterise the models and taxonomies previously described. Firstly, they have been elaborated at different stages of development of human sciences and technology, over two decades. Moreover, they have been conceived with special attention to different aspects of HMI. These factors, instead of being a drawback, offer the analyst a variety of options for choosing the most appropriate model for the problem at hand.

Table 3.1 shows a comparison of the four cognitive models and SHELL structure, described earlier in this chapter, with respect to certain modelling characteristics and types applications in HMI.

The criteria utilised for comparing models/taxonomies concentrate on the present technological systems, and can be identified in the ability of a model to account for:

- behaviour of *individuals*;
- behaviour of *teams*;

Table 3.1 Models of cognition and HMI application

	RMC	SHELL	SRK	Fallible machine	COCOM
Model Characteristics					
Individual behaviour	***	*	***	**	***
Team behaviour	*	***	*	–	*
Task/job analysis	***	**	***	*	***
Organisational effects	*	***	–	***	*
Cyclic & dynamic interactions	***	–	–	*	***
Application Types					
Accident/incident interpretation	**	***	***	***	***
Safety indicators	*	*	*	*	*

*** *Model with complete data structure* (qualitative and quantitative framework with complete data structure defined by modelling architecture).
** *Model with partial data structure* (qualitative and quantitative framework with data structure identified, but not specified).
* *Shallow model* (only qualitative framework applied for describing human behaviour).
– *Not considered or modelled.*

- *task/job analysis*;
- effect of *organisations*;
- *cyclic* and *dynamics interactions* versus static and linear models.

At the level of safety assessment and accident analysis, it is also relevant to know to which extent a model is able to cope with certain types of application. In particular, it is important to know if a model is able to:

- perform *accident/incident interpretation*; and
- evaluate *safety indicators* for auditing purposes.

Other criteria for comparison can be considered and should be developed in accordance with the needs of the analyst and in relation to the specificity of the case at hand.

From Table 3.1 it is clear that all models are well designed for representing individual behaviour, except SHELL which has already been identified more as a structure for representing HMI rather than a model of cognitive behaviour. Some modelling architectures are, however, more advanced than others, as they also contain the definition of the data and parameters that are necessary for real applications and give guidelines concerning the approaches to be applied for identifying such data in real context and work settings.

On the other hand most models are quite weak in supporting modelling team behaviour, except for the specific configuration depicted as SHELL-T that is specifically dedicated to modelling team performance.

Job and task analyses are quite easily correlated to almost all these modelling architectures. This fact makes their application and interfaces with task analysis quite simple in any type application.

The influence of organisational factors is fully accounted for in two models, which are specifically dedicated to these issues. Most other models consider such factors indirectly, i.e., they evaluate the effects of organisational factors on the parameters that affect individual behaviour and do so in a qualitative and descriptive fashion only.

The consideration of cyclic and dynamic processes is the specific focus of two models (RCM/PIPE and COCOM), which have been developed on the assumption that human performance is the result of a process implying usually several cycles of mental processes and higher cognitive functions before resulting in an explicit performance. Other models, where the procedure that governs the information processing is carried out linearly from perception to action, have not accounted for this factor.

It is noticeable that the two models that focus on organisational issues, i.e., SHELL and "fallible machine," are precisely those that do not consider cyclic and dynamic processes. On the other hand, COCOM and RMC/PIPE that concentrate on cyclic cognitive processes do not account for organisational issues, or at least they have only indirect shallow liaisons to them. This is quite reasonable as COCOM and RMC/PIPE concentrate on individual cognitive reprocesses that lead to behaviour, while SHELL and "fallible machine" models look primarily at the correlations that exist between a human being and the surrounding socio-technical environment.

In this respect, it is surprising that the model SRK does not cover with a formal data structure either of these two features of modelling issues.

All models support the analysis of accidents, even if at a different level of accuracy, according to whether they are coupled to a specific taxonomy. In practice, only RMC/PIPE has not been tied to a specific taxonomy for accident analysis. This may not necessarily be a drawback, as the model is quite simple and this gives the analyst a greater degree of freedom in selecting a classification scheme that may be better suited for the purposes of the analysis under development.

Finally, as all models do not seem have been developed for sustaining safety audits and identification of safety indicators and markers, as they only give generic and qualitative description of this type of factors. However, this aspect represents a relatively new area of concern, and models and theories of cognition that tackle this issue have not yet been developed in detail. It is necessary, in such a case, that a safety analyst applies one of the generic models existing in the literature and organise the input/output structure in the most convenient form to obtain valuable indicators for the problem at hand.

3.3 Ethnographic Studies

3.3.1 Background of Ethnographic Studies

Another fundamental contributor to (retrospective) safety studies is the evaluation and assessment of the socio-technical working environment in which HMI takes place. In particular, it is very important to acquire familiarity with the operators of a system, e.g., pilots, air traffic operators, maintenance personnel, and with the knowledge, expertise, and practices that they apply in everyday work. In particular, relevant knowledge and expertise reside in different persons within an organisation and may assume different forms of behaviour, such as formal, informal, declarative, procedural, tacit, implicit, and explicit.

Moreover, every technological setting is based on and develops specific working contexts and practices that depend on local and national habits and cultures, on the policies and corporate attitudes dictated by the organisation, and on the progressive changes that occur during the lifetime of a plant/system. These aspects are very relevant in understanding overt behaviours and facts and predicting possible trends during specific HMI instances.

In order to capture these characteristics, specific and well-defined methods and approaches must be applied.

A variety of methods and techniques are presently available for the process of knowledge identification and elicitation and for collection of information and data about real working contexts and behaviours (Brooking, 1999; Bonnie et al., 1999). In general, these techniques are contained under the umbrella of ethnographic methods or *ethnographic studies*, which may be applied for studying, understanding and coordinating distributed knowledge and for retrieving roots and foundations of organisational memory (Perry et al., 1999).

Ethnography was developed originally (Garfinkel, 1967; Turner, 1974) as a means to study lives and customs of other cultures, and it is characterised by the process of gathering data from a variety of sources that are fully immersed in a context, without disturbing the development of work settings and interactions (van Maanen, 1979; Agar, 1980; Bucciarelli, 1988).

More recently, *ethnographic studies* have been applied by the computer science and information systems community to look at social interactions around technology (Baldwin et al., 1999; Brown and Duguit, 2000; Button and Sharrock, 1997; Heath and Luff, 1991; Wood, 1996). The detailed investigative nature of the ethnography approach allows to examine specifically how groups integrate and regulate both the formal (technical) and informal (social) aspects of their daily work activities and, most interestingly, their knowledge practices and markets (Davenport and Prusack, 1998).

3.3.2 Ethnographic Studies for Retrospective and Prospective Analyses

The above rationale of ethnography implies that work settings may be analysed by a variety of methods aimed at capturing data without interfering with the work

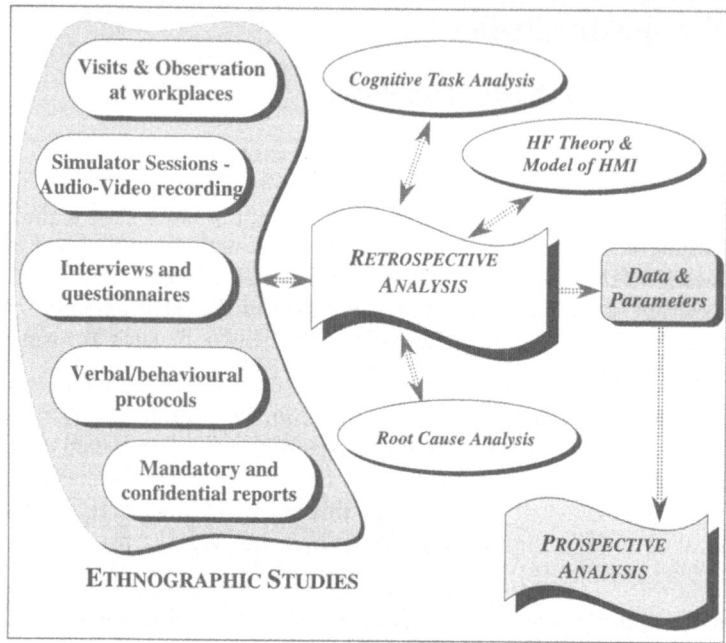

Figure 3.12 Ethnographic studies.

process, and consequently, retaining the natural complexity and the complete independence of the variables under investigation. The basic techniques that are adopted for ethnographic analyses are: *visits to workplaces, observation at workplaces, audio-video recording, verbal/behavioural protocols, interviews,* and *questionnaires* (Figure 3.12).

Conversational Analysis (CA) is originally identified as the critical technique utilised for identifying "indexical expressions" and other practical actions that characterise everyday life and practices (Garfinkel, 1967).

These data collection processes must be coupled to more analytical methods for the evaluation and assessment of working environments and real events. This can be done by task/job analysis and by the study of past experience within the organisation, which are usually contained in reports and records on accidents and incidents of mandatory or confidential nature.

All ethnographic methods are very effective in acquiring information and knowledge about practices at work and for capturing the effects of cultural and organisational factors on the way in which procedures are applied and relationships between people develop and change over time.

The role of ES in appreciating and understanding real working contexts and practices is also fundamental for *prospective* type analyses, as rarely the development of a design and safety assessment is completely novel with no reference to past experience and links to previously developed systems (Figure 3.12). Consequently, in a logical structuring of different methods, as in the case of the HERMES methodology, ES contribute to the generation of the data and parameters that are very useful in developing new HMS and interfaces.

The theories that sustain ES and, in general, the preparation and exploitation of verbal protocols, observation at workplaces, performance of interviews and questionnaires are all well known and formally established. The detailed analysis of these methods, as in the case of numerical approaches, and Fault Tree and Event Tree techniques, goes beyond the scope of this book and will not be further developed. The reader interested in performing ethnographic studies of this nature and applying these formal methods should make reference to the existing relevant literature (e.g., Garfinkel, 1967; Wixon and Ramey, 1996; Salvendi, 1997; Davenport and Prusack, 1998).

3.4 Cognitive Task Analysis

3.4.1 Background of Cognitive Task Analysis

Another essential contributor to the performance of (retrospective) safety studies is the analysis of the tasks assigned to humans. This is fundamental in understanding intentions and expectations of designers, as well as in planning future control and management operations of any HMI process.

The analysis of tasks is intended as a prerequisite for the assessment of the overall human–machine interaction, as it offers a way to structure, by formal expressions, the envelope of procedures, actions and contextual facts characterising human behaviour in working environments.

Task Analysis (TA) was first developed in the fifties, with the aim of formally describing overt human behaviour by a series of simple and elementary components (Skinner, 1957). Task analysis theory was originally applied to language learning and consequently to learning performances manifested through verbal expressions. It was then extended to the wider context of working processes (Payne and Green, 1986). Recently, as the role of operators has changed, becoming mainly of supervisory control rather than active manipulation of processes, the analysis of tasks has been adapted to these new duties. From task analysis the approach became a much debated *Cognitive Task Analysis* (CTA) (Bonar et al., 1986; Schraagen et al., 2000).

By adding the attribute "cognitive" the focus of task analysis has been placed on mental processes rather than on overt manifestation of behaviour. Thus, while the objective of task analysis has not changed, the methods of CTA have been adapted to the new locus of interest of HMI. In view of these new aims, CTA can be defined as follows: CTA is a method that attempts to specify the interaction of mental procedures, factual knowledge, and task objectives in the process of job performance.

The aspects of operator behaviour that are studied by cognitive task analysis can be summarised in terms of *contents*, *contexts*, and *mental models*. *Contents* are the technical skills, or procedures, by which the human interaction takes place, and consist mainly in the behavioural elements of task analysis. *Contexts* represent the contribution of working conditions, contextual and socio-technical factors affecting cognitive processes and overt behaviour. *Mental models* are the intermediary element between overt behaviour and cognitive process, which support the formalisation of the HMI.

3.4.2 Cognitive Task Analysis for Retrospective and Prospective Analyses

From the above definition and description of objectives and areas of application, it is clear that CTA synergistically interacts with models of HMI and ethnographic studies to characterise a set of approaches necessary for safety studies. In particular, the importance and scope of CTA in studying HMI is also well defined: CTA can affect the analysis of HMI by establishing correlations, dependencies, and links that are maintained and exploited throughout the development of design of new control systems.

Moreover, CTA is applied to perform analysis of the working environment and working practices and to structure plans, tasks, and goal setting and links between tasks and goals. CTA can be utilised for analysing people's level of expertise in order to identify training strategies and methods. This broad spectrum of application explains the important role associated with CTA both for *prospective* and *retrospective* types of analyses. In other words, CTA plays a pivotal role in a process of information acquisition about the working environment and leads to the identification of relevant data and parameters that are essential for design and safety analysis (Figure 3.13).

Following the above description of CTA area of concern the reader may be led to believe that there is little difference between CTA and ethnographic studies. These are two similar ways to analyse tasks and work processes, and a number of commonalities exist between these two contributors to retrospective studies and assessment of working contexts.

However, there is fundamental difference from the HF viewpoint that creates a clear-cut division between these two ways of analysing socio-technical working contexts, and it consists in the fact that:

* CTA focuses on the *design* of procedures and job cards and on the way in which tasks and job performance (job analysis) are planned and developed by designers in order to deal with normal working conditions, as well as dynamic and emergency situations (Figure 3.13).
* Ethnography is instead concerned with the way in which operators actually act and behave in real working environments, i.e., in ES, *practice* becomes the locus of interest for the HF analyst.

Therefore, both types of analysis are necessary to develop a clear understanding of what are the aims of tasks and how tasks should be carried out and the actual performance in everyday life.

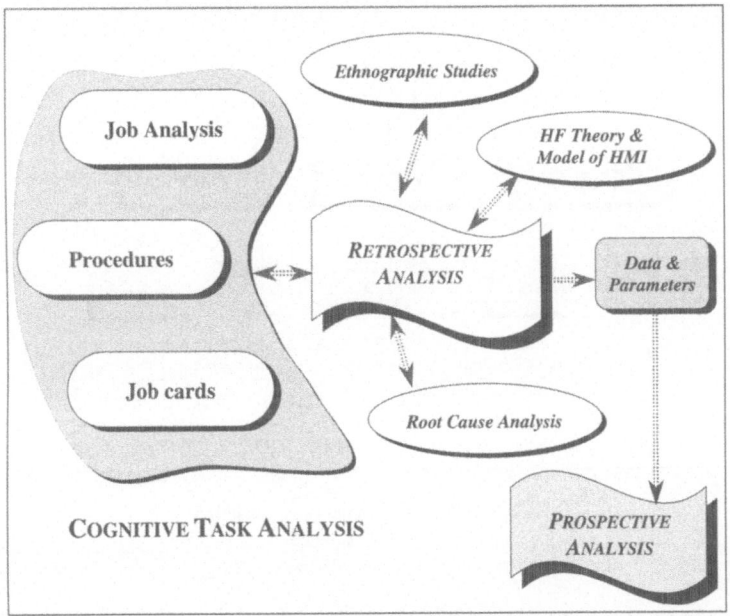

Figure 3.13 Cognitive task analysis.

The importance and role of CTA in the development of modern technological settings and, in particular, of the need to formalise CTA contribution to design and assessment of HMI, including also Human–Computer Interaction, is further emphasised by recent publications of handbooks that cover most advanced and topical developments in this area (Hollnagel, 2003; Diaper and Stanton, 2003).

Cognitive Task Analysis is based on three main contributions:

1. The *theoretical standpoint*, by which to define what is contained in a CTA and how to obtain it.
2. The *formalisms* by which it is possible to represent HMI sequences.
3. The *methods* and *approaches* to be applied in order to collect information and data in working environments.

Theoretical Standpoint and Formalism for CTA

Following the definition of CTA and considering that a *task* is always described with reference to a *goal*, it results essential to define how goals and tasks correlate in the process of job performance. In addition to goals and tasks it is important to define other basic components of CTA, such as *elementary actions, preconditions*, and *postconditions*.

Goals are the objectives to be reached by a procedure and a plan to be implemented. They are usually expressed in terms of state of the plant or physical

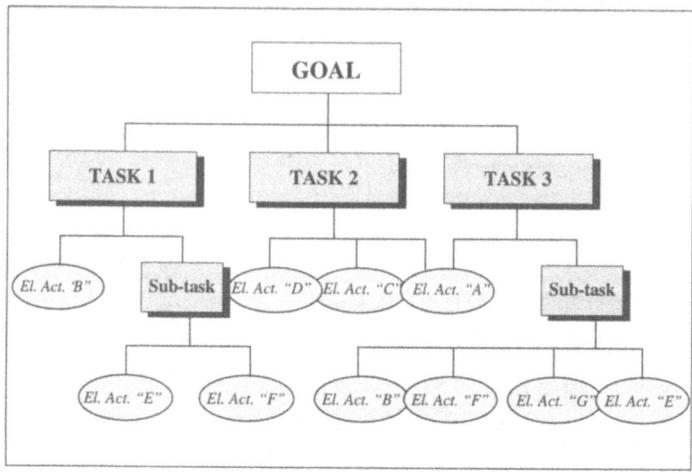

Figure 3.14 Structure of a goals–means task analysis.

process to be achieved by implementing a certain sequence of actions. *Tasks* are aggregations of *elementary actions* or procedures (*subtasks*) to be carried out in a certain sequence or combination in order to reach the goal. Subtasks are made of their own specific goals and task sequences (Figure 3.14).

Elementary actions, preconditions, and *postconditions* are other basic components of a CTA and building bricks on which the whole structure of CTA is made (Figure 3.15).

Elementary actions are the simplest forms of behavioural performance in which a task may be subdivided. *Preconditions* correspond to success conditions of previous tasks, plant state and environmental events that must be identified and permit the start-up of the task. *Postconditions* are the results of the performance of a task and affect the plant state, related tasks, context, and working environment.

The combination of interconnected goals, tasks, and elementary actions leads to a "treelike" structure which is well suited to graphically represent the development of a procedure. In task analysis of real systems, when goals and tasks are associated with many pre- and postconditions, the tree structure may become very cumbersome to develop. The use of computerised tools then becomes necessary, in combination with specific programming languages, for managing inheritance and sequencing of attributes and dependencies.

The example shown in Figure 3.15 represents a situation in which tasks 1, 2, and 3 are deeply interconnected with each other, both at the level of *pre-* and *postconditions,* as well as *elementary actions.* The situation depicted is a simplification of the real complex combination that is usually dealt with by the designer of a procedure of a real system.

In particular, in Figure 3.15 only the structure of task 1 is presented, as the whole structure would be come very complicated to show in a single figure. This explains

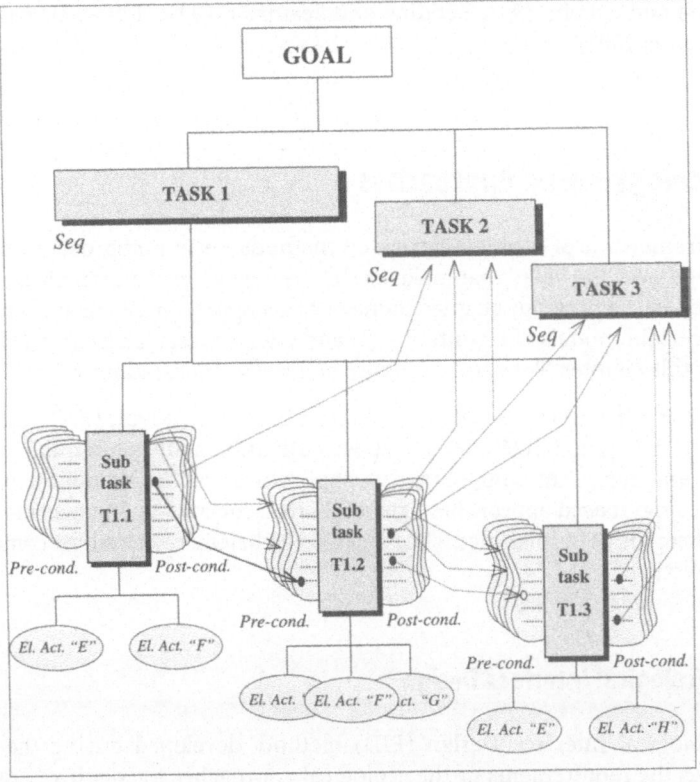

Figure 3.15 Interconnection of goals, tasks, and conditions.

why the use of a computerised support is required to design and follow the development of interconnected tasks.

Methods and Approaches

A number of approaches can be exploited to develop CTA, and to determine data relative to cognitive structures and processes underlying job performance.

These methods and techniques mainly cover field assessment of working environment and modelling techniques of logical and behavioural performances. Some of them are very formal and follow precise procedures. Other methods are less formalised, as ethnographic approaches described earlier (Scapin and Pierret-Golbreich 1990; Kirwan and Ainsworth, 1992; Seamster et al, 1997; Luczak, 1997). As for the *Ethnographic Studies* (ES) also in the case of TA and CTA, there exists a vast literature of well-developed and formalised methods. Further discussion of these methods goes beyond the scope of this book and the reader is referred to the relevant literature quoted throughout this section for the selection of specific applications of CTA (Ericsson and Simon, 1984; Hollnagel, 1991; Sebillotte 1995;

Sebillotte and Scapin, 1994; Redding and Seamster, 1994; Hollnagel, 2003; Diaper and Stanton, 2003).

3.5 Ecology-based Approaches

In the framework of ecology-correlated methods, several approaches have been developed over the years that make explicit reference to the ecological principles discussed in the previous chapter. Some of these approaches bear specific connection to certain models of cognition – ES and CTA – presented in the previous sections of this chapter. They will therefore be briefly discussed here.

In particular, the methods called "Ecological Interface Design" (EID) and "Cognitive Work Analysis" (CWA) deserve special attention as they are quite commonly applied and represent comprehensive applications of the ecological principles. Other ecology-based approaches exist and are associated to engineering psychology and cognitive ergonomics. They will also be briefly reviewed for completeness in this section.

3.5.1 Ecological Interface Design

The Ecological Interface Design (EID) method, developed during the eighties, combines the requirements of the ecological approaches to cognitive ergonomics with a well specific architecture of HMI (Rasmussen and Vicente, 1989, 1990; Vicente and Rasmussen, 1990, 1992).

The EID framework demands that three basic principles are respected every time a HMS is tackled, especially when the interfaces that govern the exchange of information between humans and machines are designed or evaluated by the HF analysts. These three principles are directly related to the concepts of ecology discussed in Chapter 2 (§2.6) and to the work of modelling and representation of HMI performed by Rasmussen (1986), in accordance with the paradigm of the information processing system previously presented in this chapter (§3.2.1).

According to the EID approach, the three principles that govern the development of a HMI system can be described as follows:

1. Apply the "means–end hierarchy" describing the work domain and the interactions between humans and machines (Figure 3.4).
2. Ensure that the *affordances* of the environment are manifested in the interfaces so as to grant human perception.
3. Apply the perception–action process depicted in the "step-ladder" modelling architecture (Figure 3.3), in order to ensure clear mapping of information display with cognitive and behavioural processes.

The application of EID to design of interfaces and analysis of work domains has bee performed in different environments, giving valuable results (Hansen, 1995).

The basic principles of application of the EID approach are at the same time very qualifying characteristics of the method, as they offer a clear and well-defined frame for application of human factors, and limiting conditions, as they bound the analyst to applying the "means–end hierarchy" and the "step-ladder" modelling architecture.

In other words, the EID approach is a specific instantiation of the general ecology principles, expressed in terms of theoretical "global perspective," and requires a step forward in the implementation of these principles for practical application. This step forward consists in the adoption of a particular model of cognition and taxonomy of human performance, based on the "means–end hierarchy" and the "step-ladder" architecture for representing HMI.

3.5.2 Cognitive Work Analysis

Cognitive Work Analysis (CWA) is a consistent methodological approach for analysing and assessing socio-technical working environments, based on ecological principles (Vicente, 1999).

The approach is essentially dedicated to the definition of the data, identification of basic information, and definition of boundary conditions for ensuring that all types of human factors applications are performed respecting the ecological validity that enables effectiveness of results.

In this framework, the identification of constrains, or influencing factors, that affect behaviour is essential. The CWA approach considers five critical factors, namely (Figure 3.16):

1. Work domain.
2. Task control.
3. Strategies.

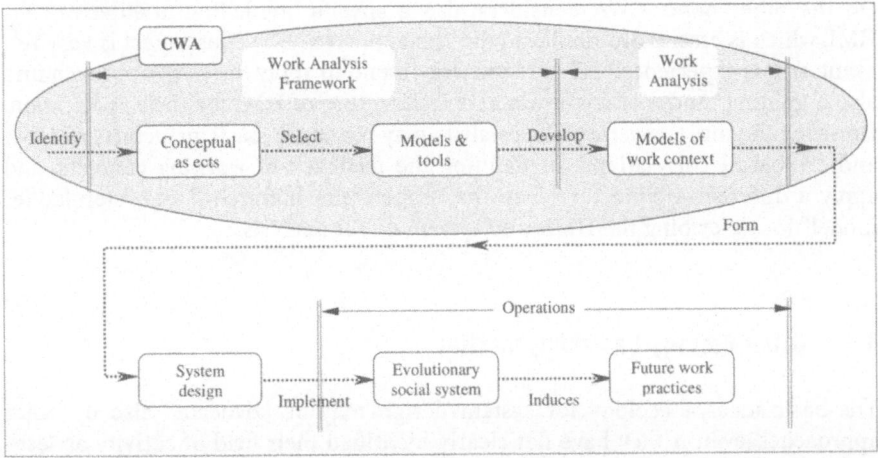

Figure 3.16 Process of application of CWA in the design process. (Adapted from Vicente, 1999.)

4. Social organisation and cooperation.

5. Worker competences.

Each factor can be associated with specific contextual conditions that demand an accurate analysis by HF analysts in order to develop the overall picture of the system under study.

The recognition of an ecological approach in order to properly account for the *affordances* of the work domain is the natural framing for CWA.

In CWA, three approaches are contemplated: *normative, descriptive,* and *formative. Normative approaches* are applied for studying or representing the designed or best procedure to perform a certain job. The typical normative approach is thus task analysis, deemed essential for any design process. *Descriptive approaches* aim at capturing the practices of performances and require techniques from experimental psychology and methods such as activity theory. *Formative approaches* attempt to view the operator, i.e., the primary actor in the control process, as an active contributor to the development of the design or at least, as the last responsible for the refining touch to the design. This process can be assimilated to the conceptual approach know as User-Centred Design (UCD).

In a more general representation of a design and implementation process, the role of CWA is essential in the initial phases of the process, as it ensures that the real working contexts and the ecology of the HMS are adequately accounted for. In this sense CWA bears strong correlations with the ethnographic approaches described in the previous section.

In any case, CWA is a very interesting way to tackle the analysis of working contexts and development of data and information that are an essential contribution for the design process. In this role, CWA is applied quite consistently in numerous applications in different working domains and supports the development of practical applications of HF methods (Naikar and Saunders, 2003).

On the other hand, CWA considers also a specific modelling architecture for HMI, which is essentially similar to the one applied in EID. This aspect is very relevant, as less experienced HF analysts who intend to apply the CWA approach find also a guiding model of cognition and architecture of HMI for their application. However, the more experience specialists may retain the basic principles of CWA and ecological correlations for defining the feedback of working contexts, and apply a different architecture than the "means–end hierarchy" and "step-ladder model" for describing the HMIs and system under analysis.

3.5.3 Other Ecology-based Approaches

The basic idea of ecology for system design may be advocated also by other approaches, even if they have not clearly identified their field of activity as "ecological" with respect to situational conditions and environment in which operations take place.

As an example, the association of design process with the analysis of operating scenarios is the conceptual base for the "scenario-based" approach to design human–computer interaction proposed by Carroll (2000).

Other approaches imply the consideration for all social and technical conditions that affect human behaviour and focus on the participation of system users and managers in all stages of system design, development and implementation. An example of this type of approaches is "participatory ergonomics" (Wilson and Haines, 1997) which implies that stakeholders are directly involved in planning and controlling a system and activity of their competence. This implies that all parties involved develop sufficient knowledge about their role and system functions for contributing properly to the work. For this reasons different "dimensions" of participatory ergonomics are identified and should be considered when following such approach to design and analysis of HMSs (Nardi and O'Day, 1999).

Another important approach for tackling HMI issues in an ecological perspective of prediction of HMIs is the use of models. As an example, the work of Mitchell (1999) in this direction is representative of a line of development which is rooted in the original work of many important specialists and pioneers of HMI (Baron et al., 1970; Sheridan and Ferrell, 1974; Stassen et al., 1990). This approach promotes the use of a model of the operator(s) for developing designs that respect the principles of ecology, including the criteria of UCD. The specific modelling architecture proposed in this approach, called the Operator Function Model (OFM) (Mitchell, 1987), is an instantiation of the conceptual standpoint that requires a clear definition of a HMI architecture to be studied.

This short review of methodologies that may be located under the wide umbrella of ecological approaches to HMS, comprehends also the techniques and technologies usually referred as Computer-Supported Cooperative Work (CSCW) and groupware (Barua et al., 1997). This domain is very large as it covers all problems and issues related to the use of computers in society. Therefore it expands beyond the technological domains toward the analysis of human–computer interactions, rather than HMI, including web activities, communication and collaboration between users. In any case, ecology principles are implicitly applied in implementations of CSCW and groupware analyses and development activities.

3.6 Root Cause Analysis of Events

The core of retrospective studies is the analysis of "what has happened" or the analysis of accidents and incidents. The aim of such analysis is to:

- reconstruct the overall scenario of accident/incidents in a temporal sequence of events, usually failures, human inappropriate behaviours, and other specific occurrences;
- link them in a causal and logical network; and then
- identify the causes of each particular event of failure, inappropriate behaviour, or manifestation, usually called "Root Cause Analysis" (RCA).

As briefly discussed earlier, RCA is different from accident investigation, as the latter is a much wider type of study that entails all aspects of retrospective assessment, including RCA. In other words, RCA tries to explain specific events, while accident analysis aims at the identification of the entire spectrum of several different events, facts, data, parameters, and indicators that have contributed to the generation and development of an accident. In practice, accident analysis is made of a number of RCA that describe and explain the sequence of events that make the accident. It is however quite common to exchange the two terms, as synonyms, and to consider RCA as the whole accident analysis.

In systemic terms, the difference between RCA and accident investigation is similar to the difference existing between Fault Trees (FT) and Event Trees (ET) techniques, as FT try to explain the causes of a specific fault or failure, while ET establish the logical links existing amongst subsequent events.

Therefore, from a conceptual as well as a formal viewpoint, the difference between accident investigation and root cause analysis is very important and should be clearly appreciated and recognised by safety analysts.

From the perspective of Human Error and Accident Management (HEAM) study, in order to consider and develop appropriate measures for prevention, recovery, and protection, it is much more important to understand reasons and causes of specific inappropriate behaviours, rather that getting the overall picture of the accident. This assertion is supported by two main considerations:

1. An accident is made of a very specific sequence of inappropriate behaviours, failures, and random occurrences that singularly would not be able to cause the accident, but when combined in a unique and specific way, generate the unwanted serious or catastrophic occurrence. It is very unlikely that an accident reoccurs with all the same characteristics.

2. On the other hand, the inappropriate behaviours and failures that make an accident are not unique events and probably have individually occurred in the past. However, this occurred in other circumstances without combining into a catastrophic sequence.

Consequently, from the HEAM viewpoint, it makes much more sense to ascertain that single contributors to an accident, such as inadequate behaviours, specific occurrences, and malfunctions are properly managed for prevention, recovery, and protection processes. In this way, the repetition of a specific accident can be avoided, but, much more importantly, many other accidents that could result from other unfortunate combinations of such inappropriate behaviours and failures can be prevented.

Some techniques for RCA based on recent theories and models will now be discussed. Later on in this chapter the issue of accident investigation will be tackled.

3.6.1 Approaches for Root Cause Analysis

In general, all RCA approaches are based on a cause–consequence tree that attempts to reconstruct the socio-technical factors, i.e., the root causes of social

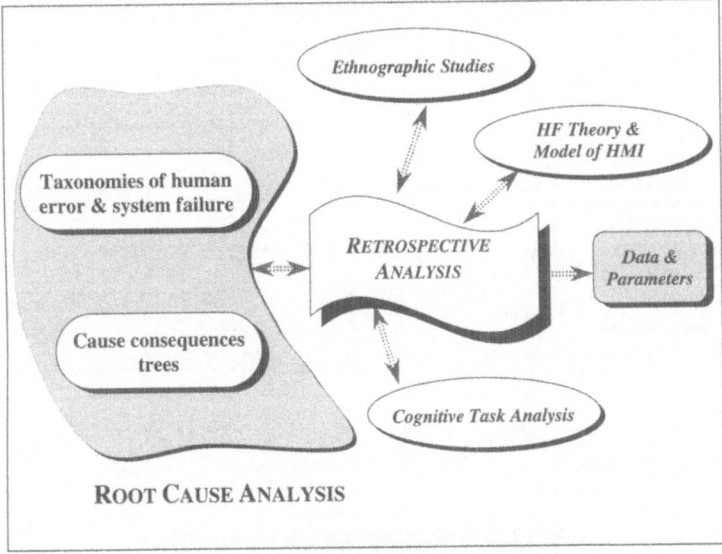

Figure 3.17 Root cause analysis for retrospective studies.

and technical nature that lie at the base of certain inappropriate behaviour or system failure and malfunction (Figure 3.17).

These approaches normally make use of models of cognition and taxonomies, such as those described in previous sections of this chapter.

RCA approaches are very useful in structuring the process of search for primary causes of inappropriate behaviour. However, they all share the same fundamental difficulty: the "stop rule" for the reconstruction of the root causes by means of cause–consequence chains is not clearly defined, and it is left up to the analyst to decide when the important primitive causes have been identified and that no further search is needed.

In the following, three different approaches of RCA will be described. They are representative of possible alternative methods that may be applied at different levels of depth. Other techniques exist for RCA, which, in general, show similarities with the three basic methods described here. As in the case of models of human behaviour and task analysis approaches, the experience and expertise of the HF analyst is essential for selecting the most appropriate method and technique that suit best the problem at hand.

Root Cause Analysis by CREAM

The taxonomy CREAM is based on the model COCOM and has been developed with the specific aim of being applied the root cause analysis of manifestation of inappropriate behaviour (Hollnagel, 1998).

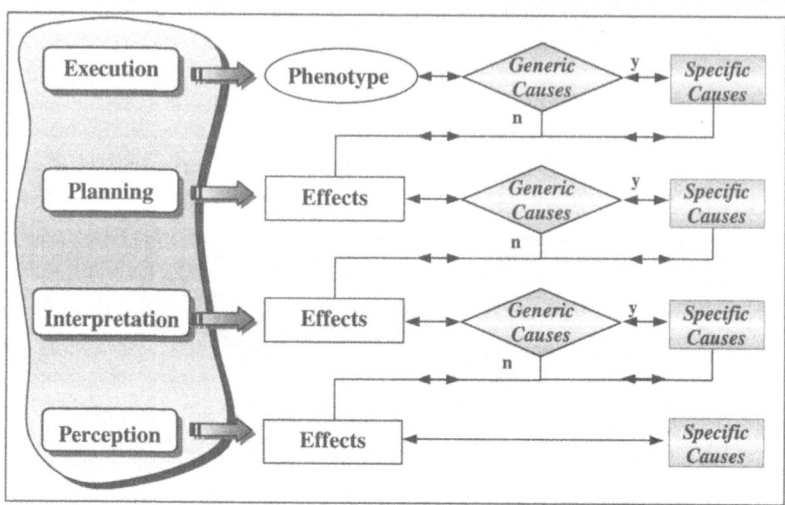

Figure 3.18 Procedure of application of the taxonomy CREAM.

In CREAM, the four *cognitive functions* of the model of competence of COCOM (perception, interpretation, planning, and execution) are considered and a logical connection of cause-effect – is applied in order to link the four cognitive functions. The aim of this process is to identify which cognitive function is at fault.

In practice, it is possible to apply the taxonomy, and therefore follow logical decision-making processes in both directions, i.e., prospectively from the perception of certain signals to the execution of an action, or retrospectively, by considering a certain manifestation of behaviour (phenotype) and searching for the causes throughout the cognitive model. In the case of root cause analysis, the process starts with a manifestation of erroneous behaviour, i.e., from a "phenotype" (Figure 3.18). Then, for each cognitive function, *specific causes*, as well as possible *generic causes* that may have originated the erroneous effect, are identified. This process applies sequentially for each cognitive function and continues until only "specific causes" are identified.

As an example, the erroneous outcome of a manoeuvre may be the result of specific causes appearing during the execution of the manoeuvre itself, or may be the result of an error in one of the previous cognitive functions leading to the execution of the manoeuvre.

From this short description, it appears clearly how the application of the taxonomy associated with the model COCOM focuses on the detailed cognitive processes that lead to erroneous behaviour and considers the environmental and contextual conditions that have generated the error. It requires a very precise recognition of the dynamic interaction between human and machine and can be applied when specific individual factors need to be discovered or defined.

On the other hand, it becomes very difficult to apply for a complete accidental sequence, which is made of several combined and interdependent events and episodes. Therefore, for the analysis of the sequence of events occurring during an accident, it may be more appropriate to apply a different methodological approach that is less precise in the evaluation of the cognitive functions involved, but favours the assessment of the accident from a wider perspective.

Root Cause Analysis by ADREP

The taxonomy Accident/Incident Data Reporting (ADREP) supports mainly retrospective analyses and classification of human-related factors associated with an accident. In particular, the ADREP reporting system codes all factors according to the findings of an accident investigation (accident report).

The ADREP method for classifying accidents was developed in the eighties (ICAO, 1987, 1991, 1993). As discussed earlier in this chapter, a new classification system has been recently developed (ADREP, 2000) with the aim of improving the way in which accidents are classified, especially from the human factors perspective.

The new taxonomy is associated with the SHELL structure (ICAO, 1997). Given that the modelling architecture of SHELL is very generic, the associated taxonomy can be considered as a living form of classification, in the sense that it is capable of accepting a new terminology and can expand according to the development of research and new findings in HF.

In ADREP, the accident is structured on a sequence of "events," organised in a "time-line" fashion that follows the order of occurrences (Figure 3.19). Usually, accidents comprise several related "events," such as "engine failure" or "under-shoot" where one event leads to another.

The RCA process that takes place in ADREP concerns the search for the primary causes of each "event." An "event" is further structured in a number of "descriptive factors" which define *what* happened during the event by listing all phenomena present, including human contributions. When a descriptive factor is expressed in relation to human factors issues, then "explanatory factors" are considered to answer the question of *who* and *why* failures occurred.

The approach applied in ADREP allows the user to be very precise and specific on the root causes, specially for what concerns individual and personality aspects, as well as social and contextual factors. However, less attention is given to specific cognitive functions.

This approach presents the advantage of being very comprehensive in all factors that may affect human performance. On the other hand, it has the drawback of being quite complicated from the psychological aspect and related taxonomy, which requires that either the user is knowledgeable in human factors and psychology, or that an "intelligent" interface for the classification of information is applied.

The application of ADREP for prospective-type studies is possible but presents some difficulties. The main reason for this lies in the fact that the method does not contain any logical structure that connects the "events" with each other. This

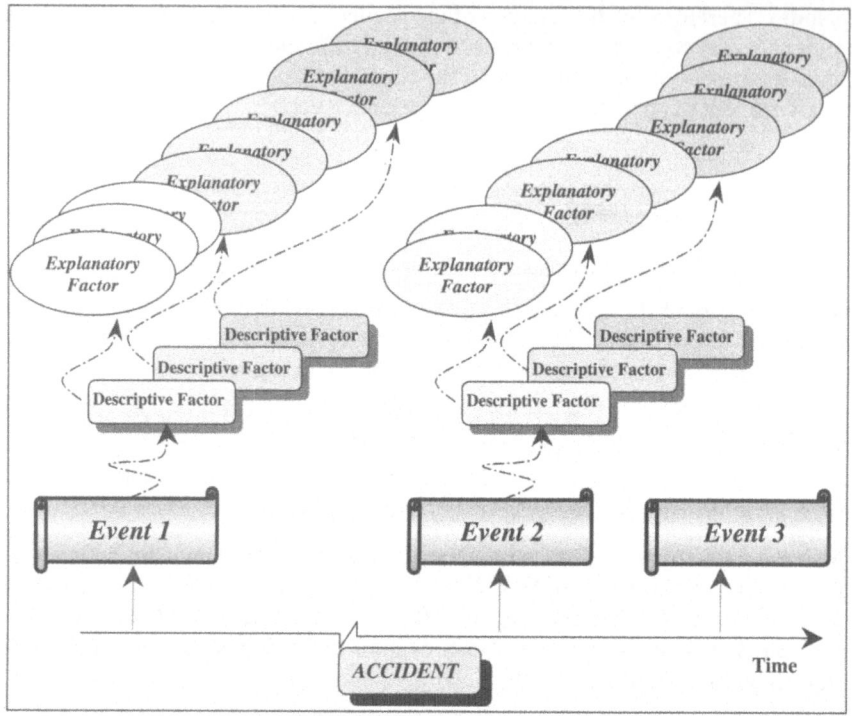

Figure 3.19 ADREP-2000 representation of an accident.

depends on the origins of ADREP, which was born with the sole aim of supporting RCA, while the sequence of events leading to the overall accident is assumed to be known from the report of the accident investigation.

Integrated Systemic Approach for Accident Causation

The Integrated Systemic Approach for Accident Causation (ISAAC) offers another framework for assessment and identification of root causes derived from combined human and plant/system-related factors.

This method, originally developed as the Simple Method for Accident Causation (SMAC) (Cacciabue et al., 1998; Carpignano and Piccini, 1999; Cacciabue, 2000a), makes reference to a model of human errors within organisational perspective and aims at the definition of active and latent errors made by different actors operating at various levels and at distributed timeframes within an organisation (Reason, 1990, 1997; Maurino et al., 1995).

As for CREAM, the ISAAC approach can be applied in both ways, i.e., it can be utilised for retrospective and prospective analysis, given the appropriate boundary and initial conditions, and the direction of the analysis that is performed. In Figure 3.20 the flowchart of ISAAC for retrospective analysis is shown.

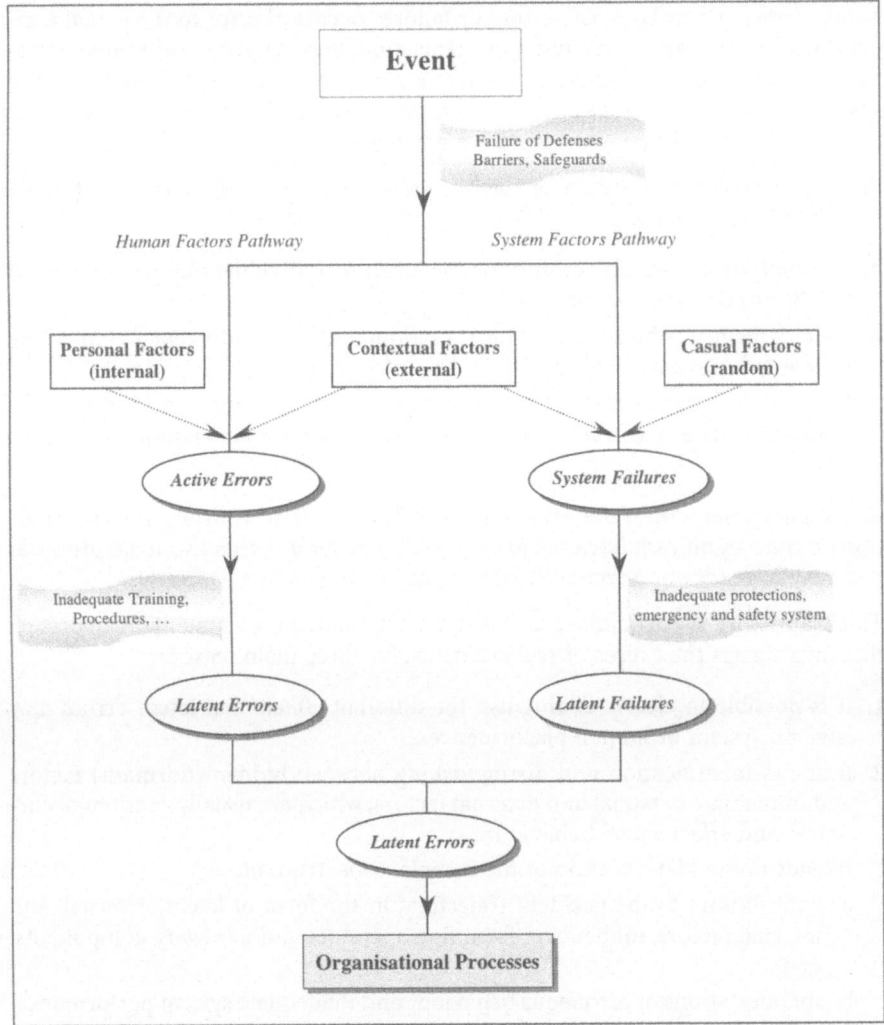

Figure 3.20 Integrated systemic approach for accident causation (ISAAC), retrospective analysis flowchart procedure.

In ISAAC, in order to explain the causes and reasons of an *event*, two pathways are considered:

1. *system factors pathway*, which considers inadequate performances of the system; and
2. *human factors pathway*, which estimates manifestations of erroneous behaviour.

In the *human factors pathway*, *active errors* are associated to inappropriate behaviour of front-line operators during the execution of a task, and, in the *system factors pathway*, *system failures* are failures of hardware and components of the plant while operating or at the time of demand.

Latent errors and *failures*, i.e., errors or failures occurred prior to the actual accident that lie dormant in the system/organisation, may influence either manifestations of inappropriate behaviour (*active errors*) or hardware malfunctions and system faults (*system failures*). Both pathways may depend or derive from latent errors, made at a different time and circumstances.

Furthermore, other causes may further enhance or generate *active errors* and *system failures*. These are:

- *Personal factors*, which account for the effect of individual physical or mental conditions on *active errors*;
- *Casual factors*, which account for random system contingencies, affecting *system failures*; and
- *Contextual factors*, which comprise physical, environmental, and local conditions, external to the individual, that may act on both *system failures* and *active errors*.

At a higher level within the organisation, it is possible to consider the contribution of more significant *latent errors*, derived from certain *organisational processes* that may engender or affect both *system* and *human factors pathways*.

This framework is particularly useful and is well suited for a systematic approach that investigates the causes of real accidents, for three main reasons:

1. It is possible to clearly distinguish the different effects that latent errors may have on system or human performances.
2. It allows identification and distinguishing between hidden (dormant) factors and immediate personal and external factors, which are usually random occurrences and affect active behaviour.
3. It leads to the identification of the complete spectrum of:
   a. contributors to the accident trajectory, in the form of latent, external, and internal factors, influencing front line operators and hardware components; and
   b. manifestations of erroneous behaviour and inadequate system performance, as they appear in the actual sequence of occurrences.

In practice, the way in which the ISAAC method is applied in a *retrospective* way requires the performance of the following sequence of steps (Figure 3.20):

1. Each *event* is associated with the failure of a certain defence, barrier, or safeguard, and is derived from a *human factors pathway* and/or a *system factors pathway*.
2. In the case of *human factors pathway*:
   a. An *active error* is identified.
   b. The analyst should then assess whether certain *personal factors* and/or *contextual factors* can be recognized that fostered the specific error at the time of occurrence.
   c. In addition to personal and contextual factors, or in the case that no such factors can be pinpointed, *latent errors* must be investigated that played a specific and unique effect on the *active error*. Examples of these *types* of

latent errors are "inadequate training," "unclear emergency operating procedures," etc.

d. It is then necessary to consider the possible contribution of more significant *latent errors*, made at a higher organisational level (*organisational processes*) that may have also engendered other failures at the system level. Examples of these *types* of latent errors are "misleading policies," "inadequate cost – benefit indications," "flawed selections of contractors," etc.

3. In the case of *system factors pathway*:

a. A *system failure* is identified.

b. Then it is necessary first to assess whether or not *casual* and/or *contextual factors* occurred that affected the specific *system failure*.

c. In addition to these factors, or in the case that no such factors can be identified, the analyst must consider the contribution of *latent failures* that affected specifically the *system failure*. Examples of these *types* of latent errors are "improper resetting after maintenance of a protection system," "erroneous design specifications for a safety device," etc.

d. Finally, as in the case of the human factors pathway, higher-level *latent errors*, made at organisational level, should also be considered as possible and relevant contributors to the *system failure* under scrutiny.

The application of the ISAAC approach for *prospective* analysis requires that the cause–consequences structure depicted in Figure 3.20 is reversed, starting with possible sets of latent errors and personal, contextual, and casual factors that may generate certain active errors and/or system failures, which then lead to hazardous events.

Formally, it is possible to describe the connections between different occurrences, i.e., errors, factors, and failures, by "backwards" and "forwards" logical links, which characterize certain occurrences as the result or the cause of correlated occurrences. These types of connections are presented graphically by "*arrowed full lines*" (backwards links) or "*arrowed dotted line*" (forwards links) (Figure 3.20).

A specific application of the ISAAC method for a complex retrospective real case study will be shown in Chapter 8, where this formalism for describing the connections between occurrences and their cause–consequence correlations has been applied in detail.

The ISAAC method is not supported by a specific model architecture and taxonomy. This presents the advantage that the approach can be applied within a less strict formalism than the approaches previously described and allows the analyst to use his/her experience in the domain. On the other hand, the absence of a predefined taxonomy may be problematic when the analyst is a novice and needs guidance for the performance of a HF study.

3.6.2 Comparison of Methods for Root Cause Analysis

The three approaches for RCA described in the previous section characterise three ways of representing the relationship between the manifestations of inappropriate behaviour and the causes that originate them. The three methods support the analyst in the process of reconstructing the RCA-tree that links a particular error to its causes, which may be distributed within an organisation at different levels and may have occurred at different points in time.

These three approaches are based on modern theories of human behaviour that focus on cognitive and organisational factors rather than behavioural aspects of HMI and share the aim of going deeper in the analysis of causes than the simple explanation of the manifestation of inappropriate performances.

In this sense, these three methods can be considered as paradigms of modern techniques for RCA and cover the spectrum of possible methodological approaches related to theories of cognition and socio-technical systems developed in the past 20 years.

Many other methods have been developed and can be found in the literature for classifying erroneous behaviour. They are mostly based on the theories and models discussed earlier and present similarities and commonalities with each other, as well as with the three methods discussed above (Rasmussen et al., 1981, 1990, 1994; Bagnara et al., 1989; Johnston and Botting, 1999; Davies et al., 2000; Paz Barroso and Wilson, 2000).

The relevant difference that exists between all these methods lies in the fact that they focus on different details and characteristics of human performance and are usually very specific for a certain technological domain, where human factors are considered safety critical issues.

In general, it is possible distinguish two main groups of classification schemes: microscopic and macroscopic type taxonomies. The RCA that may be carried out reflect this substantial difference. Microscopic-type taxonomies look at detailed aspects of the individual cognitive process and functions, and require a very accurate examination of all human interactions during an event. Macroscopic-type taxonomies focus primarily on the relationship existing between the individual human being and associated working context, and socio-technical factors.

This difference is very important and should guide the analyst in selecting the taxonomy and RCA approach that most suit his/her objectives and type of study to be performed.

Microscopic-type RCA

The RCA according to the microscopic or individual type taxonomies, e.g., CREAM, concentrates on the cognitive functions of behaviour and consequently tries to identify the specific function, in the case of CREAM either perception, interpretation, planning, or execution, that failed during a process of human–machine interaction.

When this function has been identified and the contextual conditions that have favoured the inappropriate performance have been highlighted, then the overall picture of root causes is identified. The organisational and personal causes that affect each step of the cognitive functions are also identified and the complete picture of generic and specific, context-dependent causes, is obtained.

The taxonomy CREAM can be considered the stereotype "microscopic-type" taxonomy, as it performs a detailed analysis of the individual cognitive processes (microprocesses) that lead to certain manifestations of inappropriate behaviour.

When applying CREAM, the organisational and contextual factors are still considered, but in direct connection with each specific step of the cognitive process and the temporal sequence of the HMI.

The main advantage of this type of taxonomy is that it is very accurate and goes into profound detail of RCA. The most important drawback lies in the complexity of the procedure of application and on the difficulty of finding all the precise information needed to reconstruct the cognitive processes to be examined.

Macroscopic-type RCA

The RCA according to the macroscopic- or socio-technical–type taxonomies, e.g., ADREP-2000, focuses on the links and correlations existing between individuals and organisations. The cognitive functions are not considered and only personal aspects are accounted for as possible factors influencing behaviour, in addition to all other socio-technical components.

In this sense, the classification of ADREP-2000 supports the analysis of organisational causes (macro-events) of inappropriate behaviour and hence the identification of ADREP-2000 as the stereotype "macroscopic-type" taxonomy.

The main advantage of ADREP-2000 is that the absence of search for the specific cognitive functions that failed makes the RCA very agile and rapid. This is particularly true when a computerised tool or automatic guidelines for classification support the RCA procedure.

The major drawback lies in the procedure of application and the dimension of the error taxonomy, which is very extensive as it attempts to cover all different socio-technical aspects affecting behaviour at all levels within an organisation.

The RCA according to ISAAC can be considered as a macroscopic-type analysis, as it attempts to correlate human inappropriate behaviour, as well as hardware failures, to higher-level occurrences within the organisation and focuses less on cognitive functions.

The ISAAC method is less formalised than ADREP and CREAM, as it is not strictly correlated to a detailed classification. Consequently, it does not contain an apriori and formal correlation between human erroneous behaviour and organisational and social aspects. This is an advantage rather that a drawback, as it allows the selection of the level of accuracy or granularity applied for the analysis on a case by case base. In addition, it allows the analyst to define the best-suited and specific taxonomy for the case under study.

However, it requires a bigger initial effort of definition of the taxonomy to be applied, which must consider the coupling between active errors, or failures of components, with remote causes, such as maintenance errors or management impositions, as well as with specific personal and socio-technical contextual conditions. In this sense, ISAAC enables description of the human interactions with both the micro- and macro-analysis of events, leading to the identification of the root causes within the organisation as well as individual characteristics.

The main advantage of ISAAC is related to its simplicity, which is very valuable from the application perspective, but it leads to problems when trying to formalise the results. In other words, the fact that the RCA by ISAAC is not supported by a formal taxonomy makes it difficult to organise data and collect information in a way that is then comparable with the reporting of other events. Therefore, the simplicity and rapidity of application is counterbalanced by the difficulty of formalisation of acquired information.

Comparison of RCA Based on ADREP and ISAAC

Figure 3.21 depicts the structure of the RCA model adopted by ISAAC overlapped with the structure of ADREP-2000, and shows similarities and commonalities between these two frameworks.

In particular, the definition of *event* is common to both approaches and represents the actual circumstance that has resulted from a combination of several human and systemic factors. *Active errors* and *system failures*, in the ISAAC accident causation framework, correspond to *descriptive factors* of ADREP-2000, as they represent the occurrence that led to the event and is usually represented in terms of failure of certain defences, barriers, and safeguards. The rest of the accident causation framework ISAAC, i.e., active/latent failure structure and associated factors affecting human errors and system failures, is well identified by the model and taxonomy of ADREP-2000, associated with *explanatory factors*.

In more detail:

- *Personal factors* of ISAAC are characterised in the taxonomy ADREP-2000 as *explanatory factors*, by a complete set of items referring to "physical," physiological," and "psychological" conditions as well as to "workload management," and are associated to the SHELL model element *Liveware (L)*;
- *Contextual factors* and *casual factors* of ISAAC, as well as *latent errors* or *failures*, can also be found in ADREP-2000 taxonomy at the level of contributory *explanatory factors*, classified as interactions between the SHELL modelling elements, namely *Liveware–Hardware* (L-H), *Liveware–Environment* (L-E), *Liveware–Software* (L-S) and *Liveware–Liveware* (L-L). In particular:
  —L-H interaction makes reference to design errors as well as inappropriate/poor/missing operational material that may generate system failures or human active errors;
  —L-E interaction considers an inadequate and unhealthy working environment, as well as lack of regulations, bad management, and organisational effects;
  —L-S interaction accounts for insufficient or inappropriate training and procedures;

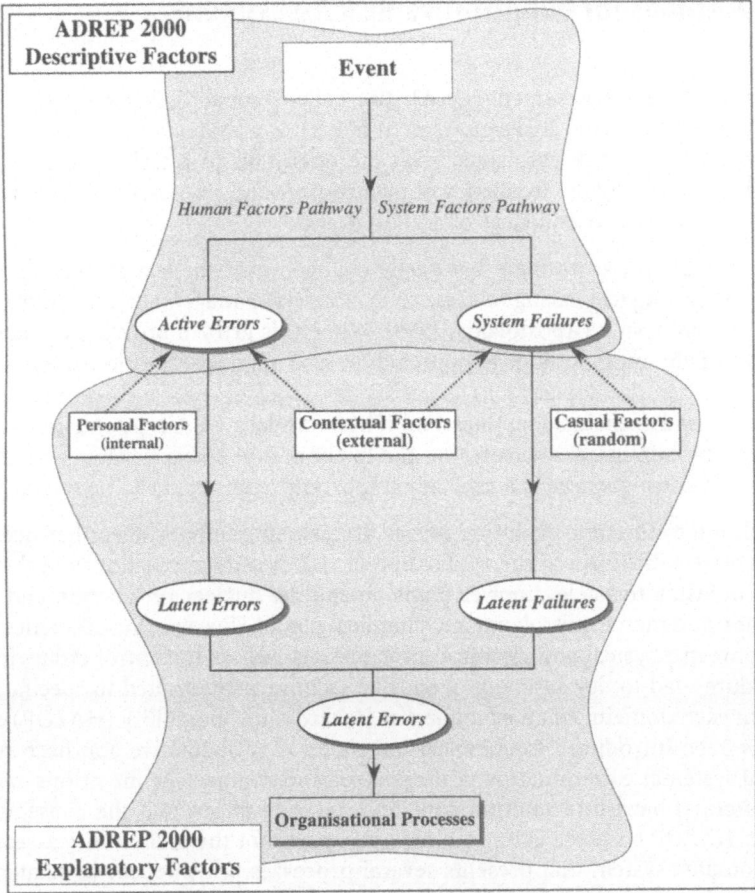

Figure 3.21 Comparison between ADREP-2000 and ISAAC.

—L-L interaction represents problems derived from the interaction of different actors during the performance of tasks in teamwork.

It can therefore be argued that the ADREP-2000 taxonomy may be applied in support of the ISAAC approach. In this way, while the drawback of lack of formalisation of ISAAC would be solved, the advantage of simplicity of analysis could be lost.

Ideally, the approach by ISAAC could be applied for a preliminary evaluation of root causes without a too strict formalisation. Then, according to the need and intention of the analyst, one could apply a more accurate and precise approach to deepen and formalise the analysis either in the direction of organisational issues or for the definition and identification of specific cognitive functions of interest to the study under development.

3.7 Methods for Quantitative Risk Assessment

Quantitative Risk Assessment (QRA), also called Probabilistic Safety Assessment (PSA), or Probabilistic Risk Assessment (PRA), is a systematic approach for the evaluation of the risk associated with the operation of a technological system, aimed at estimating the frequency of occurrence (and associated uncertainty distribution) of the consequences of certain events.

The scope of risk assessment has vastly changed over the last 15 years, progressively expanding its bearing to areas such as safety management, regulation development, and design (Apostolakis, 1998). While this growth proves the power and validity of the methodological approach, it also requires that new methods and techniques are developed so as to satisfy the requirements and specifications in new areas and domains of application. When standard QRA type analyses are performed, the numerical measures on the likelihood of certain events and of their associated consequences is a crucially important requirement to be satisfied.

In addition to the time evolution of risk assessment methods, there has been also a progressive diffusion of the application of risk-based approaches from the aerospace industry to nuclear power plants, including nuclear production and waste facilities, and then to petroleum and chemical plants. However, the difference existing between physical and chemical processes as well as in control strategies and procedures, led to development of specific techniques dedicated to specific problems in each domain. As an example, the Hazard and Operability (HAZOP) analysis has been introduced in chemical industries as a qualitative approach for the formal systematic examination of the process and engineering intentions to assess the potential hazard of malfunctions and mal-operations and the consequential effects. HAZOP has been estimated to be necessary for the preliminary assessment of a complex system that presents several processes, which occur in sequence or in parallel, each of them involving many hazardous chemical and thermodynamic reactions.

In the nuclear domain, PSA has focused on other issues, such as the evaluation of the risk associated with the release within the nuclear core container ("PSA-level 1"), or within the reactor building ("PSA-level 2"), or to the atmosphere ("PSA-level 3").

Safety regulations, and especially reactive measures imposed on industry by safety authorities following severe accidents, have an even more important impact on the evolution of risk methods than the differences due to the type of processes. As an example, quantitative risk assessment in the chemical industry, defined as the identification of potential hazards, the estimation of consequences, the evaluation of the probability of occurrence, and the comparison of the results against acceptability criteria, is nowadays part of a larger systemic approach, known as Safety Management System (SMS) (Cacciabue et al., 1994). This method is a typical proactive measure for safety assessment, which considers a plant as an integrated system, and combines standards, guidelines, procedures, auditing, safety policy, and QRA. The development and application of SMS for accident prevention followed the occurrence of a number of very serious accidents in Europe, in the late seventies in the domain of chemical and process plants. Therefore, in the chemical industry, QRA is part of a larger systemic approach and has to provide the important measures of risk, in quantitative and qualitative terms, to sustain the other elements of SMS.

In the nuclear domain, safety authorities have not retained the concept of SMS. However, the regulations of most countries demand that a "PSA-level 3" is performed, i.e., that the risk is calculated for the exposure of the population to the release of radioactive material and energy following an accident. Therefore, the probabilistic evaluation of the source term for the calculation of the effects on the population and environment becomes very important.

In this scenario of methods and approaches for QRA and SMS, the role of human factors and, more specifically, of Human Reliability Assessment (HRA) is very relevant. Adequate and substantiated methods for HRA have to be applied within any QRA or SMA application.

3.7.1 First-generation Human Reliability Methods

The human contribution to risk is an integral part of any QRA analysis and the majority of HRA methods developed and applied over the years, have seriously considered the issues of data availability and integration within a larger methodological framework.

In order to provide a substantial contribution to QRA from the human reliability side, a variety of methods were developed during the seventies and eighties. The methodological principle followed by most of these methods demands that HRA provides probabilities of success, or failure, of certain procedures involving human actions sustaining the overall QRA process. The basic data and parameters for human errors are derived from accurate field studies.

The most relevant technique and complete framework, developed in those years, for inclusion of human contribution to QRA is certainly the Technique for Human Error Rate Prediction (THERP) (Swain and Guttmann, 1983; Hannaman and Spurgin, 1984). Other methods, based on the same principles, and focused on slightly different issues, such as data collection, or semidynamic representation of HMI, were developed in the same period. These methods have been studied and compared in order to identify the most suitable ones for different types and objectives of safety analysis (Howard and Matheson, 1980; Hunns and Daniels, 1980; Seaver and Stillwell, 1982; Wreathall, 1982; Hannaman et al., 1984; Embrey et al., 1984; Dougherty and Fragola, 1988; Whalley, 1987; Williams, 1993).

All these methods represent a vast body of approaches essentially focused on behavioural aspects of human performance and may be considered as "first-generation" methods of HRA. More recently, a number of well-developed summaries and critical reviews on such methods have also been published and may serve as reference for the scope of analysis and application that first-generation methods may offer for inclusion in QRA studies (Apostolakis et al., 1990; Kirwan 1994).

3.7.2 Second-generation Methods

The first-generation HRA methods are well suited for supporting QRA, as they provide the probabilities of human errors and thus fulfil the primary requirement of reliability analysis.

However, the consideration for human behaviour included in all these techniques suffers a strong bias towards the quantification, in terms of success/failure of action performance, and there is less attention paid to the real causes of human error. These methods mostly take into account HMI if normative behaviour (i.e., behaviour as contained or foreseen by procedures) is considered, and are very superficial for what concerns cognitive processes.

Nowadays, the complexity of the system and the role assigned to human operators in the control loop, from extensive use of automation, require that special attention is dedicated to consider and tackle errors of cognition and decision making, rather than errors of overt behaviour. Indeed, the latter are usually foreseen at design level and counteracted or prevented by safety barriers or protections. This implies that errors of cognition are usually deeply rooted in the socio-technical context and become very difficult to control and contain when they infringe the engineered protection system.

Another important aspect of HMI is the dependence of human errors on the dynamic evolution of incidents. From this viewpoint, the overall picture of first-generation methods is even less encouraging. Very few methods make any reference to dynamic aspects of HMI and human–human interaction, while the static dependencies are well considered in almost all approaches.

These two issues, i.e., the ability to account for reasoning and decision-making processes, typical of cognitive behaviour, and the consideration of dynamic interplay between humans and machines, are the driving force of most methods of recent development. The solution of these two issues represents a real breakthrough that leads to the development and application of a new body of techniques and justifies their grouping into a "second-generation" HRA methods. Some of these methods will now be briefly described as examples of "second-generation" approaches. However, a rich literature on this matter exists and the reader is referred to it for identifying the spectrum of methods that can be considered for solving specific problems related to risk assessment and human factors (Williams, 1993; Kirwan, 1994; Wreathall et al, 1996; Kirwan et al., 1997; Bieder et al., 1998).

Method ATHEANA

ATHEANA (A Technique for Human Event Analysis) is a fully structured methodology developed for the U.S. Nuclear Regulatory Commission to respond to the needs of new HRA approaches oriented on cognitive behaviour of operators as well as performances and actions (Barriere et al., 1998).

As this method is to be integrated within a QRA-type approach, it must respond to some fundamental requirements, mainly: (a) produce probabilities and uncertainty boundaries associated with erroneous performances or erroneous imple-

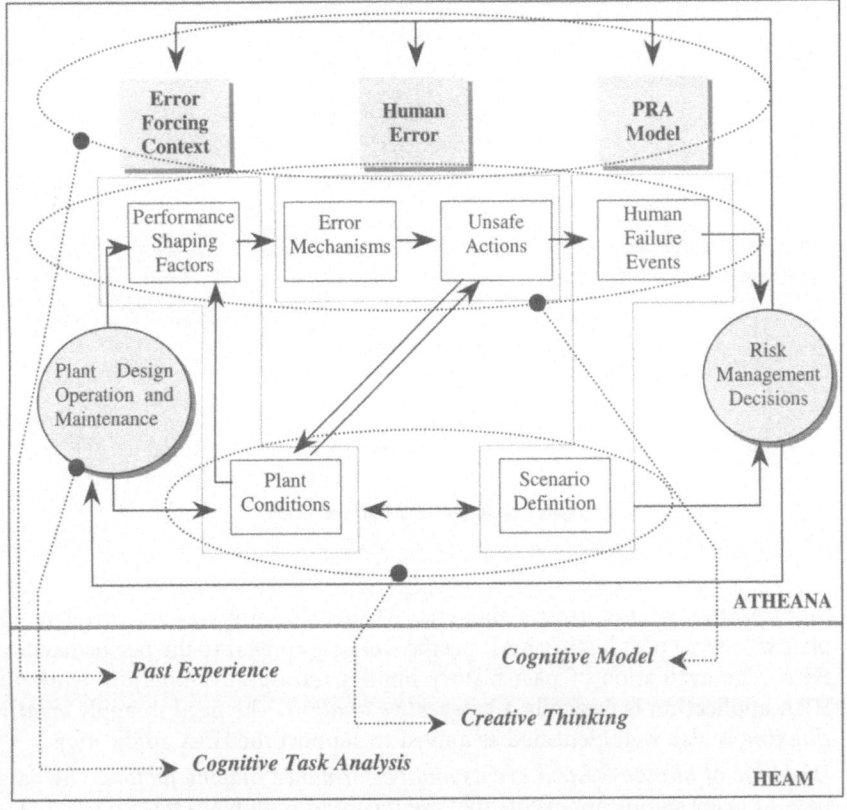

Figure 3.22 ATHEANA framework and human error and accident management requirements.

mentation of procedures; (b) use consolidated sets of data; and (c) ensure validation of method. For these reasons, the application of ATHEANA goes through a
relatively lengthy process, based of five main steps:

1. *Preparation* (select overall scope of analysis; collect background information;
   establish priorities for examining initiators/event trees; prioritise plant func-
   tions/system).
2. *Identify* human failure events (HFE) and unsafe acts (UA).
3. *Identify causes* (mistakes and circumvenctions; slips and lapses).
4. *Quantify HFEs.*
5. *Incorporate* into PSA.

The method that is proposed in ATHEANA tackles the most important open questions and issues of second-generation methods and follows a process by which root
causes of different nature, including organisational issues, and plant conditions are
combined to produce probabilities of erroneous behaviour (Figure 3.22).

In ATHEANA a number of important aspects and requirements of HEAM are
considered:

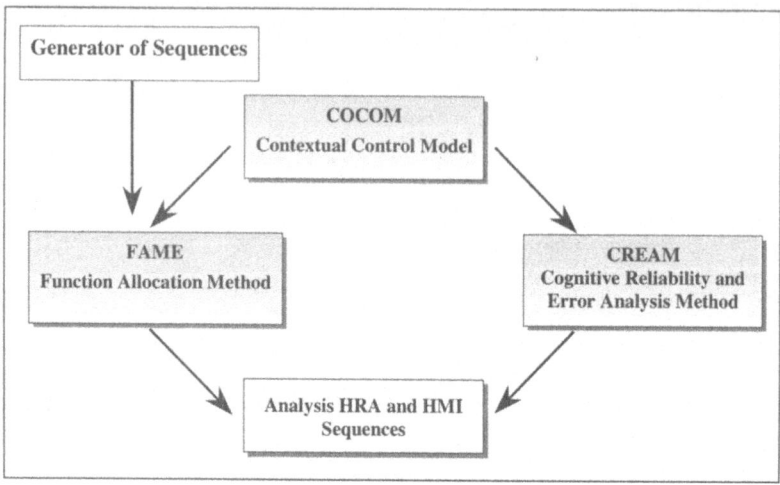

Figure 3.23 COCOM-CREAM-FAME methodology.

- *Retrospective vs. prospective analysis.* ATHEANA proposes to correlate the plants/systems past history and specific working context to the performance of HRA. The evaluation of past history implies *retrospective analysis,* while the HRA application is basically a *prospective analysis.* The need to apply *creative thinking* is also well identified as a must to support the HRA application.
- *Database of human–system events and performance shaping factors.* The database of prior significant events that are recorded in human–system event classification scheme is a very important feature of the methodology and represents one of the distinguishing elements of the method.
- *Model of cognitive activities.* The model applied in ATHEANA is based on the information processing system (IPS) paradigm, so as to retain the most important feature of cognitive activities.
- *Task analysis.* The use of CTA is advocated in ATHEANA for collecting information on tasks and procedures. The possibility of applying formal approaches or more advanced analyses, specifically focused on cognitive processes, is left to the choice of the analyst.

Method COCOM-CREAM-FAME

This approach (Hollnagel, 1998) considers human cognitive behaviour and environmental factors for HRA assessment and is based on three elements (Figure 3.23):

1. a mechanisms for analysing the allocation of functions (FAME);
2. a model of human information processing a mechanism, which accounts for contextual control factors (COCOM); and
3. a cognitive reliability and error analysis method (CREAM).

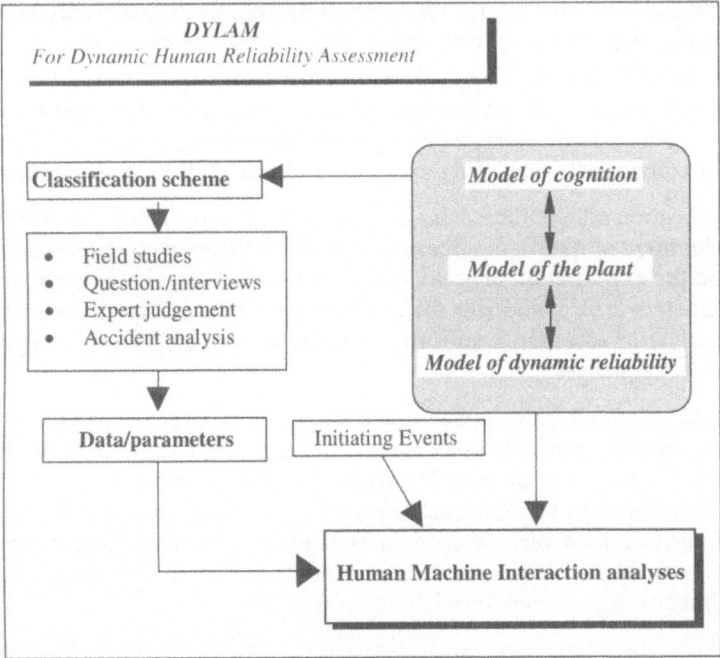

Figure 3.24 DYLAM for HRA framework.

The methodology comprises of a model of cognition COCOM by which human interactions can be simulated focusing on the cognitive aspects, considering the effects of context and different control behavioural modes. The human cognitive aspects are then combined and evaluated in order to generate the data structures that allow to classify human errors for inclusion in a formal reliability analysis.

This methodological approach for Human Machine Interaction is then imbedded in a process that generates the sequences of events according to predefined criteria and dynamic considerations, which allows to simulate plant and operator in an integrated fashion.

Method DYLAM

The Dynamic Logical Analytical method (DYLAM) for human error risk assessment (Cacciabue, 1997) is based on four main components, two of a theoretical nature, i.e., a *cognitive model* of human behaviour and a taxonomy or *classification*, and two of a methodological and practical nature, a *database* and the *formal method* for applications (Figure 3.24):

1. A *cognitive model* is selected. The model must account for the fact that the operator interacts dynamically and interactively with the plant.
2. A *classification* scheme of erroneous behaviour is strictly correlated to the cognitive model.

3. A *database* of inappropriate behaviours is derived from past events, containing both functional response of the plant and human performances.

4. A *formal method* is then applied for structuring the interaction of the models of cognition and plant and for controlling the dynamic evolution of events. This method generates the normal or inappropriate dynamic behaviour of operators and/or components, according to logical and probabilistic considerations.

The classification scheme, based on the model of cognition, guides the field studies, the development of questionnaires and interviews, the extraction of expert judgement, and the examination of accidents/incidents, with the aim of estimating data and parameters to be included in the analyses. A series of initiating events, selected by the analyst in accordance with the objective of the safety study, triggers off a whole set of sequences.

The DYLAM methodology, applied uniquely for the prospective type of studies, is quite complicated, especially as a consequence of the dynamic reliability module. Therefore, it is mostly applicable for detailed studies of selected subsystems with limited complexity and operational impact. On the other hand, it offers the possibility of performing a very detailed analysis of the system and interactions with humans for many different combinations of faults and errors. However, it is limited by the simulation power and availability of data.

3.8 Recurrent Safety Audits Throughout an Organisation

Recurrent Safety Audits (RSA) throughout an organisation are studies carried out at regular and recurrent intervals to evaluate and measure specific safety indicators by which it is possible to ascertain safety standards of an organisation, and detect possible risky conditions.

In addition to these two safety critical features, RSA must be correlated to methods that follow the initial findings of risky conditions for discovering their reasons and causes. Moreover, it can be applied for identifying where and which changes are necessary for fulfilling new safety requirements, in order to keep adequate safety levels throughout the whole lifetime of a plant.

Safety audit is obviously a *retrospective* type of analysis which relates directly to the identification of data, critical system functions, specific human–machine characteristics that require particular attention and that need to be evaluated in order to assess possible hazards. However, it can be argued that such types of analyses are also *prospective* in that they attempt to evaluate the safety state of an organisation with respect to a variety of hypothetical initiating events and possible failures/errors by measuring selected safety indicators.

Safety audit is the critical step in all safety management control cycles, as they constitute the feedback that enables an organisation to maintain and develop adequate ability to manage risk and retain appropriate safety standards (Byrom, 1994).

All systems tend to deteriorate over time or to become obsolete as result of change. In addition, usually, the focus on efficiency induces a relaxation of the safety measures, implemented at design and construction stages. This calls for systems to be regularly audited. Auditing provides information on the implementation and effectiveness of plans and performance standards and indicates a way forward to improve or re-establish adequate safety levels.

The ISO standards 10011–1, 1990 defines audit as "a systematic and independent examination to determine whether (quality) activities and related results comply with planned arrangements and whether these arrangements are implemented effectively and are suitable to achieve objectives."

The fundamental sources of information on which an audit may be based are typical ethnographic approaches by which to examine a working environment and observe people at work, and reviews of documentation and data, which offer a factual and objective measure of the adequacy and completeness of the systems dedicated to the control of risk and management of safety (Cacciabue et al., 1994).

The outcome of this process of preparation for the actual audit process consists in the definition of a number of parameters and indicators that are then actually measured and evaluated during the activity at plant site.

In practice, the assessment of the overall safety level of an organisation requires a sort of methodological framework where methods and approaches for considering HMI are combined and integrated (Wilson et al., 2001).

The importance and role played by recurrent safety audits of organisations has been identified many times as a key issue to preserve systems from accidents. Furthermore, it has been shown that organisations with poor or nonexistent safety audit processes are much more exposed to hazards and have experienced serious accidents (Cacciabue et al., 2000).

The most consistent way of performing an accurate analysis of the safety level and trend of an organisation is to examine each of the sociotechnical elements that characterise the dynamic evolution and processes of an organisation. As discussed earlier in Chapter 2, these elements are: *organisational processes, personal and external factors, local working conditions,* and *defences–barriers–safeguards.*

In summary, it can be argued that the process of RSA of an organisation is a quite complicated and extensive process that requires the application of several methods and field analyses. Usually, for real applications a framework like the HERMES methodology is utilised for the specific purpose of analysing an organisation and deriving safety indicators. In the next chapter, a much more accurate account of practical implementation guidelines for RSA will be developed. Moreover, in Chapter 7, a specific application of RSA will be performed according to the HERMES methodology, selecting amongst existing methods those that most suited the safety study.

The TRIPOD Method

The method called TRIPOD was developed by research teams from the universities of Leiden and Manchester, and aims to identify the root causes of accidents and unsafe acts involving operators (Reason, 1997).

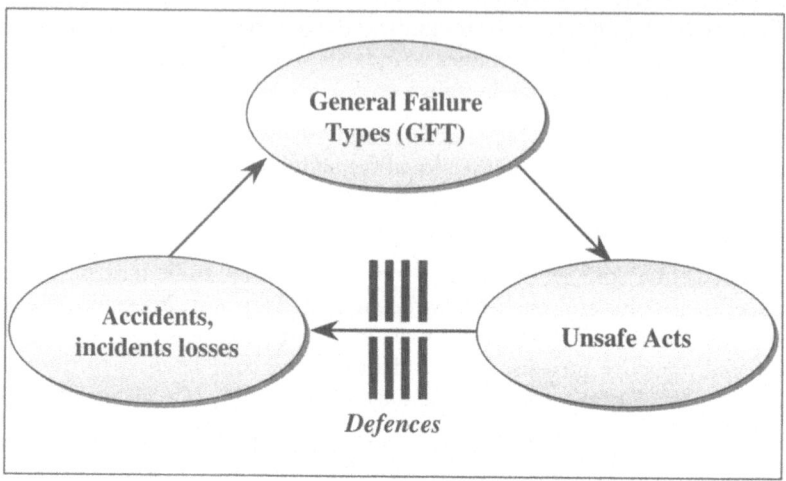

Figure 3.25 TRIPOD method.

The underlying idea is that to reduce human contribution to accidents/incidents it is necessary to measure regularly the safety state of an organisation and identify the offspring of local triggering factors or of General Failure Types (GFT) and apply appropriate remedial interventions.

The basic philosophy of TRIPOD considers, as the name implies, three basic elements on which the method rests (Figure 3.25):

1. The *unsafe acts*, i.e., the performance of inappropriate or hazardous operations that may break the system defences and barriers.
2. The *accident*, i.e., *incident losses* that are the result of unsafe acts and failure of the defences.
3. The *general failure types* that represent those latent conditions, either engendered from past events or produced as result of erroneous decisions at management and organisational level that affect substantially all performances of front line operators.

In order to define the GFT the authors of the methodology have developed a more detailed list of failure types that need to be studied by the analyst. This list of factors, although generic, should help in looking for GFT specific to different domains by asking/interviewing personnel involved in the everyday management of a system, or by obtaining the necessary information from task specialists and/or by direct observation of workplaces.

The list of GFT that have been identified can be associated with the different processes involved in the overall historical development and management of a plant that may foster and promote the unsafe acts. These general failure types are shown in Table 3.2. However, the analyst should devise this list on the specific characteristics of the organisation being studied.

Table 3.2 Organisational processes, general failure types, and error-forcing conditions

Processes	GFT
Statement of goals	• Incompatible goals
Organisation	• Organisational deficiencies
Management	• Poor communications
Design	• Design failures
	• Poor defences
Build	• Hardware failures
	• Poor defences
Operate	• Poor training
	• Poor procedures
	• Poor housekeeping
Maintain	• Poor training
	• Poor procedures
	• Poor maintenance management

Where the GFT can be further specified as:
- *Hardware*: concerns the available quality and quantity of tools and equipment.
- *Design*: implies failure by the designer to provide sufficient guidance or give sufficient evidence to his/hers goals, or simply to properly dimension/design parts of plant.
- *Maintenance management*: focuses on the maintenance planning and management rather that on actually performed maintenance activities.
- *Procedures*: relates to the quality of procedures and their workability in practice.
- *Housekeeping*: concerns the working conditions of personnel availability, responsibilities, inadequate investment, etc.
- *Incompatible goals*: refers to conflict either at individual level, or at group level, or at organisational level.
- *Communications*: covers all forms of communications.
- *Organisation*: considers all forms of organisational deficiencies, namely, organisational structure, organisational responsibilities, and management of contractor safety.
- *Training*: concerns all problems associated with training, such as insufficient and poor training, inexperience of tutors, poor task analysis, etc.
- *Defences*: comprise failures in detection, warning, protection, recovery, and containment.

The TRIPOD method is a typical *bottom-up* approach that starts from the observation of the actual working environment and human–machine system and looks for the factors and indicators that may enhance a specific risky condition that requires intervention.

Other approaches, especially safety management systems, are very different, as they tend to define the rules and characteristics that a system should have in order to be classified as safe. These other types of approaches are usually called *top-down* methods and are necessary as prior analysis tools to be put into practice before implementing a design.

3.9 Summary of Chapter 3

This chapter has aimed to present a short overview of the variety of methods and models that can be found and applied to support the performance of HF analysis.

The review of approaches is not, and did not intend to be, exhaustive of all existing techniques. However, it is sufficient to show the spectrum of available and applicable means to perform a HF analysis.

Much more important is that the analyst understands and accounts for the complexity of the HF problem and tackles it with the appropriate perspective. This implies that the analysts, and the management of any organisation under scrutiny, are prepared to spend the necessary effort and time for developing knowledge and experience of real working context before dwelling in the application of formal approaches. This process ensures that adequate and validated data and parameters are identified for performing the safety studies.

This way to deal with the problem of HF analysis is usually associated with a *bottom-up* approach, as it requires that working contexts be analysed without a fixed modelling and methodological structure in mind. From these initial conditions, analysts are able to select the most appropriate techniques for studying the system at hand and develop the necessary analytical tools for discovering the required information.

On the other hand, *top-down* approaches advocate as essential starting condition the existence of a clear and definite methodological approach. This can then be followed by a much more focused analysis of the field and context for the identification of relevant data and parameters.

This discussion has not been tackled in this chapter, as it is assumed that the real issues do not lie in the starting point of the analysis, but rather on the attitude that the analyst undertakes with respect to the system being studied.

In reality, it is impossible for the *expert* analyst to carry out a field observation totally based on a bottom-up approach, i.e., with no prior model or expected paradigm in mind concerning the organisation. Similarly, it is not possible to expect that even a well-developed and comprehensive approach fits any organisation and can fully guide the safety study without adequate consideration for the peculiarities and practices that are imbedded in the everyday life and working contexts.

In other words, an effective HF analysis cannot be developed uniquely as a *bottom-up* or *top-down* approach. Rather, the integration of these two approaches is the key to generate valuable and effective HF studies.

For these reasons, the issue of top-down vs. bottom-up approaches has been given little importance in this chapter, in favour of the analysis of the available methods that can be utilised to carry out retrospective and prospective integrated analyses in a HERMES perspective.

3.9.1 Theories and Methods

The models and paradigms that can be applied for studying HMI have first been analysed, focusing on the information processing paradigm, as the leading framework to describe HMI. In addition to the IPS paradigm, another generic structure (SHELL) has been reviewed that is commonly applied for considering wider sets of interactions existing in working contexts, and includes socio-technical, personal, and teamwork factors affecting human behaviour.

The approaches for performing ES and CTA have been considered. These two areas of development are, on their own, very extensive and relevant, and demand the application of structured and formalised methods, which are the subject of specialised monographic works. Consequently, this chapter has only considered the underlying theory on which these approaches are based. The reader interested in getting more insight in these two areas has been referred to the relevant literature. In any case, the HF specialist concerned in design and assessment of HMI and HMS must apply and develop adequate CTA and ES concerning the system under study in order to ensure the effectiveness and validity of the HF analysis.

The issue of ecological validity, already discussed in the previous chapter, has been considered from the perspective of methods and models that have been associated to this concept. Some of them bear strong links with the models of HMI and methods for ES and CTA.

The RCA methods that support primarily accident investigation processes have then been reviewed. The links and connections that exist between these RCA methods and the theories and models of HMI and HMS, in accordance with the HERMES application, have been highlighted.

The human factors contribution to risk assessment has also been considered. A number of specific approaches and techniques to deal with the performance of human reliability studies, from the probabilistic viewpoint, have been investigated. The most recent methods for dealing with cognitive aspects to human errors and dynamic reliability issues have been revised.

Finally, the safety audit problem and the integration of human factors consideration in RSA for assessing the safety levels of organisations have been tackled. This is a typical approach where the HERMES methodology becomes particularly useful. The existence of integrated approaches for performing safety audit is only considered in the wide domain of SMS. However, SMS consider whole aspects of a plant, primarily, the technological ones. In the case of specific HF safety audit, a method has been reviewed. In general, however, the performance of a safety audit based on HF issues requires the application of a more comprehensive approach, typical of the HERMES methodology.

All in all, at this point of the book, the reader should have developed sufficient knowledge of models and theories for HF analysis to enable him/her to plan in detail a specific study relative to a plant/organisation, in relation to a well-defined area of application, namely, design, training, risk assessment, and accident investigation. In the next chapter the development of specific procedures for applying HERMES and HF theories to these four areas of applications will be developed, before tackling four real applications of the HERMES methodology to real plants/organisations of different level of complexity and in different technological domains.

4
Guidelines for the Application of HF Methods in HMS

The objective of this chapter is the development of guidelines for considering Human Factors (HF) within the areas of application of *design, training, safety assessment*, and *accident investigation* for a Human–Machine System (HMS).

The rationale is to offer the reader clear stepwise procedures for applying and integrating, in practice, various consolidated and existing methods according to the overall frame of the HERMES methodology. This process of integration and merging of methods and models in a logical and balanced sequence depends primarily on the experience and expertise of the designer or safety analyst. Moreover, in certain cases, it is not necessary to use the most complicated methods to reach the goals of the analyst, simpler approaches being quite adequate to deal with the problem at hand. In some cases, the selection of the most suitable methods depends on boundary conditions and data, while in other cases, the scope of the study and the need of integration of human factors and system assessment methods guide the level of accuracy required.

The following sections will start with a discussion on the impact of HF on any technological assessment and, in particular, in the four areas of application. Detailed guidelines and stepwise procedures will then be developed for each one of them.

4.1 Impact of Human Factors on Areas of Application

The inclusion of Human Factors in design, training, safety assessment, and accident investigation usually aims at improving friendliness and efficiency of operational environments or at tackling safety-related issues, such as accident prevention and management of emergency situations.

Focusing on safety measures, the consideration of human error and accident management is essential in achieving really effective human–machine interactions during accidental conditions and designing safety systems that are actually useful and utilised during emergencies. However, human factors methods must be consistently applied even when studying normal operations to improve the operating

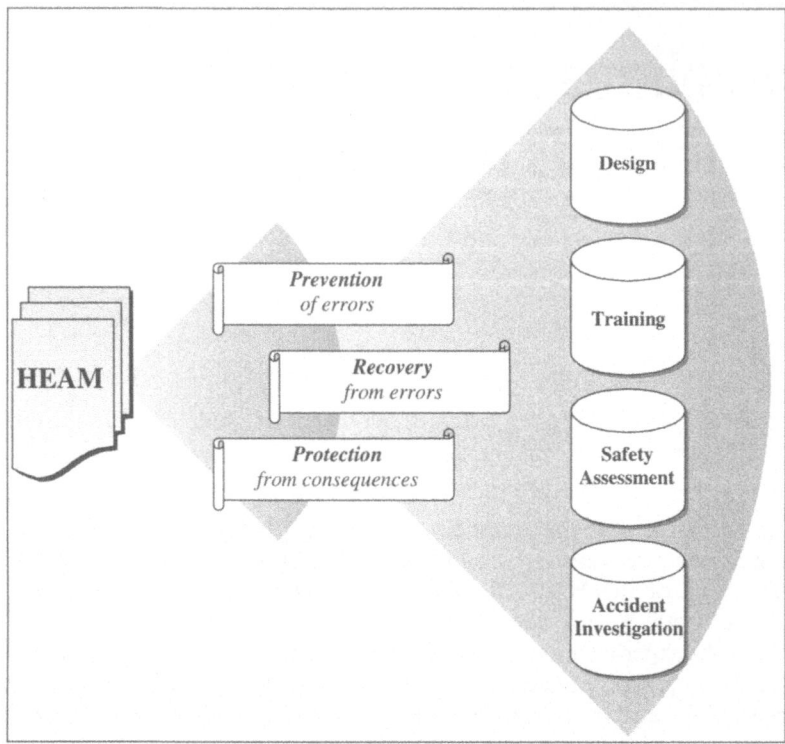

Figure 4.1 Aims and impact on safety of human error and accident management.

environment, thereby creating the optimal conditions for accident prevention and emergency management.

The availability of specific stepwise procedures for assisting the analyst in selecting and applying relevant Human Factors methods for each area of application, is therefore essential to ensure consistent and consolidated safety measures. The well-developed and formalised methods described in the previous chapter will be framed in such procedures, in order to offer the analyst a set of approaches and techniques from amongst which the most suitable ones can be selected for practical applications.

As this book focuses on safety and on Human Error and Accident Management (HEAM) measures, the following three basic objectives will be considered, either independently or in a combined way (Figure 4.1):

- *prevention* of human errors and system failures, in order to avoid incidents and accidents;
- *recovery* from errors and failures, once they have occurred, so as to re-establish normal conditions; and
- *protection* of humans and environment, when neither prevention nor recovery is possible, and an incident or accident has developed.

Once the goals of the HMS and safety measure under study are defined, it is necessary that the most suitable and qualified existing methods be identified, selected, and implemented. This requires good knowledge of literature, understanding of boundary conditions, limitations, and of data required by the selected methods. Moreover, the actual socio-technical context and organisational structure must be evaluated and analysed in order to derive specific data and knowledge that lead to the detailed application for the case at hand.

During the implementation of the various methods, the analyst needs to monitor and revise the application in order to ensure that the goals of the study are attained and that the selected methods are applied in a consistent way.

Finally, once the study is completed, the overall evaluation of the design/safety assessment must be carried out in order to ascertain whether the original objectives were attained, or what must be repeated and revised.

The analyst must follow a process by which the objectives of the HEAM measure and analysis are clearly identified, selected, and monitored during the development of a specific study, and the results finally evaluated in terms of quality and quantity of outcome with respect to the original goals.

In practice, four main stages should be followed that are quite typical of project management and may be described as the DSME process, i.e., Definition, Selection and Implementation, Monitoring, and Evaluation (Figure 4.2).

In more detail, the four steps of the DSME process are:

1. *Define* the aims of the HEAM analysis. This implies that either one of the three main objectives of HEAM measure, i.e., prevention, recovery, or protection is considered on its own, or that they are tackled at the same time, in some form of combination.

2. *Select and implement* specific methods. This requires that either a fully formalised approach that suits the analyst's need is identified, or a methodological framework that allows reaching the goals by applying a sequence of well-defined methods is built. This step is normally quite complex and requires field assessments and retrospective analyses for the definition of real data and evaluation of real socio-technical contexts.

3. *Monitor* attainment of goals. This requires that, during the application of the selected approach or methodology, the analyst constantly verifies that the HEAM measure under design or analysis focuses on the expected goals, and that the selected methods are consistently applied.

4. *Evaluate* results. This final step aims at evaluating the effectiveness of the results with respect to the overall objectives of the HEAM study.

4.2 Guidelines for the Application of HF Methods in Design

From the human factors perspective, the design of a specific piece of technology concentrates on the interfaces and controls, and on the procedures for imple-

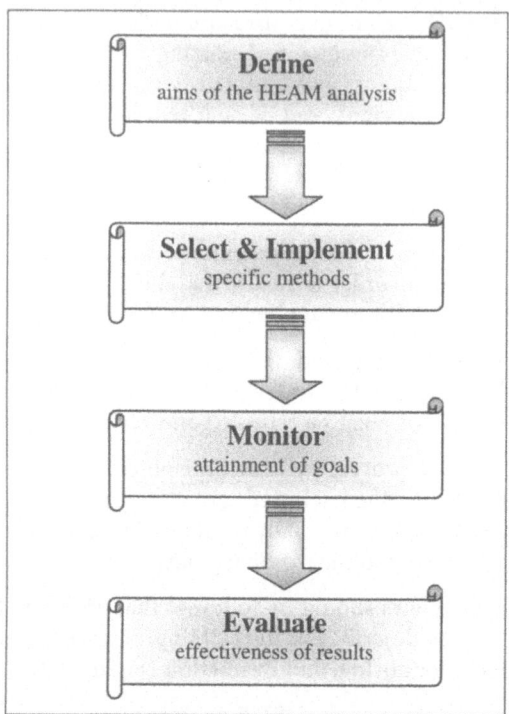

Figure 4.2 Four stages for implementation of human error and accident management measures – the DSME process.

menting the expected functions, in different working contexts, such as normal, transitory, and emergency conditions.

Guidelines and procedures for designing HMSs have been developed already several times in the past by well-known authors (Sanders and McCormick, 1976; Rouse, 1980, 1991; Sheridan, 1992; Salvendi, 1997) and they follow the general principles of the human-centred design approach and joint cognitive modelling, as they have evolved over time, staring from the initial IPS paradigm, which stipulates that the human being can be represented as an information-processing system. Following these original ideas for designing HMS, a series of improvements have been generated in accordance with the development of technology, especially in the case of rapid diffusion of automation (Billings, 1997; Sage and Rouse, 1999; ISO, 1999; Piccini, 2002).

This section does not intend to propose a new or different way to consider Human Factors in design processes, but rather to develop guidelines that can be followed by a designer for implementing the principles of ecological and human-centred design processes and associated methods, so as to utilise properly and exploit existing models.

The only requirement taken into account, as a conservation principle, is that the structure proposed in the generic methodology for integrating retrospective and prospective application, i.e., HERMES methodology, is preserved and followed at all times. This ensures that consistency and continuity exist between (a) the data and methods applied to collect information; and (b) the input data and approaches that are applied to make inferences and predictions in safety analysis.

In order to clearly define the objectives and functionality of a control system or safety measure, the designer or safety analyst must combine the needs of prevention, recovery and containment of errors, faults and consequences, together with the development of a modern technological system. In particular, three fundamental principles for designing HMI systems must be accounted for and merged. These are the principles of Supervisory Control, System Usability and User-Centred Design (UCD). They all rotate around the same concepts, which are examined from slightly different perspectives. The designer must effectively keep them in consideration and merge them in a balanced and effective manner.

The supervisory control principle implies a clear understanding of functions and roles of humans and automation (Rouse, 1990; Sheridan, 1992). System usability has been defined by the International Standard Organisation with the objective of enabling designers to measure the effectiveness, efficiency, and satisfaction associated with the use of a product (ISO, 1993; Bevan and Mcleod, 1994). UCD is an iterative design approach to usable systems development that, based on a clear modelling architecture of HMSs and user's feedback, produces more and more improved design solutions (Rouse, 1991; Billings, 1997; ISO, 1999).

These three principles must be combined in such a way that the designer can consider the role of the user from the initial stages of the design process, and then can verify and improve effectiveness, efficiency and usability by direct feedback and by field assessments with adequately selected users.

These three principles will be described briefly before developing the stepwise procedure to be followed in designing HMS.

4.2.1 Supervisory Control Principle

A "classical" way to perceive automation and supervisory control in modern technology was developed in the early nineties (Figure 4.3) (Sheridan, 1992, 1999).

In brief, five different levels of control can be envisaged. In the case of *manual control*, whether the process of controlling a system is performed by the operator alone (configuration *a*) or is supported by a computerised system (configuration *b*), the operator is in complete control and acts almost directly on the system. In *manual control* configurations, the support given by a computerised system represents only a prosthetic form for improving operator strength and working ability.

In the case of *supervisory control*, the computer becomes the major actor in controlling the system. The human supervisor defines goals of computer opera-

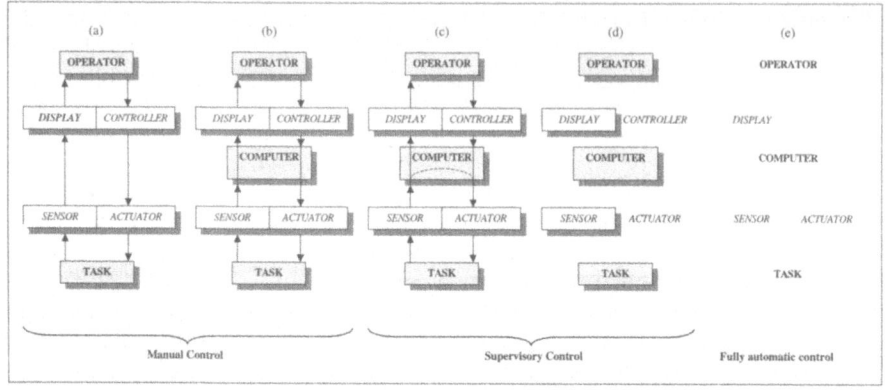

Figure 4.3 Paradigms of control process in technological settings (adapted from Sheridan, 1992).

tion, objective tradeoffs, models, plans and "if-then-else" procedures. Then, the human supervisor may assume direct control ("trade control") (configuration *c*) or may share control of some variables and processes with the computer (configuration *d*).

In *fully automatic control* (configuration *e*), the human operator is totally detached from the control of the system/plant and can only observe the implementation of the activities planned and decided during a preliminary phase. He/she can, in practice, only "pull the plug out" as an ultimate measure in influencing the process when "something goes wrong."

The last configuration of system control can no longer be defined as *supervisory control* and represents a condition in which systems are completely autonomous and should be designed and developed only in situations that are very well known and recognised by designers as well as by users. This is the case in some domains of application such as, for example, the automotive environment or guided transport systems, where certain systems operate in complete autonomy and the driver or operator can only observe their performance. Examples of such systems in automotive environment are "electronic traction control," "electronic stability control," and "electronic suspensions." It is interesting to note that the latter systems present an autonomy that is even greater that the highest level of automation envisaged by Sheridan. Indeed, as an example, the electronic suspensions cannot be made inoperative during operation, i.e., the driver cannot simply "pull the plug out" while driving. To do so it is necessary to perform a laborious and complicated intervention on the electronic control system, that can be done only in a specialised workshop (Cacciabue et al., 2002).

4.2.2 System Usability Principle

The International Standard Organisation has defined system usability as "the extent to which a product can be used by specified users to achieve specified goals with effectiveness, efficiency, and satisfaction in a specified context of use" (ISO, 1993, 1999).

Even if this definition may seem very clear and complete, the concept and scope of usability has been very much debated in the past years and is still the object of many research projects and initiatives, mainly related to the development of information technology (EC, 1998; Bevan, 1999; Cacciabue et al., 2002a). In principle, according to the guidelines of the ISO standards, the three principle requirements of effectiveness, efficiency, and satisfaction may be defined as follows:

- *Effectiveness* is the "accuracy and completeness with which users achieve specified goals" and allows the execution of all tasks for which a system was developed.
- *Efficiency* is a measure of the "resources expended in relation to the accuracy and completeness with which users achieve specified goals" and considers performance with respect to required temporal and economic resources.
- *Satisfaction* is the "freedom from discomfort, and positive attitudes to the use of the product" and reflects the user-friendliness, stress and frustration in usage of a system.

These definitions, however, do not represent a precise instrument to measure the usability level of a system, as they do not exactly define what features make an object usable.

Nevertheless, the limits of these definitions can't be exceeded, as usability is a relative concept depending on three factors, namely *task*, *user*, and *environment* (Bevan and Macleod, 1994). These three factors combine in generating the particular context of use for the system, and, in practise, it is impossible to establish in advance universal rules for different systems development. An evaluation of the real working context is necessary by adequate field examination of the interactions between task, user, and environment.

As a consequence, in order to design usable systems, these three factors (task, user, and environment) should be studied first. They constitute the system context of use. Information gathered on these factors permits definition of the system requirements and users' needs, necessary to make the system satisfy the usability definition, at least for what can be achieved by preliminary study and field assessment on the part of the designer.

4.2.3 Human-centred Design Principle

The idea of Human-Centred Design (HCD) was developed in the same years of the information processing system metaphor for maximising and exploiting the role of the human being in the interactions with "modern" machines (Norman, and Draper, 1986; Rouse, 1991; Sheridan, 1992). With the progress of automation and information technology, the HCD concept has been further refined and adapted to different working contexts and domains of application (Billings, 1997; ISO, 1999). However, the basic idea of HCD has remained constant and, as the name says, it relies on the fact that whichever system, product, tools is designed for a user, the role, needs, and peculiarities of the latter must be accurately considered.

In current technology development, the relevant role of automation and the supervisory control tasks assigned to operators demand that the basic idea of HCD embraces and considers the various levels of supervisory control and the usability principles. In practice, the HCD, or Human-Centred Automation (HCA), approach consists in the fact that the role of supervisory control assigned to the human operator is maintained, while planning the interfaces, procedures, and tasks, avoiding the trap of fully automatic control system. At the same time, the active contribution of end-users must be ensured at all steps of the design process, in the form of "participatory design" providing continuous feedback to the design process, to grant user friendliness, and maximising the exploitation of all features and potentialities of the control system.

The way to account for the user in the early phases of design is to consider a model of operator/user behaviour that interacts with the system performance. In this way, it is possible to generate an overall simulation of the "joint cognitive system" (Hollnagel and Woods, 1983) that enables the designer to perform predictions and calculations of possible interactions and control processes and to consider optimal system features and best control strategies. In such dynamic simulations, both the technical process and the human operator are modelled as equivalent parts of the joint cognitive system.

According to ISO standards (ISO, 1999), the main features of HCD can be summarised as active involvement of users; appropriate allocation of functions; iterations of design solutions; multidisciplinary design.

4.2.4 Integrated Approach to Design HMS

The three principles of supervisory control, system usability, and user-centred design must combine in such a way as to provide an integrated approach to the design of HMSs that retains and exploits all their essential features (Figure 4.4).

It is possible to develop an iterative procedure that is able to support the designer's activity and integrates these three principles (Figure 4.5).

From the very initial phases of the design process, the designer can consider the user role and tasks, by applying *models* of behaviour of operators/users in conjunction/interaction with system/supervisory control models. The selection of these models is of paramount importance for the whole design process, as it defines the context for simulations and experimental tests and for the taxonomy of possible inappropriate or erroneous behaviours to be evaluated during the whole design process.

In the early phases of the design process, the designer should also acquire maximum knowledge and *experience* on the environment and working context in which the system will operate.

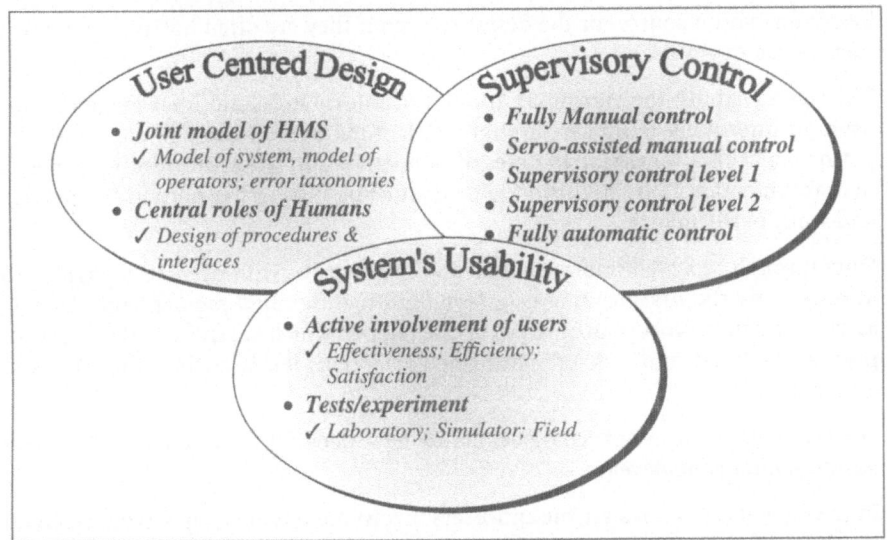

Figure 4.4 Peculiarities of UCD, system's usability, and supervisory control.

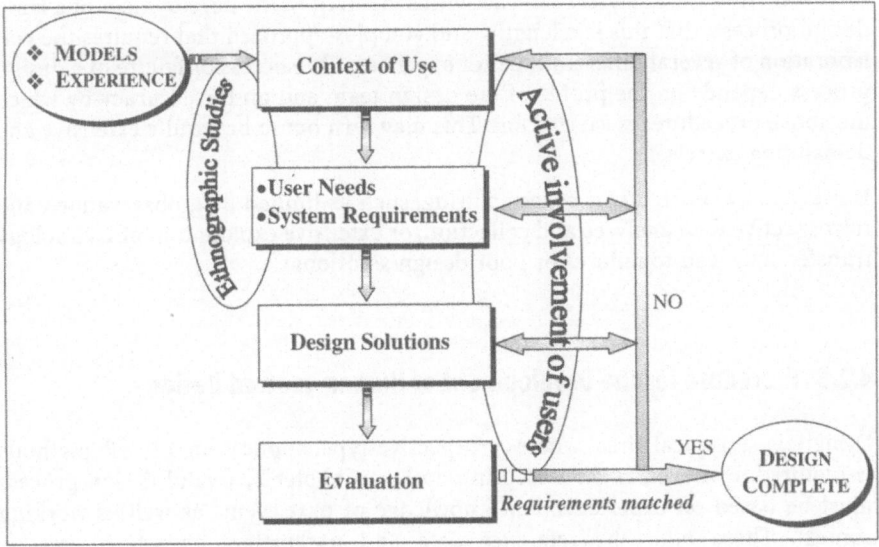

Figure 4.5 Human–machine interaction integrated approach.

The collaboration of the end-users is already important at this stage as it allows familiarisation with the context and tasks to be performed. From this initial process, the *context of use* of the system can be developed.

After this initial stage, observations on working context, interactions with users, and analyses of tasks and allocation of functions are necessary. End-users are the best information source for the designer, even if they are often not able to express their needs in words.

The way to exploit the enormous amount of information and knowledge of end-users in improving design is through *ethnographic studies*, which consist in the performance of a variety of analyses of familiarization with real working contexts that are intended to be noninvasive or disturbing during the normal procedures and work performance.

Only through repeated improvements and interactions with users it is possible to know exactly the *user needs* and *system and organisational requirements*. During such an iterative process, prototypes are developed which are more and more complete and refined, and tests are performed at various levels, such as in laboratory and in simulators.

Only when these issues of ecological nature have been fully resolved can the actual *design solutions* be developed.

In the final phases of design, the end-users involvement is also important and could result in another process of iteration. In particular, the *evaluation* of the implemented design solutions requires tests and experiments with prototypes imbedded in real systems.

It is clear from the analysis of the activities involved in the development of a HMS design process, that this is a lengthy and complex approach that requires the collaboration of several different types of expertise. The success or failure of a design process depends on the profile of the design team and on the accuracy by which the above procedures is carried out. This may turn out to be a quite extensive and demanding exercise.

However, any shortcuts or corner cutting, such as limited field observations and retrospective data analyses and collection, or extensive exploitation of technology transfer, may lead to failures or poor design solutions.

4.2.5 Procedure for the Development of Human-centred Design

Design is a typical area where prospective-type application of HF methods is required. However, as discussed in depth in Chapter 2, a valid design process must be based on experience and knowledge of past events as well as working context. These offer the reference data and parameters on which creative new design development is based. This process of well integrated prospective and retrospective analysis is typical on the HERMES framework shown in Figure 2.6.

The need to incorporate the principles of supervisory control and usability in HCD approaches matches very well with the HERMES requirement of integration of prospective and retrospective analysis.

The basic ideas described in the previous sections can be summarised in the following requirements to include human factors in design processes of HMS:

- *Active involvement of users.*
  Users must be involved in all steps of a design process, including the preliminary design, when a model and simulation of user behaviour can be accounted for simulating possible behaviours and interaction processes.
- *Appropriate allocation of functions.*
  Allocation of functions consists in the specification of which tasks and responsibilities are to be assigned to the user and which, instead, should be performed by the system. It is wrong to assign to the system all that is technically feasible. This allocation should be based upon several issues, such as abilities, attitudes and limits of the user with respect to the system in terms of reliability, accuracy, speed, flexibility, and economical cost.
- *Iteration of design solutions.*
  It is impossible to define user needs and system requirements accurately from the beginning. They need to be taken into consideration by an appropriate modelling, and only through repeated improvements and iterations is it possible to know exactly which requirements the system should have and which are the real needs of users. In order to accomplish such an iterative process, prototypes are developed which are more and more complete and refined.
- *Multidisciplinary design.*
  Human-centred design is a complex approach requiring the collaboration of several professionals with expertise in different domains. Usually a design team is based on: end-user; purchaser, user manager; application domain specialist, business analyst; system analyst, system engineer, programmer; marketer, salesperson; user interface designer, visual designer; human factors and ergonomics expert, human–computer interaction specialist; technical author, trainer and support personnel.

Finally, the guidelines for the development of HCD approaches must reflect the basic steps of implementation of human error and accident management measures, i.e., the DSME process discussed earlier in this chapter.

In practice, the specific process of designing a control and safety system and/or an interface and the associated procedures may make reference to the integrated approach for designing HMI systems discussed above and represented in Figure 4.5. However, a more complete and detailed procedure and guidelines for developing designs of HMI systems should include all steps considered in a HERMES-type approach. These include also the preliminary phases of definition of basic goals, selection of models, and data associated with past experience, as well as the activities following the actual design process that include training the end-users.

A detailed stepwise procedure for developing a design may be summarised in the following seven steps (Figure 4.6).

Step 1 Select goals and objectives for the system/safety measure to be designed, including supervisory control, user-centred design, and system usability principles (DSME stage *define* aims):

☐ Objectives of HEAM measure, or HMS, to be designed (prevention, recovery, protection).

☐ Level of automation to be attained.

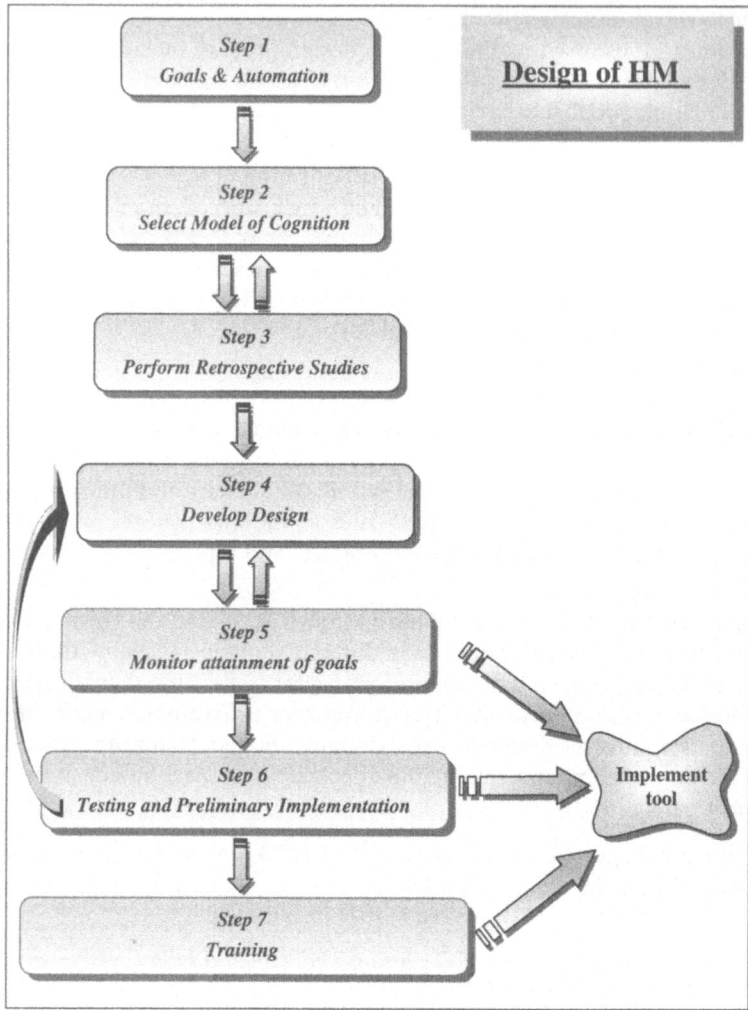

Figure 4.6 Stepwise procedure for designing HEAM measures.

☐ Cognitive functions involved in the HMI, e.g., perception, interpretation, planning, and execution.

Step 2 Select the model of cognition and associated taxonomy to be coupled with system models for simulating expected human–machine interactions and consequently designing system's features and relative control procedures. The model should focus on the relevant functions of cognition identified in step 1 of the procedure (DSME stage *select and implement* specific models and methods):

☐ If the goal is *prevention*, then:
  • privilege perception and interpretation of information received;
  • ensure that preventive manoeuvres and execution of actions can be easily and promptly carried out.

☐ If the goal is *recovery*, then:

- make sure that perception and interpretation of information can be performed by existing tools and instruments; and
- privilege planning and execution of manoeuvres and actions for re-establishing system to its normal performances.

☐ If the goal is *protection*, then:

- privilege the execution of escape processes; and
- operation of the protective equipment.

Step 3 Perform retrospective studies (DSME stage *select and implement*):

☐ Evaluation of the social and working context and user needs by ES.

☐ Performance of CTA for the use of the specific technology and its integration with the rest of the equipment.

☐ Evaluation of past experience, with particular reference to the system under development, by RCA and accident/incident investigation methods;

☐ Consider the possible role of the new tool with respect to its hypothetical contribution in incidents and accidents previously occurred.

Iterate with Step 2.

Step 4 Develop the design of HMS, including interfaces and procedures for control and management (DSME stage *select and implement*):

☐ Consider data and parameters derived from the analysis of past experience and apply engineering creativity for the evaluation of possible new initial and boundary conditions.

☐ Identify benefits and possible hazardous consequence of system performance.

Step 5 Monitor that goals of the system and HEAM measure are considered during the implementation of the design process (DSME stage *monitor* attainment of goals).

Iterate with Step 4.

Step 6 Testing and implementation (DSME stage *evaluate* effectiveness of results):

☐ Develop a test plan (experimental design) for the execution of experiments on the prototype tool.

☐ Perform theoretical and computerised simulations of HMI using models and simplified numerical methods.

☐ Implement HMS and interfaces on simulators and on board of test plants.

☐ Perform tests for:

- assessing functionality and operations according to design expectations;
- validation of user effectiveness, efficiency, and friendliness of operations, control procedures, and interfaces of HMS;

If need be, return to Step 4 for modifications and improvements of design.

Step 7 Ensure appropriate training of operators on objectives, use, and implementation of new technology.

A case study of a real application of the above stepwise procedure for designing the interface of an anticollision system in the automotive environment will be discussed in detail in Chapter 5.

4.3 Guidelines for the Application of HF Methods in Nontechnical Training

The need to train operators and users of technological systems for the management of human resources and for the coordination of activities in the working environment has received different levels of attention in different domains of technology.

This type of training is usually called "nontechnical" training and differs substantially from "technical" training that aims at developing operational and manual skills of operators so as to exploit and maximise the use of the properties and functionalities of a system. Nontechnical training, instead, aims at developing the ability of operators to integrate human and social aspects of the working environment within the specific technology and tasks.

The relevance of this different and new type of training was firstly recognised in the aviation domain in the early seventies, when the Federal Aviation Administration (FAA) began to demand that pilots, beside their airmanship skill and ability, be evaluated also on their ability to manage the variety of nontechnical duties on board, such as communicating with other pilots in the cabin and air traffic controllers on the ground, or managing workload derived from long haul flights, or coordinating crew operations. This request of the FAA started a process of development of methods and approaches that focused initially on the individual crew member. Already in the early eighties, however, the need to consider the whole crew as a team was recognised and, in the second half of that decade, several commercial airlines developed and implemented seminars on their own of "cockpit or Crew Resource Management" (CRM). In this period, the role and importance of CRM was also recognised by the International Civil Aviation Organisation (ICAO) (Taggart, 1987; Wiener et al., 1993).

Nowadays, the function of nontechnical training in aviation extends well beyond the simple classroom training and is fully integrated with the technical training, both at the level of simulator (Line-Oriented Simulation, LOS) and on-board training, in what is usually called Line-Oriented Flight Training (LOFT) (FAA, 1978, 1996a,b, 1997).

The success of CRM for pilots and the recognition of its relevance in the overall training process for improving safety, led to the extension of this type of training to all other aviation environments, involving cabin assistants, air traffic controllers, and maintenance personnel. Moreover, the recognition of the role played by organisational and cultural aspects in many accidents and incidents led to the development of courses and training practices that focus on organisational aspects, and people began to talk of "company resource management," extending nontechnical training to the whole organisation, including management at all levels.

In the domain of aviation, it is possible to find in the literature significant theoretical studies and complete frameworks for the development and adaptation of CRM to a specific airline or organisation (Foushee and Helmreich 1988; Wiener et al., 1993; Taggart, 1994; Johnston et al., 1995; Smallwood and Fraser, 1995; Jensen, 1995).

The amount of research and development that has been performed on nontechnical training for aviation is certainly extensive and the most recent publications are focusing more on aspects of implementation and correlated issues to CRM rather than the actual training itself, such as, for example, organisational issues (Telfer and Moore, 1997), teamwork (Orlady and Orlady, 1999), debriefing and instructions (Hunt, 1997; Prince et al., 2001), error management (Masson and Koning, 2002), course evaluation (Goeters, 1998), etc.

In technological domains other than aviation, the application and exploitation of nontechnical training is much less recognised and considered. This is also due to a specific deficiency in current norms and standards required by certifying authorities.

In the field of nuclear energy production, a move towards the implementation of concepts and approaches similar to those in place in aviation is in progress (CNSC, 2003). In other areas, such as maritime and rail transport as well in chemical and process plants, the need to train operators has been recognised but not yet formalised (Wilson et al., 2001).

In general, the need of developing and applying nontechnical training in all domains where human activities are safety critical remains an issue of minor importance and is usually "hidden" in the overall training programme of an operator. On the contrary, it should represent, as in aviation, a specific, well-recognised and essential contributor to the figure of the operator and in the certification process of different qualifications.

The need of performing nontechnical training at the same level of technical training will certainly be recognised in the near future as an essential contributor to safety and the development of nontechnical training will become a regulatory requirement for certification in all technological domains.

4.3.1 Procedure for the Development of Nontechnical Training

As in the case of design, the development of nontechnical training, or a "human factors course," requires the evaluation of all sociotechnical aspects existing in the specific organisation for which the training is developed.

This initial standpoint embraces an implicit assumption: the nontechnical training must be adapted to the specific characteristics and peculiarities of the organisation to which is dedicated. In practice, while a number of basic subjects represent the core base of training, the most important elements of a "human factors course" should focus on the specific socio-technical aspects and actual working environments of an organisation.

Moreover, as already discussed in previous sections of this book (Chapter 2), the process of training is an activity that requires the consideration of the outcome of accidents and incidents occurred in the past within the organisation (retrospective assessment) and the evaluation of possible safety critical situations that may occur in the future (prospective assessment).

These factors have to be evaluated and studied in depth before developing a "human factors course."

In order to do so, it is necessary to carry out all the basic steps of implementation of the HERMES methodology, already extensively discussed in Chapter 2 (Figures 2.4 and 2.6).

In practice, the stepwise procedure for developing nontechnical training within an organisation can be summarised in the following six steps (Figure 4.7):

Step 1 Consider features of nontechnical training programme/course (DSME stage *define* aims):

  ☐ Set goals of training programme in terms of:
  - prevention of errors and recognition of critical conditions;
  - recovery from errors and management of emergencies; and
  - protection for humans and environments in case of accident;

  ☐ Identify the level of automation and technology applied in everyday work;

  ☐ Select functions and processes applied during human interactions, e.g., team work, co-operative and collaborative type work, individual performances, etc.

Step 2 Select reference model(s) of cognition and taxonomies to be utilised as paradigms throughout the whole training programme (DSME stage *select and implement* specific models and methods):

  ☐ Develop trainers (facilitators) guidelines.

Step 3 Perform the analysis of organisation and attitudes of personnel, e.g., operators, assistants, maintenance personnel, (air) traffic controllers, management, etc., according to the principles of retrospective studies (DSME stage *select and implement*):

  ☐ Ethnographic studies, i.e., visits and observation at workplaces, audio-video recording, verbal/behavioural protocols, interviews, and questionnaires;

  ☐ Examination of tasks and procedures by cognitive task analysis;

  ☐ Review of the past experience of a system, and, in particular, analysis of incidents and accidents previously occurred.

Iterate with Step 2.

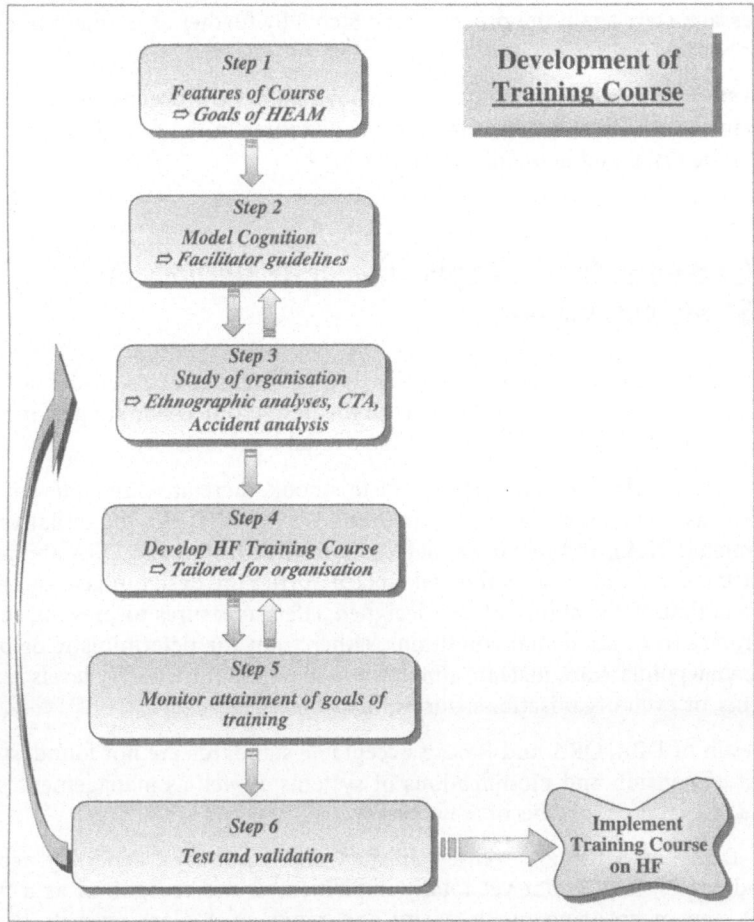

Figure 4.7 Stepwise procedure for developing nontechnical training course.

Step 4 Develop HF training programme based on general frameworks of nontechnical training courses and tailor it on the organisation, i.e., consider needs and goals and results of retrospective/ethnographic studies (DSME stage *select and implement*).

Step 5 Monitor that goals and objectives of training are maintained during the development of the programme (DSME stage *monitor* attainment of goals).

Iterate with Step 4.

Step 6 Test and validate effectiveness of training (DSME stage *evaluate* effectiveness of results):

☐ Develop and implement a test plan (experimental design) based on appropriate questionnaires, interviews, and indicators of effectiveness;

☐ Evaluate detailed content of programme/course.

If necessary, start again the process from Step 3 by further assessment of organisational needs.

A case study of a real application of the above stepwise procedure for developing a human factors Crew Resource Management course in the domain of civil aviation will be discussed in detail in Chapter 6.

4.4 Guidelines for the Application of HF Methods in Safety Assessment

The process of safety assessment implies the performance of a well-defined set of analyses which are usually combined with the design process or are performed on an existing plant for the evaluation of the overall level of safety.

As discussed in the previous chapters of this book, there are three relevant types of safety assessment, namely: Design Basis Accident (DBA), Quantitative Risk Assessment (QRA), and Recurrent Safety Audit (RSA). DBA and QRA are integral part of the design process as they take place during the design phase and aim at the evaluation of the ability of the designed safety measures to prevent, control, and protect from accidental conditions, either from the deterministic or probabilistic viewpoints. RSA, instead, aim at the evaluation of the safety levels reached by plants, or even organisations, during their lifetime.

As a result of DBA, QRA and RSA, if acceptable standards are not found, specific request of revision and modifications of systems as well as management procedures and practices may become necessary.

While DBA and QRA are well-established and, nowadays, generally accepted methodologies, RSA is not yet totally implemented and recognised as a fundamental step in ensuring the integrity and safety of a plant over its lifetime. Nonetheless, RSA probably represents the most efficient way to ensure that the required levels of safety associated to a plant and an organisation are ensured at the beginning of their life and are maintained while aging and changes and modifications occur during operations.

DBA and QRA are typical prospective type analyses, while RSA are much more focused on the existing and required safety levels of a system, and therefore, are essentially based on retrospective-type approaches.

The following sections will develop a set of specific guidelines for application of DBA, QRA, and RSA, and, as in the case of design and training, will consider the basic steps of implementation of HEAM measures, according to the principles of the HERMES methodology, discussed in Chapter 2 (Figure 2.6).

4.4.1 Procedure for the Development of Design Basis Accident Analyses

Design basis accident analyses consist in studies of specific accidents believed to represent the set of worse possible accidental scenarios, and serve the purpose of designing safety measures, and, in particular, defences, barriers, and safeguards (DBS), so that they offer adequate support for prevention, control, and protection under these "maximum credible accident" conditions. DBA studies imply the evaluation of all engineered safety devices, standards, procedures, and training, including human interactions and plant performances.

There are no specific methods that can be applied for DBA, but rather a methodological sequence of steps can be identified for obtaining the expected outcome. The goals of DBA analysis are to consider specific features of the designed safety systems and evaluate their effectiveness during the worse possible conditions, i.e., the most dangerous combination of failures, malfunctions and inappropriate behaviour are considered while evaluating operability and effectiveness of safety systems.

The human role in a DBA study is quite obvious. The HEAM analysis associated with DBA aims at ensuring that human interventions during these worse conditions remain doable and are effective for safety purposes. At the same time, a set of erroneous and inadequate performances is considered and, in line with the goals and hypothesis of a DBA, the overall safety of the system is evaluated.

This approach has been proven insufficient for fully assessing safety, as in many circumstances the unsuccessful operation and performance of a system has derived from the combination of different conditions and situations that are per se irrelevant, or at least not critical for safety, but become catastrophic when brought together in a certain configuration. However, the DBA remains an essential approach to evaluate the safe operation and functioning of a system in severe and extreme working conditions.

A detailed stepwise procedure for developing DBA analyses associated with the design of a plant may be summarised in the following six steps (Figure 4.8):

Step 1 Consider features of system design (DSME stage *define* aims):
- ☐ Goals of HEAM analysis, i.e., prevention of errors/failures, recovery from errors/failures, protection for humans and environments in case of accident.
- ☐ Level of automation.
- ☐ Cognitive functions to be exploited during the HMI.

Step 2 Select the model of cognition to be used for DBA. This model must not necessarily be the same as the one utilised for designing the system (DSME stage *select and implement* specific models and methods).

Step 3 Develop a set of data and parameters for DBA study based on retrospective analyses (DSME stage *select and implement*):
- ☐ Evaluation of working context by ethnographic studies.
- ☐ Review of past experience of the system, in particular, analysis of incidents and accidents previously occurred.
- ☐ Examination of tasks and procedures for the use of the specific system by cognitive task analysis.

Iterate with Step 2.

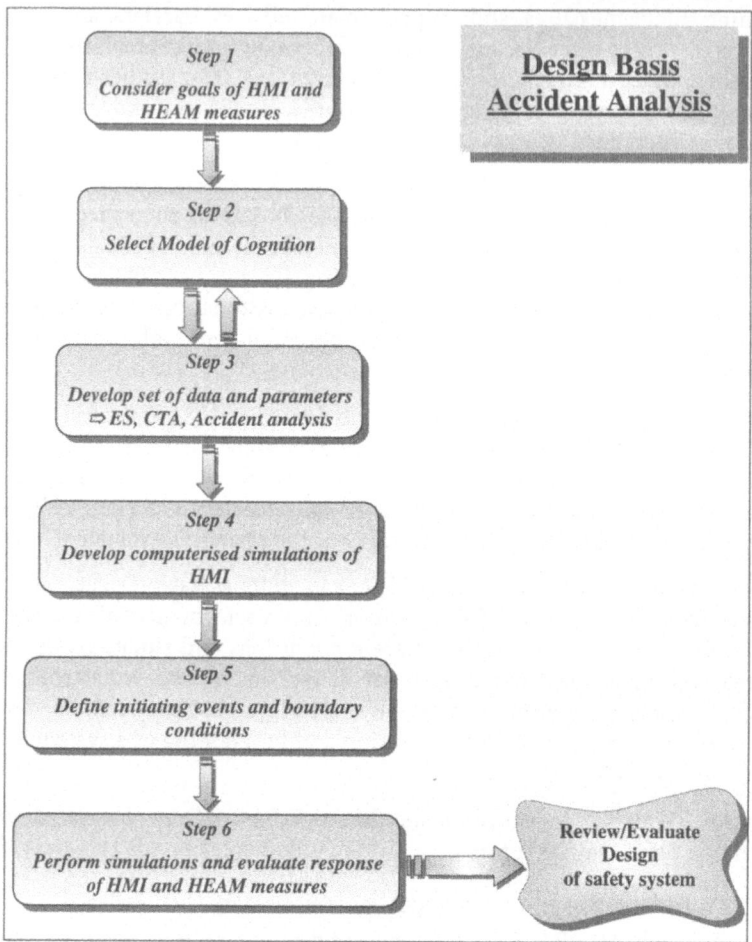

Figure 4.8 Stepwise procedure for design basis accident analysis.

Step 4 Develop computerised simulations of HMI, using selected models of cognition and system, and applying adequate numerical methods (DSME stage *select and implement*).

Step 5 Define, by creative thinking and experience, a body of initiating events and boundary conditions that represent the worse possible condition of operation for safety systems (DSME stage *monitor* attainment goals).

Step 6 Perform simulations and evaluate response of system and objectives of the HEAM analysis, in terms of expected goals and performances (DSME stage *evaluate* effectiveness of results).

If results are successful, then the design is validated. If not, then a revision of the design may be required and the overall process of design and safety assessment may need to be repeated.

4.4.2 Procedure for the Development of Human Reliability Assessments

Human Reliability Assessment (HRA) is an essential contributor to quantitative risk assessment that characterizes the evaluation of the overall risk associated with a plant or system.

In practice, all human reliability methods, and specially the methods described in Chapter 3, follow similar procedures for application that integrate retrospective and prospective approaches in order to develop a database of human errors as realistic as possible with respect to the system under study.

From the safety perspective, it is once again very important to recognise the need to clearly identify the goals of the system that is studied, i.e., prevention, recovery and protection, and to follow the main steps for development and implementation of measures for Human Error and Accident Management (HEAM), according to the basic structure of the HERMES methodology.

The detailed stepwise procedure for the development of a HRA analysis, within a quantitative risk assessment, is based on six main steps (Figure 4.9):

Step 1 Consider features of the system under study (DSME stage *define aims*):
- ☐ Goals of tool, i.e., prevention of errors, recovery from errors, protection for humans and environments in case of accident.
- ☐ Level of automation.
- ☐ Cognitive functions to be exploited during the HMI.

Step 2 Select the model of cognition to be used for HRA (DSME stage *select and implement* specific models and methods):

Step 3 Develop a set of data and parameters for the HRA study based on retrospective studies (DSME stage *select and implement*):
- ☐ Evaluation of working context by ethnographic studies.
- ☐ Review of past experience, and, in particular, analysis of incidents and accidents previously occurred.
- ☐ Examination of tasks and procedures for the use of the specific safety system by cognitive task analysis.

As QRA analysis concerns probabilistic assessment, this set of data contains mostly modes, types, and frequency of occurrence of systemic failures and inappropriate human behaviours.

Iterate with Step 2.

Step 4 Develop or apply an appropriate reliability method that combines causes and effects of failures and/or inappropriate behaviours, and leads to the evaluation of the consequences with associated frequencies of occurrence and uncertainty distribution (DSME stage *select and implement*).

Step 5 Define, by creative thinking and experience, a body of initiating events and boundary conditions for the reliability assessment (DSME stage *monitor attainment of goals*).

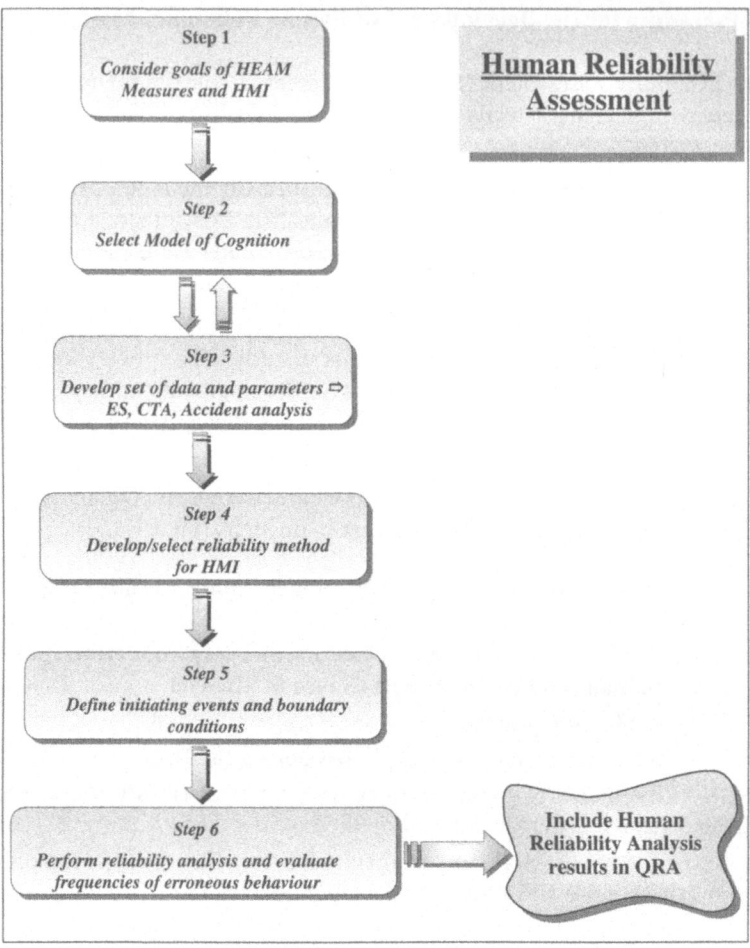

Figure 4.9 Stepwise procedure for human reliability assessment analysis.

Step 6 Perform calculation and evaluate the response of the safety system under study in terms of expected frequencies of occurrence of human errors (DSME stage *evaluate* effectiveness of results).

Include HRA results within the overall QRA analysis.

4.4.3 Recurrent Safety Audits Throughout an Organisation

Indicators of Safety: A Fundamental Approach to Safety

A fundamental standpoint sustained all throughout this book is that "the evidence from a large number of accident inquiries indicates that bad events are more often the result of error-prone situations and error-prone activities than they are of error-prone people" (Reason, 1997). The management of errors made at all levels within an organisation, by anticipating or preventing their occurrence and eliminating those contributing factors that are fundamental elements in sustaining accident generation and development, is an essential means of improving organisations and ensuring safer performance of operations.

In complex technologies, many elements contribute, in addition to the mere technological aspects, to the management of a system and its human machine interaction. In particular, organisational and cultural traits and personal and external factors are strongly interconnected with working conditions, and, as already discussed in Chapter 2, they must be integrated in the overall safety assessment of defences, barriers, and safeguards.

In the case of DBA and QRA, the integration of these factors aims at the identification of the safety level of a plant at the design stage, i.e., before the actual implementation of the plant with all its hardware and software. Sometimes, during the lifetime of a plant, QRA are revisited or partially redeveloped when a new safety evaluation is required, e.g., after important design modifications following a major accident.

However, a complete safety assessment of an organisation requires that the factors identified and considered by DBA and QRA be evaluated at recurrent intervals during the lifetime of a plant, i.e., after its implementation and during operational conditions, by a specific type of safety approach called RSA.

In a RSA specific safety indicators are identified and measured, screening throughout the entire organisation, making it possible to:

1. Ascertain that the standards reached or existing at the time of the original design and certification are kept at the same level during the lifetime and practical operations of the plant.
2. Find out whether new standards introduced by novel regulations are satisfied by the current state of the system and/or identify where changes are necessary to fulfilling such new requirements.
3. Discover potential evolution of the organisation towards different levels of safety/risk conditions, and identify the reasons and causes of such a change. If these are not acceptable, propose changes to re-establish adequate level of safety.

As already discussed, the performance of a valuable and effective RSA may only be achieved by an appropriate integration of retrospective and prospective analyses. In particular, in the case of RSA this requirement becomes even more important, as the analysts must:

● Acquire complete knowledge of the safety characteristics (safety levels, indicators, DBA, and procedures) of the systems at beginning of operations and how these have evolved or have been modified during the operational life of the plant. This is a typical retrospective-type analysis.
● Perform an evaluation of the safety level of the plant in its current conditions and evaluate whether the plant is sufficiently "safe" and shows acceptable safety

Table 4.1 Generic Indicators of Safety (*IoS*) of a plant and an organization

☐ *Organisational Processes (OP)*	☐ *Local Working Conditions (LWC)*
✓ Unconscious beliefs	✓ Workplace design
✓ Values	✓ Automation
✓ Norms	✓ Tools and actuators
✓ Fundamental assumptions	✓ Protective equipment
✓ Unwritten rules	✓ Job planning
✓ Etc.	✓ Etc.
☐ *Personal and External Factors (PEF)*	☐ *Defences, Barriers, Safeguards (DBS)*
✓ Coordination ability	✓ Engineered safety devices
✓ Physical characteristics	✓ Protective equipment
✓ Mental conditions	✓ Training
✓ System performance	✓ Policies, standards, controls
✓ Random system contingencies	✓ Procedures, instructions, supervision
✓ Etc.	✓ Etc.

standards, or whether improvements and changes are needed to re-establish acceptable levels of safety.

All RSA methods consider HF aspects that combine socio-technical elements with the dynamic properties of the plant and organisation (Byrom, 1994; Wilson et al., 2001). In Chapters 2 (§2.2.2) and 3 (§3.6) these elements have been identified as: *Organisational Processes (OP)*, *Personal and External Factors (PEF)*, *Local Working Conditions (LWC)*, and *Defences–Barriers–Safeguards (DBS)*.

Each element can be further subdivided into a number of more specific constituents that can be measured and evaluated for the assessment of the safety state of an organisation (Table 4.1). These are the *Indicators of Safety (IoS)* of a plant and an organisation and are the quantities that are selected, identified, and then studied during a RSA. As a general guideline, the structure of the *IoS* shown in Table 4.1 can be utilised for a more precise definition of the *IoS* specific to the plan/organisation under audit.

These elements of a human–machine system are strongly interdependent and, as soon as a structured analysis is performed on any of them, the relation to others becomes immediately visible (Reason, 1997). As an example, after examining DBS and their definitions, it becomes clear that they are connected to, for example, higher decisional processes of an organisation, i.e., OP, and to local working conditions.

RSA Matrix

The process of development of a RSA, with the focus on human error and accident management, is then based on the evaluation of each element in the socio-technical system, i.e., OP, PEF, LWC, and DBS, and associated constituents with respect to the three basic objectives of a safety system, namely, *prevention* of human errors and system failures, *recovery* from errors/failures, and *protection* from consequences.

It is therefore possible to develop a *RSA-matrix* for each element of the socio-technical system that enables the analyst to evaluate the essential features of safety and to assess levels of safety. Examples of the *RSA-matrices* are shown in Table 4.2a–d, which contain:

- Rows that consider the basic constituents of the socio-technical elements of a plant/organisation, i.e.:
  —for *Organisational Processes (OP)*: unconscious beliefs, values, norms, etc.
  —for *Personal and External Factors (PEF)*: coordination ability, physical characteristics, mental conditions, etc.
  —for *Local Working Conditions (LWC)*: workplace design, automation, tools and actuators, etc.

Table 4.2.a RSA-matrix for OP (organisational processes)

Indicators of Safety	Prevention	Recovery	Containment
Unconscious beliefs	x	–	–
Values	x	x	x
Norms	x	x	x
Fundamental assumptions	x	x	–
Unwritten rules	x	x	x
....	–	–	–

Table 4.2.b RSA-matrix for PEF (personal and external factors)

Indicators of Safety	Prevention	Recovery	Containment
Coordination	x	x	–
Physical characteristics	x	x	x
Mental conditions	x	x	–
System performance	x	x	–
Random system contingencies	x	x	x
....	–	–	–

Table 4.2.c RSA-matrix for LWC (local working conditions)

Indicators of Safety	Prevention	Recovery	Containment
Workplace design	x	x	x
Automation	x	x	x
Tools and actuators	x	x	–
Personal equipment	x	–	x
Job planning	x	x	x
....	–	–	–

Table 4.2.d RSA-matrix for DBS (defences, barriers, and safeguards)

Indicators of Safety	Prevention	Recovery	Containment
Engineered safety devices	x	–	x
Protective equipment	–	–	x
Training	x	x	x
Policies, standards & control	x	x	x
Procedures, instructions, supervision	x	x	x
....	–	–	–

—for *Defences–Barriers–Safeguards (DBS)*: engineered safety devices, protective equipment, training, policies, standards and control, procedures, instructions and supervision, etc.

- Columns containing the three basic objectives (*prevention*, *recovery*, and *protection*) of the safety system or HEAM measure under audit.

The values that are associated with each element of the *RSA-matrices* can be of quantitative or qualitative nature, and represent the actual value of the *indicators of safety*. In Table 4.2a–d, the "x" indicates that, in general, the selected IoS may be considered as relevant for the specific RSA-matrix. However, the analyst must develop the actual matrices and IoS to be applied in real cases by means of appropriate field studies and ethnographic analyses of the socio-technical contexts.

In practice, given that the elements are strongly interconnected as mentioned above, the *RSA-matrix* for the evaluation of DBS is in many cases sufficient for developing a quite accurate picture of the overall safety state of an organisation. In general, RSA matrices for all elements of a socio-technical system may still be developed for a more precise and accurate estimate of specific aspects of safety.

RSA-Matrix and IoS

During an RSA process, the indicators of safety are evaluated in order to assess whether the plant and organisation involved are operating within acceptable safety margins, and the safety measures of the system conform with current norms and standards.

In order to reach these objectives, it is necessary that three sets of specific values of *IoS* for each *RSA-matrix* are considered and compared during an RSA analysis:

1. The first set contains the *Indicators of Safety* (IoS_t) evaluated at the time of the audit, and represent the current safety state of the plant/organisation.
2. The second set of *Indicators of Safety* (IoS_N) contains the values that are required by norms, regulations and standards at the time of the RSA, in order to ensure that the plant and organisation are operating within acceptable safety conditions.
3. The third set of *Indicators of Safety* (IoS_o) contains the values as resulting from the original design.

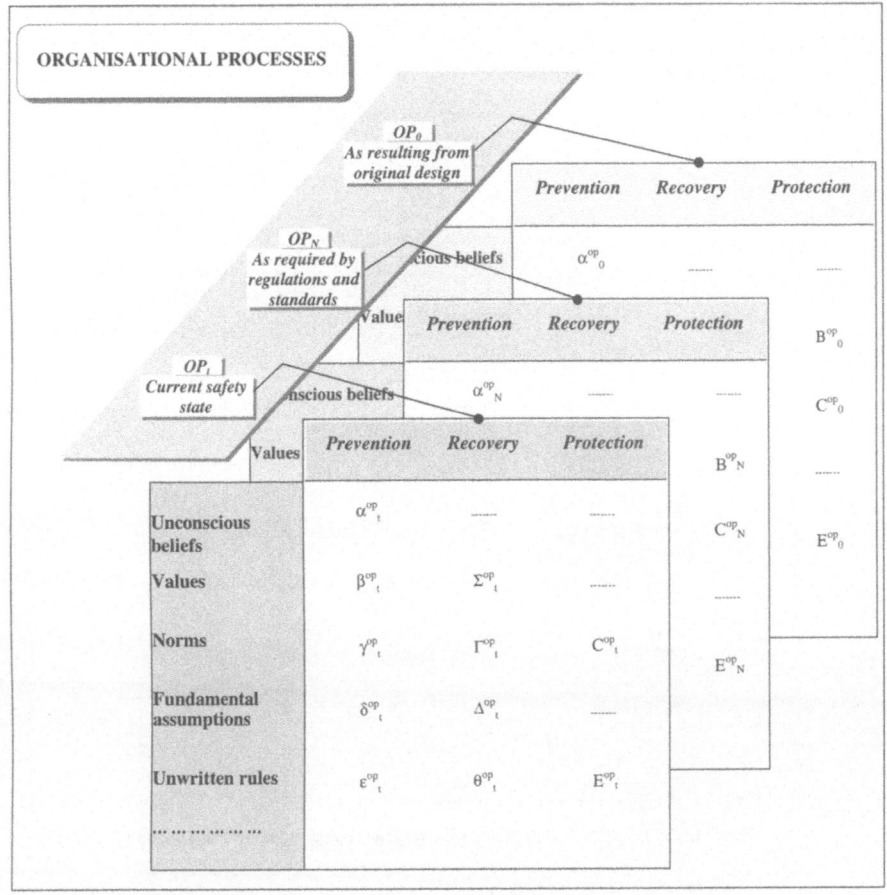

Figure 4.10.a Indicators of Safety (*IoS*) and *RSA-matrix* for Organisational Processes (OP).

In practice, in the case of the *RSA-matrix* for DBS, the protective and mitigating functions provided by defences, barriers, and safeguards are evaluated comparing the current state of the system with respect to the expected level of safety, as well as the original design. Similar considerations may be developed with reference to the other *RSA matrices* concerning more focused socio-technical aspects, namely, OP, PEF, and LWC.

The *IoS* identified in Figure 4.10a–d are only indicative of the variables that need to be evaluated for defining safety levels, as well as their impact on different aspects and features of HEAM measures and safety systems. The exact structure of *RSA-matrices* and *IoS* must be defined by the safety analyst on a case-by-case basis and derive from the retrospective analysis (ethnographic assessments, CTA, and root causes analysis) that must be carried out as a preliminary step in any systematic safety study, as thoroughly discussed in Chapter 2.

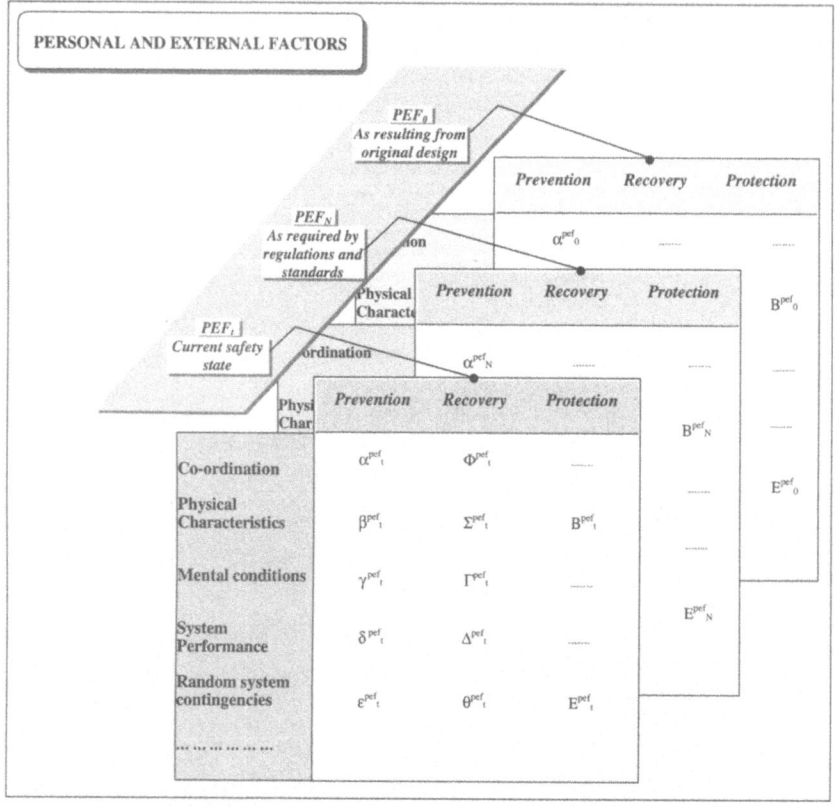

Figure 4.10.b Indicators of Safety (*IoS*) and *RSA-matrix* for Personal and External Factors (PEF).

The comparison of the values associated with the *IoS* for the different *RSA-matrices* allows the final assessment of the safety measures that exist within an organisation and of its overall safety level. On the basis of the results of RSA, it is then possible to assess whether operations should continue as they are, or whether improvements of structural and socio-technical nature are needed in order to raise the safety margins of the system to acceptable and adequate levels.

In the latter case, the outcome of RSA allows the safety analyst to identify the areas of concern and, in most cases, also the type of amendments, changes, and activities that must be implemented in improving safety and re-establish acceptable values of the *IoS*.

In other words, a RSA enables the evaluation of "whether" problems exist, and, if so, "where" they may be located, and gives indications about "what" should be done to solve them.

In practice, a comprehensive approach to management of errors/failures and assessment of an organisation should look at, and be directed to, different levels

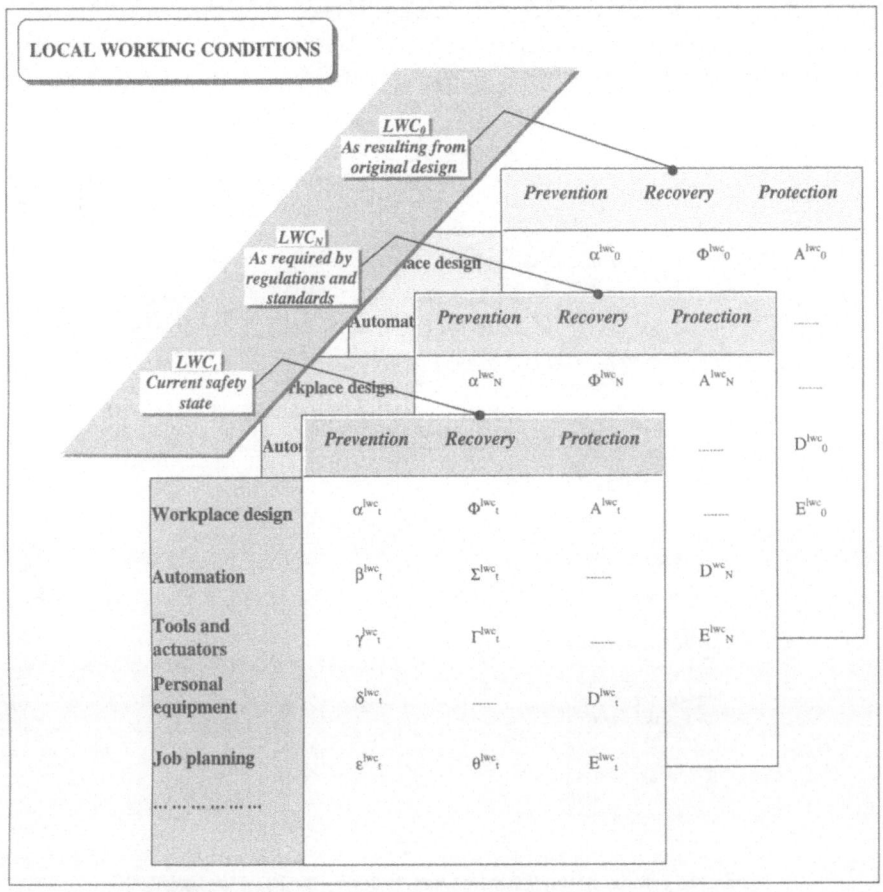

Figure 4.10.c Indicators of Safety (*IoS*) and *RSA-matrix* for Local Working Conditions (LWC).

of the organisation, namely, the individual, the team, the task, the workplace, and the organisational and technical processes. (Reason, 1997).

Historically, many organisations, especially in aviation, already target their error management resources at the individual level, both by performing training and recurrent training, evaluations and proficiency checks. Similarly, the measures to examine team performances and working conditions have been quite extensively developed. A typical example of these measures in civil aviation are the crew resource management courses and advanced qualification programmes which have nowadays reached a very high level of quality and are part of modern training processes, as well as being recognised by formal norms and regulations, even if a complete standardisation has not yet been reached (FAA, 1990; Wiener et al., 1993).

The evaluation of tasks and job assessment is also an area where much work has been carried out, for it concerns both the analysis of human–machine interaction

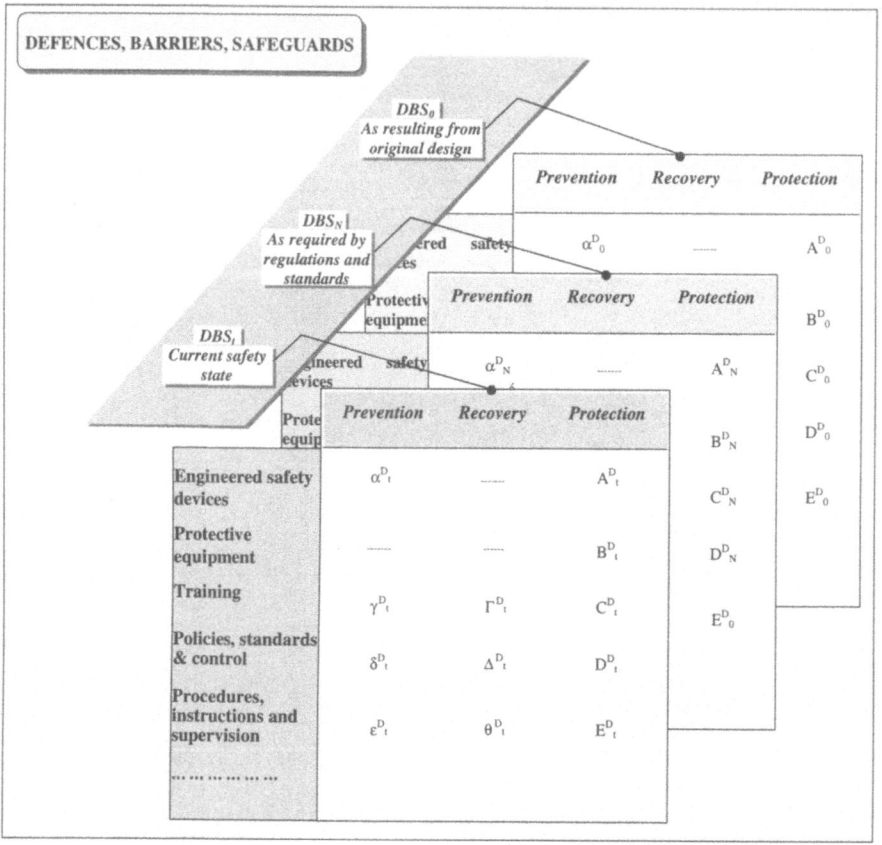

Figure 4.10.d Indicators of Safety (*IoS*) and *RSA-matrix* for Defences, Barriers and Safeguards (DBS).

and the development of computerised tools. The domains of aviation and petro-chemical systems are well advanced and lead these developments. There are also domains where RSA, or a general need to assess safety levels, is recognised. They are the railway and maritime transportation systems, energy production, and manufacturing. As an example, in the domain of railway, the most recent methods developed for error management and safety assessment of organisations tend to focus on the higher levels of the organisation and on workplaces, so as to identify less apparent factors and indicators of safety margins and risk (Wilson et al., 2001).

Following the above discussion on the definition and identification of IoS of an organisation and technical system, it is now possible to proceed to outline a precise guideline and procedure that can be applied for developing a consistent and successful RSA.

Procedure for the Development of Recurrent Safety Audits

The methods for RSA follow procedures that are very similar to those applied for other safety assessment approaches like DBA and QRA. This generic procedure is well integrated in the general framework HERMES for retrospective-prospective types approaches as already discussed in Chapter 2.

However, while DBA and QRA are usually focused on systems or more specific parts of a plant and on the quantification of parameters and frequencies of occurrences, a RSA involves the whole socio-technical organisation and aims at evaluating whether the safety standards originally considered and attained at the design stage, and conform with regulatory requirements, are maintained during the lifetime of a plant. With this major objective in mind, a RSA approach usually looks for safety indicators of quantitative (numerical, e.g., physical properties, coefficients of strength, etc.) as well qualitative nature (non-numerical, e.g., fuzzy measures, attitudes, etc.) that can be measured and evaluated compared to appropriate standards.

From the system safety and HEAM perspective, it is once again very important to differentiate clearly between prevention, recovery, and protection. In the case of RSA, this implies that various safety measures must be evaluated with respect to their planned objectives and their capability to respond timely and according to design during the lifetime of the plant.

Finally, in developing a RSA approach it is essential that the four stages of the DSME process (definition, selection and implementation, monitoring, and evaluation) are followed, in order to ensure a complete coordination between the attainment of the goals of the HEAM measure under examination and the requirements of human machine interaction.

In practice, the detailed stepwise procedure for developing of a RSA within an organisation can be summarised in the following six steps (Figure 4.11):

Step 1 Consider features of the system under study (DSME stage *define* aims):

    ☐ Goals of system under evaluation, i.e., prevention of errors, recovery from errors, protection for humans and environments in case of accident.

    ☐ Level of automation.

Step 2 Select the models of organisation and HMI to be used for RSA. These models should lead to (DSME stage *select and implement*):

    ☐ Generic *IoS* of the system in terms of prevention, recovery and protection for humans and environments. These indicators are, at this stage, quite generic and can be derived from the *RSA-matrices* discussed in the pervious subsections (Table 4.2a–d).

Step 3 Perform retrospective analysis in order to develop (DSME stage *select and implement*):

    ☐ Specific *IoS* for OP, PEF, LWC, and DBS. These *IoS* are precise quantities and parameters that pertain to the organisation and system under scrutiny.

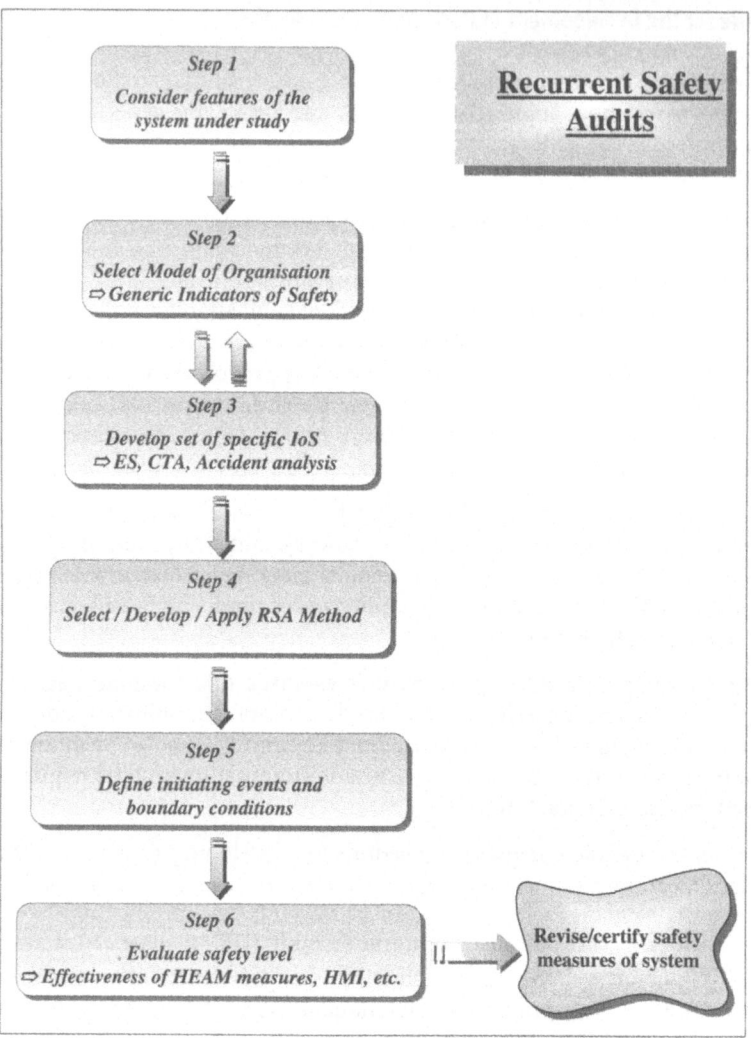

Figure 4.11 Stepwise procedure for Recurrent Safety Audit analysis.

Iterate with Step 2.

Step 4 Select, or develop, and apply an appropriate RSA method that combines causes and consequences of failures/errors, and leads to the quantitative and qualitative evaluation of the specific *IoS* (DSME stage *select and implement*).

Step 5 Define, by creative thinking and experience, a body of initiating events and boundary conditions for safety assessment (DSME stage *monitor* attainment of goals).

Step 6 Perform calculations and evaluate IoS and effectiveness of existing safety system and HEAM measures to offer adequate means of prevention, recovery and protection from human errors and system failures (DSME stage *evaluate* effectiveness of results).

If results are successful then system and organisation are validated and safety levels confirmed. If not, a revision of the design or an improvement of safety measure may be required.

A case study of a real application of the stepwise procedure for performing an RSA in the domain of railway transportation systems will be discussed in detail in Chapter 7.

4.5 Guidelines for the Application of HF Methods in Accident Investigation

The study of accidents, and the classification of their outcomes, is the most straightforward application of retrospective analysis. The results of such studies offer valuable and essential data, information, and insight that are utilised for (prospective) evaluation and designing of new safety systems and HEAM measures, as well as for training.

As this book is concerned with HF, the focus of this section will be on the human contribution to accident causation. However, the guidelines that will be developed are general and may also be applied to the evaluation of systemic root causes of accidents.

Accident investigation is a well-established area of application of safety in modern technological environments. In the aviation domain the procedures and practices of performing accident investigation have been formalised at international level for many years (ICAO, 1984, 1986, 1988). In particular, the Annex 13 of the Convention of the International Civil Aviation Organisation (ICAO, 1988) prescribes a suitable format for accident reporting, which allows considerable flexibility in order to account for factual description of systemic and human-related occurrences. In more recent years, this format has been further elaborated focusing, in particular, on human factors (ICAO, 1991, 1993, 1997). In this new format, all deviations from normal and expected performance can be structured in a detailed and articulated taxonomy.

In domains other than aviation, especially in surface and maritime transportation systems, energy production (nuclear), petrochemical and process plants, accident investigations also play a very important role in the assessment of overall safety scenarios. The procedures that are implemented in these domains follow quite closely those defined in aviation and are formalised in similar ways, at the level of national authorities (Gow and Otway, 1990; Byrom, 1994).

The significant relevance assigned to accident investigations is quite obvious, and depends on the consequences in terms of plant destruction, number of lives lost, and environmental damage caused by a single accident in certain domains. Moreover, accidents have a considerable impact on public opinion and concern, and may have a drastic effect on the overall development of a certain system or even a whole technology. The negative, sometime disastrous, consequences of severe accidents for the economy of a company or even for the survival of an entire field cannot be mitigated by an accurate and thorough accident investigation. Typical examples are the bankruptcy of the airline Easyjet following the crash in Miami, even though the airline was shown to be extraneous to the causes of the accident following the enquiry, and the actual ending of development and exploitation of pacific use of nuclear energy following the accident of Chernobyl.

Unfortunately, the time that public opinion takes to respond to the consequences of a major accident and the time demanded for performing an adequate accident investigation are very different. The former takes only days or weeks, while the latter usually requires several months and can take years.

The only field that seems somehow immune to a strong negative feedback of public response to accidents is the automotive and road transport domain, even if more than half a million people are killed worldwide each year in traffic crashes and about one person in 200 of the world's population dies from injuries received in traffic accidents (European Parliament, 1998). The reasons for such a low impact on public concern are many and differ greatly from country to country. In general, this is due to the relatively low number of deaths caused by a single accident and by the enormous diffusion and popularity of this means of mobility and transportation. Other subtler reasons that may vary strongly from nation to nation, are cultural and personality traits, risk perception and communication, and individual attitudes such as invulnerability or machismo.

The importance of accident investigation lies in the richness and quantity of information and data that can be generated. The results of an accident investigation usually demonstrate the rarity and singularity of the specific events. They show the combination of factors and root causes that led to the accident and propose new safety measures that ensure that the specific sequence of events and combination of factors will not happen again.

However, in modern technological domains, serious accidents are very rare, and occurrences of minor importance, also called "incidents" or near misses, are much more frequent. Therefore, the richness of information that may be included in a data reporting system, based on reports of incidents and near misses, is much more relevant than that contained in a database on severe accidents. Unfortunately, in the case of incidents and near misses the data reporting is not mandatory, as in the case of accidents and serious incidents. The collection of information from individuals and operators in these cases becomes much more difficult and it is developed on a voluntary basis and requires many complicated features of confidentiality, trust, and nonliability that need adequate regulatory control.

What is important is that the methods for analysing reports on accidents and incidents or near misses are all integrated into a common methodological approach and modelling formalism that makes all these different types of data compatible and comparable. In this way it is possible to exploit the outcome of accident/incident investigations in prospective type of studies for designing new safety measures or for improving safety assessment or training.

The existence of databases on near misses and voluntary reporting represents thus a crucial source for preliminary investigation and assessment, as already discussed in previous sections on recurrent safety audits.

4.5.1 Practical Approach to Accident Investigation

In practice, the development of an investigation of an accident, or an event that occurred or was reported, entails the application of the methods and techniques discussed in previous chapters.

Consequently, before analysing the guidelines for developing accident/incident analysis, it is interesting to see how some of the methods and models discussed earlier can be combined in the process of accident investigation. These can be summarised in a number of parallel and interactive steps that describe the methodological framework and outcome of an accident analysis (Figure 4.12). These steps are: collection and organisation of data about an accident; study of the organisations involved; analysis of data; evaluation of the effectiveness of DBS and the development of recommendations; and storage of data. They are all contemplated in the HERMES methodology, and will be hereafter briefly discussed.

Collection and Organisation of Data About an Accident

The collection of data about an accident is a critical and normal process of acquisition of evidence and outcome of an accident that takes place in the immediate aftermath of an occurrence. The obvious requirement, in this step, is that data should be affected as less as possible by the process of collection and should give the most realistic picture of the consequences of the accident. In particular, in the case of human factors analysis, recordings of verbal interactions between people involved, such as pilots, air traffic controllers and comments by observers of the events are essential information for understanding what happened.

The acquired data must be structured and organised in a logical sequence. This implies that a number of *events* should be identified, each of which may be considered as a specific element to be analysed in more detail on its own. The logic by which events are structured is usually a temporal sequence, generating an Event Time Line (ETL) that gives the overall picture of the sequence of events. There often exist an obvious correlation between two or more events, in the sense that one determines the other and makes the latter a direct and unavoidable consequence of the former. This logical correlation must also be identified and pinpointed in the ETL and accurately considered in the following steps and analysis.

Study of the Organisations Involved

A necessary and crucial phase of accident analysis is the collection of information and data concerning the organisations involved. This phase must be carried out in parallel to the initial phase of data collection and offers very important traces

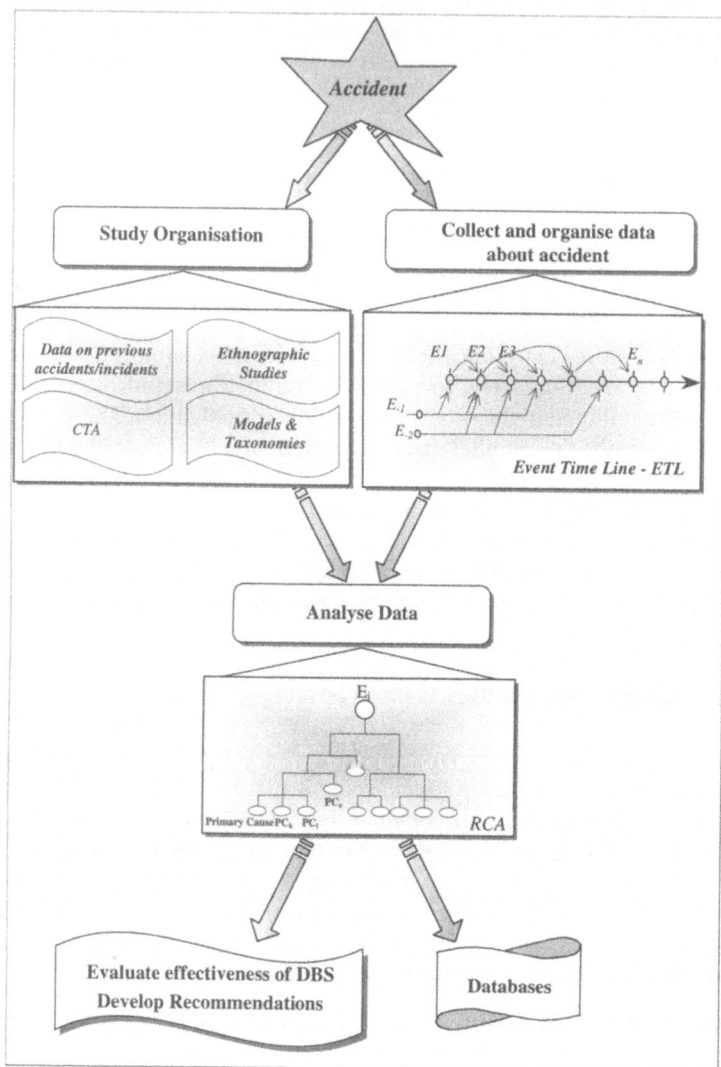

Figure 4.12 Methodological framework for an accident analysis.

and hints in understanding "why," "how," and "what" has happened during the accident.

The methods described earlier for performing retrospective studies must be applied in order to develop a useful and fruitful review and audit of an organisation. The outcome of these studies contributes to the overall accident analysis and to the understanding of behaviours and performances that otherwise would not be immediately clear or explicable in generic or unbounded contexts.

A mixture of formal and practical approaches, supported by theoretical standpoints, sustains the review and examination of an organisation. The essential theoretical standpoint is the *joint cognitive model* of human—machine interaction and associated taxonomy. By selecting a reference model, the analyst defines depth and scope of the study, as well as contexts over which the study will span. By applying the model and taxonomy it is possible to generate the connections between the different manifestations of behaviour and guide the classification of performances, errors, and malfunctions.

The use of other formal methods, i.e., CTA and the study of accidents and incidents previously occurred within the organisation, offer means of acquiring further insight about the accident under study. On the one hand, they show how general philosophies and policies established by management are implemented in procedures and are applied in practice by front line operators (Degani and Wiener, 1994b). Moreover, they offer a picture of past (negative) experiences and give a measure of the counteractions that such previous events generated within the organisation. Rarely does data exist about positive occurrences, i.e., the near misses and their management, in terms of successful operations and interventions performed by front line operators and managers that contributed to the recovery of risky situations. This information, when available, offers a very valid ground for understanding behaviours and performances.

Finally, the performance of ethnographic studies, with extensive field observation, leads to the appreciation of the company climate and the ecological dimension of the working environment in which the accident developed.

It is quite clear that this form of analysis of an organisation and the assessment of its socio-technical context at the time of the accident are fundamental aspects to investigate in order to reach an understanding of facts and behaviours. This analysis is as important as the collection and structuring of data for the description and understanding of the accident and cannot be overlooked, in the overall process of an investigation.

The risk of not performing this part of an accident analysis with the appropriate accuracy and attention is enormous. In such a case, on the one hand it is almost certain that the real root causes of what has happened may not be identified. The front actors would probably be blamed as being solely responsible for the accident, as their behaviour would be easily and immediately detected, with no other evidence of conditioning contexts and priorities being considered. But, more important, the possibility of developing remedies based on a wide span and fruitful advice about management of errors would not be obtained. In other word, no lesson would be learnt from the analysis of the accident, producing results of minor relevance that possibly could be dangerous.

On the other hand, a well-performed study of the organisations involved represents a solid base for understanding behaviours and for the whole accident investigation.

Analysis of Data

For each *event* RCA must be carried out in order to identify its real and initiating causes. A RCA usually consist of a simple cause–consequence tree that looks for the interplay of factors leading to overt inappropriate behaviour. These are normally considered the *primary causes* of the accident and are expressed as human- or systemic-related factors that combined to generate the sequence of events.

However, a number of logical and temporal correlations as well as dependences and interplays between events and primary causes exist. These must be accounted for in the development of the RCA. Moreover, some events may be very relevant to the causal path of the accident and play a pivotal role in the entire sequence. They should also be pinpointed and analysed with particular care.

Evaluation of Effectiveness of DBS and Development of Recommendations

Following the three initial steps of analysis, the accident consequences need to be evaluated with respect to the existing Defences, Barriers, and Safeguards (DBS).

In particular, the analyst should be able to measure or estimate DBS' performance and effectiveness during the accident, in accordance with the original plan and design specifications. In many accident cases, the degradation of DBS and the relaxed attitudes towards safety in favour of production and cost efficiency, led to the reduction of quality and effectiveness of DBS and to their inadequate performance during the accident (Cacciabue et al., 2000).

DBS must be considered in view of their ability to sustain the three main features of HEAM, namely, *prevention* of system failures and/or human errors, *recovery* from hazardous or critical conditions, once errors and failures could not be prevented or avoided, and *protection* for people and the environment from the consequences of an accident, when normality could not be recovered.

The goal of the analyst at this stage of the accident investigation is to develop recommendations for safety improvement. These recommendations should focus on areas of intervention so as to avoid the reoccurrence of the same accident or critical events, and, more importantly, to prevent the repetition of the same root causes that may favour new accidents. This usually concentrates on the improvement of existing measures of prevention, recovery and protection and/or on the suggestion of new and more effective measures.

However, when recommendations for new protection and intervention means are made, they need to be integrated within the current and existing plant. Therefore, a wider study is required that involves designers and the whole organisation that manages the plant, in order to ensure that the implementation of such new measures is consistent with the recommendations of the accident analysis and, at the same time, is adequately integrated with the overall set of existing procedures and practices for plant management.

Storage of Data in Databases

The storage of data on accidents in databases is a very important and difficult process. It demands that the analysts in charge formalise the results of an accident investigation in formats that are specific for the taxonomy associated with the underlying accident model and database structure.

The way in which data are normally recorded in databases is by analysing accident reports, and selecting from such reports, usually very long and detailed, the relevant information to be stored in the formats of the classification scheme of the database. Most accident reports contain a final section that summarises the most relevant results and describes them in the form of "findings." These findings are usually sufficient for data-recording purposes. However, when uncertainties about the classification of some facts occur, detailed sections of the report may be considered for specific consultation and more precise data recording.

It is clear that the process of storing data from accident reports is a quite difficult process and demands a very diversified competence on the part of the analyst. Particular requirements include:

- Knowledge of the technological domain under scrutiny. This enables the analyst to understand the terminology adopted in the accident report.
- Knowledge and expertise in HF and HMI. As many root causes of accident are related to socio-technical factors, the taxonomies of modern databases allow quite accurate classification of individual, organisational, and social aspects. Without adequate expertise in this domain it is impossible to exploit the potentiality of such a classification scheme and give proper account of root causes.
- Knowledge of computerised structures of databases. This enables the analyst to navigate rapidly and efficiently through the database to select the specific domain of classification for the recording of data.

These three types of experience and expertise are rarely found in a single analyst in charge of data storage and classification. On the other hand, it is quite obvious that an inadequate classification of data may be even more damaging that no classification at all, as it would lead to erroneous interpretation of accidents and would contribute to false statistical results.

Consequently, in order to reduce the number of errors in classification of data, while retaining the accuracy and detail modern taxonomies, several systems have been developed to support the analyst in data recording. These techniques aim to guide their user to exploit all aspects of the classifications and to ensure that the real root causes identified by the accident investigation are properly recorded for future use and analysis (Cacciabue, 2000b).

4.5.2 Procedure for the Development of Accident Investigations

From a practical point of view, following the steps described in the previous section can successfully develop a process of accident investigation. However, as the accident investigation is one of the four fundamental areas of application of a general methodology that combines retrospective and prospective approaches, it is important that, as already done for design, training, and safety assessment, also in the case of accident investigation the process is framed within the HERMES approach.

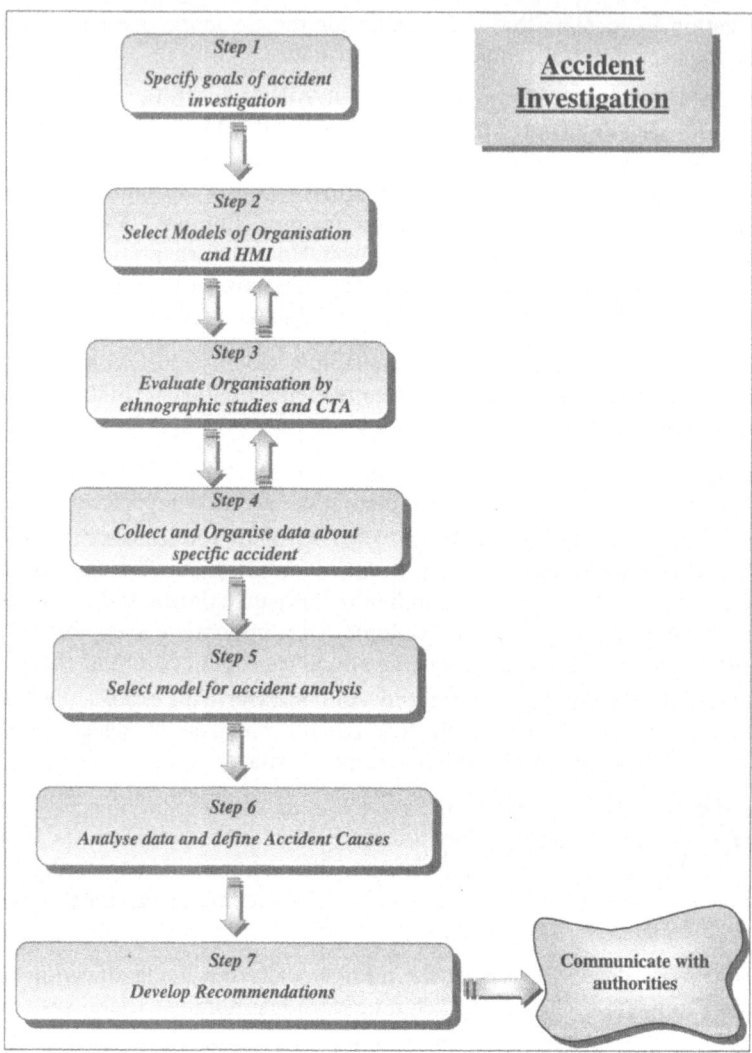

Figure 4.13 Stepwise procedure for accident investigation.

As many different elements contribute to its formulation, the study of an accident is not normally based on a single method, but usually follows a framework, where the contribution of different converging and integrated methods are applied. Moreover, the guidelines of accident investigation follow the principles of the DSME approach, discussed earlier in this chapter.

In summary, the process of accident analysis develops in seven steps that may be summarised as follows (Figure 4.13):

Step 1 Identify main objectives of accident investigation, i.e., identification of root causes, prevention of future failures/events/accidents of the same nature, recovery from similar system failures and/or human errors, protection for humans and environment in analogous cases (DSME stage *define* aims).

Step 2 Select the model of organisation and HMI to be used in accident investigation for (DSME stage *select and implement* specific models and methods):

☐ Selecting the taxonomy of system failures, error types, and modes in relation to contextual, socio-technical, and personal conditions.

☐ Defining the logical and sequential connection between errors and cognitive functions, systemic failure modes, and effects.

Step 3 Perform the analysis of organisation and attitudes of personnel, according to the principles of retrospective studies (DSME stage *select and implement*):

☐ Ethnographic studies, i.e., visits and observation at workplaces, audio-video recording, verbal/behavioural protocols, interviews etc.

☐ Examination of tasks and procedures by Cognitive Task Analysis.

☐ Review of past experience of system, and, in particular, analysis of incidents and accidents previously occurred.

Iterate with Step 2.

Step 4 Collect and structure data about the accident in association with the selected models of the organisation and results of studies of step 3 (DSME stage *select and implement*).

If necessary, iterate with Step 3.

Step 5 Define/select model for accident analysis and root cases evaluation (DSME stage *select and implement*).

Step 6 Define specific root causes and basic factors that affected the accident and the sequence of events (DSME stage *monitor* attainment of goals).

Step 7 Once the investigation is complete, the identification of causes is really proactive and effective if it contains relevant recommendations about the areas of change and possible modifications and ameliorations that are deemed necessary to improve the overall safety level of the organisation (DSME stage *evaluate* effectiveness of results).

The amendments identified by the accident investigation should aim at avoiding the repetition of similar circumstances and contextual conditions that favoured the accident. Their actual implementation and transfer to regulatory and normative measures depends on the authority in charge of safety of the relevant domain. This goal goes beyond the objectives of the accident investigation.

Another very relevant and delicate issue that needs clarification is the role and boundaries of the accident investigation with respect to the judicial enquiry. As already was clearly pointed out, the accident investigation aims at the identification of root causes that engendered and fostered an accident and its development. The judicial enquiry is set to identify responsibility and blame. There is an obvious connection and correlation between these two objectives. However, their standpoints and bases are totally different. Consequently, while the judicial enquiry should utilise the results of the accident investigation, the conclusions that may be drawn from a legal viewpoint must go beyond the technical and safety-related aspects dealt with in the accident investigation.

These legal and judicial matters go well beyond the scope of this book and therefore are not developed further. The only relevant issue that is underpinned here is that the technical accident investigation, to be performed as described in this chapter needs to be totally independent from the judicial investigation and must be granted access to all information that may be relevant for its development and for reaching its objectives. Lacking these conditions, the results may not be totally consistent and complete, and may also affect to a large extent the development of the judicial enquiry.

A case study of a real application of the stepwise procedure for performing an accident investigation in the domain of energy production systems will be discussed in detail in Chapter 8.

4.6 Summary of Chapter 4

This chapter represents the liaison between the theoretical description of models and methods for considering HF in system analysis and the development of a number of real applications in different industrial environments and organisational contexts.

The architecture HERMES remains the reference methodological frame that guides the development of any HF application. However, in this chapter, specific stepwise procedures of application of HF methods have been developed for the areas of application of design, training, safety assessment, and accident investigation.

These procedures must follow, in addition to the fundamental requirements of HERMES, the principles of project management, which are based on four steps to be preformed in sequence, namely: first defining the aims of the human error and accident management system under study; then selecting and implementing specific HF methods in the analysis of the HMS under evaluation; third, monitoring the attainment of the goals defined in the first step of the procedure while performing the HF analysis; finally, evaluating the effectiveness and efficiency of the results with respect to objectives and user friendliness.

In this chapter, the procedure for including HF in the area of safety assessment, and in particular, safety audit has been developed in great detail. This is due to the fact that this specific type of safety assessment is relatively novel with respect to the others, and very relevant for its potential impact on safety and on future applications. The concepts of IoS and RSA matrix have been discussed at length.

5

Application of HERMES for the Design of HMI Systems: Anticollision Warning in Automotive Environments

This chapter will consider the application of the HERMES methodology for the design of a system with strong HMI characteristics.

As the application concerned is a real case, the reader must consider that the overall study and development required a consistent period of several months of work. Moreover, a number of thorough and dedicated analyses were carried out in order to collect the necessary information for tackling different aspects of HMI relative to the system under design and development. A team of experts made up of designers, human factors specialists, and technology developers was engaged in the overall process.

Focusing on the human factors contribution to the design process, the stepwise and methodological details of the implementation of HERMES will be discussed, while the aspects concerning the technological development of the system and all the related technical tests and evaluations will be covered in a less accurate way, as these are common and well-known methods for engineering design and development.

Moreover, the reader must remember that the HERMES methodology does not contain guidelines for the implementation of specific HF methods. HERMES represents a framework for applying consistently and fruitfully a set of techniques that integrate retrospective and prospective analyses, with particular reference to human factors, in the four main areas of application, namely design, training, safety assessment, and accident investigation. The way in which these techniques and data are actually implemented depends on the specific methods selected by the designers and analysts, and on the accuracy and granularity required by the design/analysis carried out.

The following sections will commence with the problem statement of the specific design objective of this case study. Then the process of selection of the basic methods, models and data utilised for the HMI design process will be discussed. Finally, the progressive application of the selected methods will be shown, according to the HERMES framework and respecting the procedure for designing Human Error and Accident Management (HEAM) measures described in Chapter 4 (§4.2).

5.1 Problem Statement

In the domain of automotive environment, i.e., road transportation systems, the number and severity of accidents occurring throughout the world results in huge numbers of deaths or severely injured people every year. In most cases, accidents have more than one cause. However, environmental conditions, such as reduced visibility caused by darkness and/or bad weather, are a major contributor to these occurrences.

Supporting drivers in such adverse conditions, by enhancing perception of obstacles and possible imminent collisions with other vehicles, objects and/or persons, becomes critical for the prevention and reduction of accidents, and contributes to containing driver workload.

The design of such a system presents several major difficulties from both the technical and human factors perspective. In particular, the effectiveness of the system depends on two major features of the system, namely: the functionality of different sensors that enable the early identification of moving obstacles and possible imminent collisions; and the ability of the Human–machine Interface (HI) to mitigate driver workload and convey well-defined warning strategies in risky situations.

The European Commission has supported a project to deal with these issues, based on the development of an integrated system that merges two types of sensors derived from two different technologies, namely, the far infrared sensing and the microwave radar (Andreone et al., 2002). The system should cover a wide variety of adverse environments conditions, such as fog, rain, darkness, etc., and should develop a HI that supports the driver with an effective warning strategy in the case of danger of collision, and respects the requirements of international standards on usability and human-centred design (ISO, 1993, 1999).

In particular, from the HMI perspective, such an anticollision system should support the driver in a continuous mode and operate in the cases of reduced visibility conditions and driver-impaired or erroneous behaviour.

The design and development of such a tool has required the collaboration of a team of numerous experts and the contribution of car and original equipment manufacturers, in particular, developers of infrared and radar technologies. The designers of the tool dealt with a variety of complex technical problems and managed to solve them by a number of innovative solutions in accordance to the above requirements. Similarly, car manufactures implemented the instrument on board existing vehicles, merging its characteristics with the already very high number of instruments and tools on board modern vehicles (Hancock et al., 1996; Carsten, 1999; Carsten and Nilsson, 2001).

This chapter will focus only on the development of the interface of such a system and will concentrate on the way in which the HERMES methodology has served as a framework for the application of different approaches during the phases of conception, definition, and evaluation of the HI.

The technological problems and solutions adopted, especially for the integration and data fusion of the infrared and radar sensors, and for the implementation of the instrument on board vehicles will not be discussed, as they are outside the scope of this book.

Moreover, a number of results, formulations, and design choices have been generalised or slightly modified from the original ones made within the project, in order to preserve the necessary confidentiality of the technological development. This does not affect the objective of this chapter to demonstrate the application and effectiveness of the HERMES methodology for designing HMS.

5.2 The HF Design of an Anticollision Warning System

The design process of a safety device and relative interface is a well-established and very extensively studied procedure. The spreading of computer technology and the vast exploitation of automation for controlling systems of different complexity have required specific attention by the designers in the interaction processes between the user and the available automation.

The technological issues and problems related to the introduction of computerised systems and cybernetics for controlling systems (Wiener, 1948; Ashby, 1956) and more recently to the use of automation (Sheridan and Ferrel, 1974; Sheridan, 1992; Rouse, 1980, 1991; Billings, 1997) are known. Similarly, the human factors problems associated with the use and exploitation of such powerful technology have also been pointed out (Kantowitz and Sorkin, 1983; Bainbridge, 1983; Dreyfus and Dreyfus, 1986).

The most appropriate solution for designers of Human–Machine Systems (HMS) with extensive exploitation of automation consist in a process of application of best practices that combine technical engineering implementations and human factors assessments, in a sort of iterative process that involves users and progressively leads to the final development of the system. This approach has been identified since the early years of human computer interaction and, nowadays, is a well-established and standardised process (Sanders and McCormick, 1976; Gould, 1988; Helander, 1988; ISO, 1993; Bevan and Macleod, 1994; Salvendi, 1997; Billings, 1997; Sage and Rouse, 1999; ISO, 1999).

In Chapter 4 (§4.2) this iterative process was already extensively described, and can be summarised essentially in a sequence of seven steps leading to the implementation of the system in the market. During all these steps specific attention is given to the *context of use, user needs,* and *system requirements.* Following the design solution, special tests are preformed to monitor attainment of goals. Before, implementing the system, special training is devised for ensuring user friendliness, ability, and adequate exploitation.

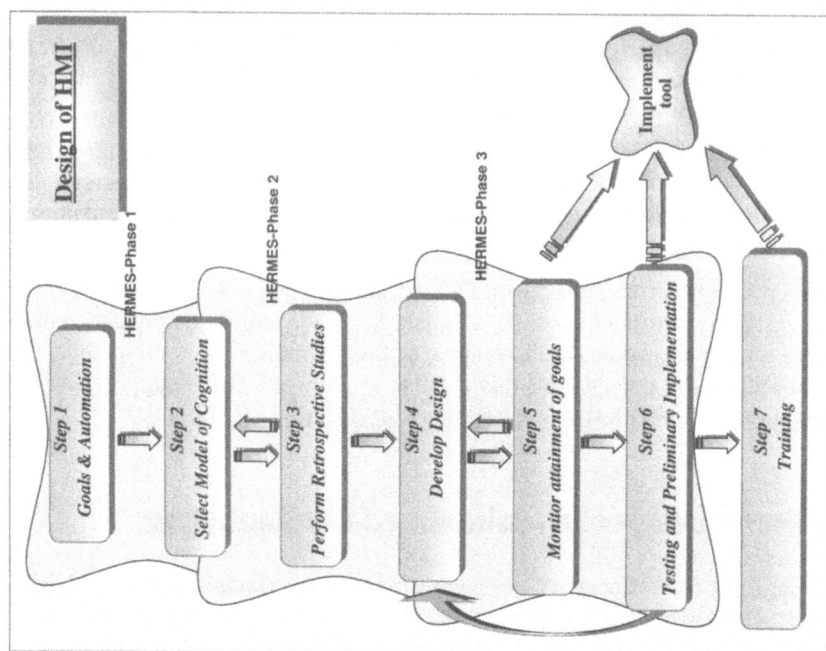

Figure 5.2 Design and implementation of a HEAM system and HERMES phases.

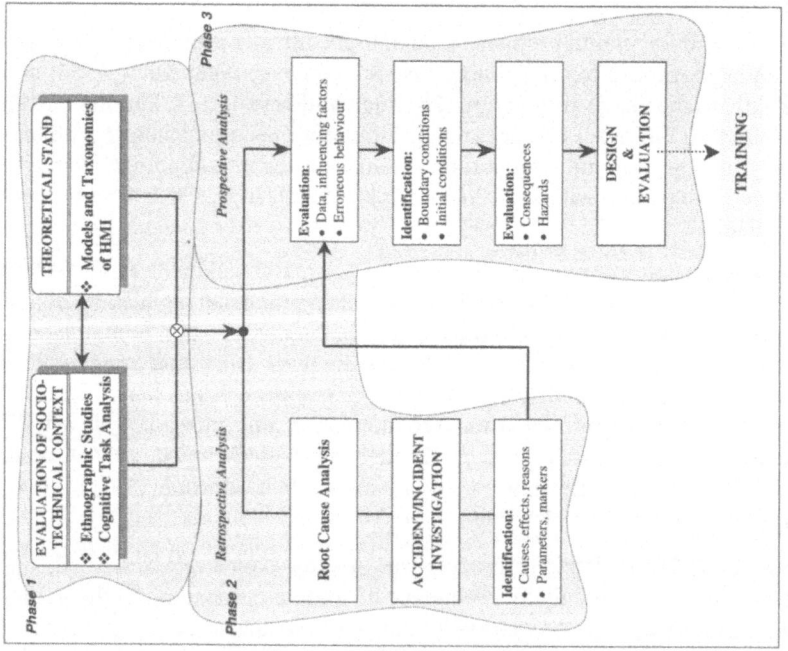

Figure 5.1 Application of HERMES for the design of an Anticollision Warning system.

In all these steps or phases the continued focus on the users of a system is granted by direct contact through interviews, observations, surveys, and participative design. The aim of such studies is to understand cognitive, behavioural, attitudinal, and anthropometrical characteristics of users and their tasks.

The methodology HERMES was applied throughout the design process and, in particular, in supporting the development of the HI and for the definition of the plan of tests and experiments of different design solutions. The role of the HERMES approach and the effect that the selection of the most suitable methods for HF analyses play on the overall design process are therefore quite obvious.

In practice, the activity of the HF specialists was performed in three correlated phases of application of HERMES that were associated to the steps of design development as follows (Figure 5.1):

Phase 1. Evaluation of socio-technical context of work and definition of theoretical stand for design development and testing (*context of use*). This phase can be associated with the steps of definition of the goals of the system and selection of level of automation to be implemented.

Phase 2. Study road accidents and performance of laboratory and simulator experiments for defining user psychophysical characteristics (*user needs*). This led to the definition of scenarios for further testing the system to be developed in real traffic conditions. Moreover, during this phase, contextual data and parameters for the most suitable HF design solutions were developed (*system requirements*). This phase can be linked to the steps of retrospective studies and preliminary design development.

Phase 3. Support the final design process of the anticollision system and associated interface, by performing a sequence of experimental tests and field evaluation for the assessment of efficiency and user friendliness of the system on board real vehicles. This phase completed the process of testing and experimental validation of the system, mainly from the HF perspective.

Figure 5.2 shows how the stepwise procedure for designing HEAM measures and the HERMES framework can be integrated. In reality, the whole process of design and implementation of a system for HEAM goes beyond the design and testing of the preliminary implementation of a technological solution (§4.2 and Figure 4.6). Two more steps should be carried out before the complete insertion of the system on the market, namely, the *training* process of the users, either by the development of adequate manuals and/or of on-board active instructions, and the *implementation* of the system in the construction process of the entire vehicle. These two steps were not performed in the case presented here, as the whole project was of research nature and therefore only an exploitation plan was developed with the design of the prototype system. Consequently, they will not be discussed further. However, they represent the industrialisation part of a design process and the essence of application of the HERMES methodology is fully covered by the case herewith presented.

Moreover, as already mentioned, the technological aspects and problems dealt with by the manufacturers of equipments and vehicles will also not be discussed.

We will focus here on the way in which HF issues have been dealt with and solved in the HERMES methodological architecture. We will refer to the technical issues and solutions that have been adopted, simply by presenting the conclusions reached after discussion and evaluation of different technological and human-related alternatives, at different stages of development of the design.

5.2.1   Phase 1: Context of Use and Theoretical Stand for Design, Testing, and Evaluation

The first phase of work is usually very important in the application of HERMES, as it represents the selection of the basic instruments for performing the human factors analysis and offers the analysts the possibility to study the organisation in which the methods are applied and exploited.

In this case, phase 1 was essential for the entire development at technological and human factors level. In particular, two important aspects of HF methods are relevant in this application, because of the specific area and domain of application. First, the principles of UCD approach were fulfilled by considering a model of the driver, interacting with the environment and vehicle under control, throughout the entire design and testing process. The identification of this model is very important, as the selected paradigm serves as reference for the design of interfaces and affects the whole plan of experiments and tests. The selection, and preservation throughout the whole design process, of a reference cognitive model of the driver granted consistency and coherency in the application of methods.

Second, the fact that the design process concerned a tool in the automotive environment reduced the relevance of the analysis of the cultural and organisational aspects affecting the sociotechnical context of the "organisation" in which such technology was to be implemented. Indeed, nowadays cars and road vehicles are utilised throughout the world, and the need to develop specific design solutions, related to particular context and organisational aspects, is less relevant, with respect to other factors, such as the variety of age, expertise of users, and physical and environmental characteristics of use.

The major peculiarities of the automotive environment can be summarised as follows:

- *Type of User*. A large variety of people, from teenagers to very old people, including those with physical, sensory or cognitive deficits, can obtain a driving license. In other domains, the types of users are much less diversified. Moreover, drivers receive minimal initial training, often little or no refresher training at all, and no training with specific new vehicles or devices. In contrast, the operators of plants and users of systems in other domains, e.g., pilots, air traffic controllers, control room operators, medical personnel, aviation technicians, and so forth, are highly selected, relatively homogenous, well trained, and regularly supervised or monitored.

- *Temporal Demands.* The temporal demands for in-vehicle warning practises are typically much more time-critical than for other domains. For a potential crash situation, there may be only a second or two between the recognition of conflict and the moment of impact. In contrast, emergency situations in other contexts may be measured in tens of seconds, or even minutes.

- *Working Context.* Most technological domains are based on specific and dedicated workplace environments. This is not the case for the automotive domain. This difference has many implications in terms of acceptability of constraints, practices, and assumptions about the user. As an example, the acceptable level of intrusiveness or annoyance and tolerance of false alarms in a car may be quite different than in an aeroplane cockpit. In addition to this context internal to the vehicle, there is also a context that is external to it, namely, the roads and traffic. This is also a very specific peculiarity of automotive transportation, as it represents a domain where there is an extremely high density of traffic and the rules for traffic management are almost completely left to the decision of drivers rather than a specific traffic management authority.

- *Social Context.* There is a social context in driving that is substantially different than other environments. The presence of other passengers in the vehicle means that various driver warnings may be public. Consumer acceptance of an in-vehicle warning device, or behavioural reactions to the warning, can be influenced in various ways by the social context, e.g., embarrassment, feelings of competence, need to appear daring or skilled to peers. Social context is also present in other domains (aviation, air traffic control, train-driving, etc.), but it is more affected by organisational factors and national cultures.

- *System Integration.* In other domains, the workplace is often designed as an integrated system, so that the requirements can be well defined from the beginning. For vehicle applications many devices may be aftermarket add-ons, not integrated into the original design of the vehicle. In these cases, the environment is evolving and therefore is not fully predictable and it is not consistent from vehicle to vehicle.

All these factors affected the ethnographic studies of phase 1 of the work, minimising the study of organisational factors, in favour of the identification of user needs covering a wide range of the population.

In practice, phase 1 of the work aimed at the identification of the user needs and the selection of the cognitive model of the user and was structured into three steps:

Phase 1 – Step 1. Definition of the characteristics of the tool and warning strategies from a safety perspective.

           Step 2. Definition of user needs by task analysis and initial ethnographic studies associated with the warning strategies.

           Step 3. Selection of model of user and identification of cognitive functions and processes associated to warning strategies.

Definition of the Characteristics of the Tool and Warning Strategies

The main goal of this design process was to develop an integrated driver assistance system, merging the functionality of two different sensors, namely, a far infrared camera and a microwave radar. The system should operate in the case of:

- driver lack of obstacle perception, due to reduced visibility conditions and adverse weather conditions; and
- driver impaired or erroneous behaviour under different traffic scenarios.

Existing Driver Support Systems

Driving support systems for vision enhancement can be subdivided in nonintervening systems and intervening systems.

A nonintervening system has no influence on the vehicle dynamics and only displays information and warnings by means of an acoustical, haptic, or optic signal. An example of this is the thermal imaging, or far infrared technology night vision system by General Motors, which was introduced onto the market in the year 2000 (Figure 5.3). This "thermal imaging" device creates pictures based on heat energy emitted by objects in the viewed scene. The virtual image, designed not to obstruct the view and to keep driver attention on the road, is projected by a head-up display (HUD) in the driver's peripheral vision. Drivers can glance at the virtual image without refocusing and continue to keep their eyes on the road. The system powers up when the key is in the "on" position, a dedicated photocell indicates that it's dark outside, and the headlamps are on. Drivers can turn the system on and off, and can adjust the image intensity and vertical position. The image, generated by a liquid crystal display (LCD), is magnified and mirrored onto the car windscreen. The image seems to float below the horizon at a distance of few meters and requires

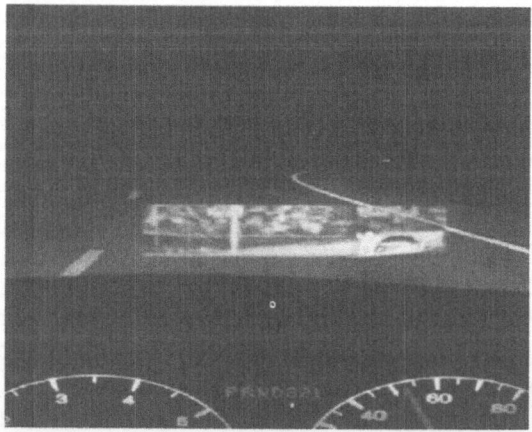

Figure 5.3 Far infrared technology night vision system by General Motors.

Figure 5.4 Adaptive cruise control by Daimler-Chrysler.

only a short time to get used to using. Due to the limited brightness of matrix LCDs, the image can only be seen during darkness.

An intervening system uses a processed sensor signal to influence the vehicle dynamics, like the "Adaptive Cruise Control" (ACC). The human–machine interface (HI) is reduced to an optical interface in the instrument cluster and to some control buttons. Although the radar sensor signal of the ACC is highly dynamic, the interface is designed to be simple and requires no permanent attention. An example of intervening system is the Daimler-Chrysler Adaptive Cruise Control (Figure 5.4).

Characteristics of the Tool Under Design

The anticollision warning discussed here will be called Road Collision Warning System (RCWS). The first *design option* selected for the RCWS was that:

- *The RCWS should be designed as a nonintervening warning system.*

The system uses a combination of complementary sensors in order to provide an enhanced warning function for the driver in case of a dangerous situation on the road ahead. The system includes the following components (Figure 5.5):

- Far Infrared (IR) camera
- Forward collision warning radar sensor
- Yaw rate sensor (vehicle angular rate)
- Vehicle sensors (i.e., light status, vehicle speed, steering angle sensor etc.)
- Sensor merge platform and human–machine interface presentation
- Image processing platform

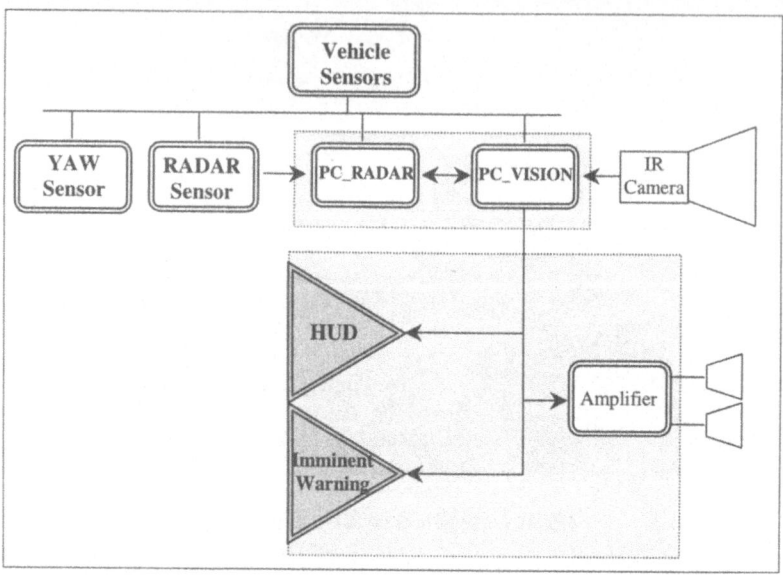

Figure 5.5 Road collision warning system (RCWS) architecture.

- Human–machine interface units (HUD and visual warning)
- Loudspeaker (for the acoustic warning)

Warning Strategies

As the RCWS was developed as a nonintervening supportive and nondisturbing "warning system," the definition of the warning strategies was one of the critical points of the design process.

Aspects such as "eye-off-road" time and the possibility of providing visual warnings directly integrated with the visual representation of the road scenario on the windshield were accounted for in the definition of the warning strategies. Four aspects of the overall warning strategy were considered in detail:

- *Warning information content*
- *Warning timing and distance*, relative to the hazard
- *Warning levels*
- *Warning channels*, i.e., visual, acoustic, haptic

These four aspects are not independent of each other, and the selection of an alternative for one of them affects all the others. Moreover, a vast literature exists on this subject (NHTSA, 1993, 1996; EC, 1999). This was widely explored before making a complete plan of warning strategy.

As for the *warning information content* transferred to the user, the physical entities that should be detected by the RCWS are the following:

- vehicles, such as cars and lorries;
- pedestrians, and cyclists;
- motorbikes; and
- road infrastructures and fixed or moving obstacles.

The *warning timing and distance* that affect the safety performance of the system can be associated with the distance from the obstacle and the expected time required to collide with it. In particular, three measures can be considered:

1. The *time to collision* (*TTC*), which is the time that results from the distance Δx between the leading vehicle's rear bumper ($\text{car}_{\text{leading}}$), or obstacle, and one's own car ($\text{car}_{\text{system}}$), divided by the difference of velocity ($v_{\text{system}} - v_{\text{leading}}$) between them:

$$TTC = \left| \frac{\Delta x}{v_{\text{system}} - v_{\text{leading}}} \right| \qquad \text{Eq. 5.1}$$

2. *Headway* (HW), which is the time that results from the distance Δx and the velocity v_{system} of one's own car system, i.e., the TTC, when the obstacle is steady or almost steady, with $v_{\text{leading}} \approx 0$, or $v_{\text{leading}} \ll v_{\text{system}}$:

$$HW = \frac{\Delta x}{v_{\text{system}}} \qquad \text{Eq. 5.2}$$

3. *Predicted minimum distance* (PMD), i.e., the minimum distance between a vehicle and a potential obstacle predicted in real time, for the warning system to become active. If $PMD = 0$ then the impact is forecasted; if $PMD > d_{\text{threshold}}$, then the obstacle is not to be considered dangerous.

Because the time that is available for a driver to react to hazards by changing the driving parameters, decreases the danger of a the situation, the warning of the collision avoidance system must be activated in *warning levels*. In accordance with the guidelines for crash avoidance warning devices of different road and traffic safety authorities (NHTSA, 1996, EC, 1999), the following *design option* was chosen with respect to warning levels for the RCWS:

- *The RCWS would be based on two levels of indications, i.e., <u>cautionary warning</u> and <u>imminent warning</u>, to distinguish between potentially dangerous and highly probable situations.*

Two corresponding distances and times are associated with the two levels of warning:

1. The *cautionary warning* occurs when the detected obstacles are potentially dangerous, as the *TTC* and *PMD* are significantly low, but not below given thresholds.

2. The *imminent warning* occurs when the obstacle detected has a high probability of being hit, and is located at a distance below defined thresholds ($TTC < t_{threshold}$; $PMD < d_{threshold}$).

Finally, the HI of the RCWS should be capable of effectively supporting the driver by strongly reducing the data stream together with the appearance of intuitive, simplified messages/displays, on different *warning channels*, namely:

- "visual warnings," by presenting text messages, icons, virtual images, or other signals on specific locations, e.g., dashboard or windscreen;
- "aural warnings," by sound or speech messages of various levels and directional effects; and
- "haptic warnings," by exerting forces, vibrations, pulses on different vehicle controls systems such as the accelerator, break pedals, steering wheel, or seats.

Definition of User Needs

After having defined the generic operational and safety characteristics of the RCWS system, the remaining steps of phase 1 focused on the specific definition of the design variables and on the human factors methods to be applied for designing the interface with the user.

The first of these steps consisted in the analysis of the user needs (Cacciabue et al., 2001). The automotive domain is one of the most studied environments. Cars and vehicles are extremely diffused means of transportation and are utilised by practically the entire spectrum of people, independent of their age, gender, profession, culture, or social context. For these reasons, a thorough study of the literature in the domain was performed. Moreover, the evaluation of the working context had to expand beyond the organisational analyses of the working context, according to the HERMES methodology, covering a wide variety of possible users, rather than a specific subset of persons with very well defined corporate culture and training background.

A detailed study of past and ongoing projects and research initiatives was carried out, and the guidelines provided by national safety authorities were reviewed (FHWA, 1999; NHTSA, 1996; ACAS, 2002; AWARE, 2002; SAVE, 2002; COMMUNCAR, 2002; EUCLIDE, 2002).

At the same time, a questionnaire and associated guidelines for an interview were developed to examine the impact of the specific type of warning system on the widest possible variety of drivers. In order to capture a wide variety of people from different national, cultural, and social conditions, the questionnaire and interview were distributed to the passengers of an international European airport (Cacciabue et al., 2002b).

A high number of passengers were interviewed and filled in the questionnaire during several weeks of intensive data collection.

The results of the analysis of the literature and of the questionnaires/interviews have been very useful in defining the user needs from the environmental and human viewpoints, and for highlighting some open issues that need special consideration in the development of the interface and functions of the system. The following subsections describe the findings of these analyses from different viewpoints.

User Needs from an Environmental Viewpoint

The environmental factors that affect the performance of an anticollision warning system, as the RCWS, can be analysed according to the *type of roads*, *speed range of action*, *traffic*, and *visibility conditions*.

A first important remark is that in the case that the system is not reliable in a type of road/situation, e.g., urban environment, down hills, near bridges, etc., it should automatically turn off and should warn the driver in order not to offer a notion of false reliance on it.

Users seem to like the idea of having such a support in all speed ranges. However, the maximum speed to be supported by the system would be limited for technical reasons, leading to a more robust and inexpensive system.

Concerning traffic conditions, users prefer the support of a vision enhancement system in cases of low and medium traffic density:

• low traffic density, with the front vehicle at *headway* over 3 sec.;
• medium traffic density, with front vehicle at *headway* 2–3 sec., for more than 2 min.; and
• high traffic density, with front vehicle at *headway* 1–2 sec., for more than 2 min.

In very high traffic densities, with mean speed below 20 km/h for at least 2 min, the system may not be useful at all, or it can be turned into a simple collision avoidance warning mode, rather than supporting an enhanced environmental perception. Thus, a multiple target handling system, also able to work in high traffic densities, did not seem to be really necessary.

From the external environmental point of view, users prefer the system mostly in reduced visibility conditions, which may be defined as:

• medium visibility between 50 and 150 m;
• low visibility, when being able to discern targets in less than 50 m; and
• extremely low visibility, when one is able to discern targets in less than 10 m.

A major issue is if the system should be automatically or manually turned off in the case of good visibility, to reduce driver workload. On the other hand, if the driver becomes familiar with the system, he/she might find it useful even under such conditions. From the analysis of the questionnaires/interviews it was derived that users prefer the use of the system at night or during dusk/dawn. Thus, the development of a complex and expensive system, also being able to work on a sunny day, was not supported by the actual user needs. Also, it is preferable from the user's point of view to use the system mostly in fog and on wet road conditions, as opposed to rain or snow and on dry or icy roads.

User Needs from a Human Viewpoint

The analysis and classification of user needs and preferences from a human factor perspective was very important for a first screening of the types and modes of the warning system.

The level of system acceptance, and the ability/need to operate the system were rated on a scale from 0 (not at all accepted/operable) to 5 (full need/interaction). The drivers were grouped according to the typology of use of the vehicle, and their age group.

The analysis of the questionnaires led to the same results observed in literature. In particular, truck and ordinary car drivers showed very high levels of acceptance (levels 4–5), while taxi drivers showed lower scores both for what concerns acceptance and ability/need to operate. This can be easily explained by the specific type of activity carried out by taxi drivers, which is predominantly on town roads, where visibility is rarely a problem, speed is relatively low, and traffic density is usually very high. These are the environmental conditions for which the system was considered less useful.

With respect to age group, the level of acceptance and the need of the system have been recognised by all age groups, namely 18–35, 35–65 and over 65, with a particular need identified by those over 65.

In terms of HI, the user preferences were studied in order to discover which type of warning was most suitable for the system under development. In particular, the three different types of *warning channels*, i.e., aural, visual, and haptic, were studied to decide about their implementation in the system. The relative importance of different aural systems and image displays were investigated to identify the most effective ones.

Another important aspect of the human–machine interface to be analysed from the user need and human factor perspective is the type of image display. This can be substantially of two types, namely: head-down or head-up display (HDD or HUD). For each type, the image can be either virtual or real. Therefore, four types of HI configurations for vision enhancement systems can be considered (Figure 5.6):

- virtual image display, head-up mode (VID-HU) (Figure 5.6, option 1);
- virtual image display, head-down mode (VID-HD), (Figure 5.6, option 2);
- head-up display, real image (Figure 5.6, option 3);
- image displayed on the instrument panel (Figure 5.6, option 4).

Findings from User Needs Analysis

According to the survey and analysis of the literature it was possible to define a number of priority scenarios for using and testing the RCWS system. These scenarios were also developed by expert opinion of the HF specialists engaged in the

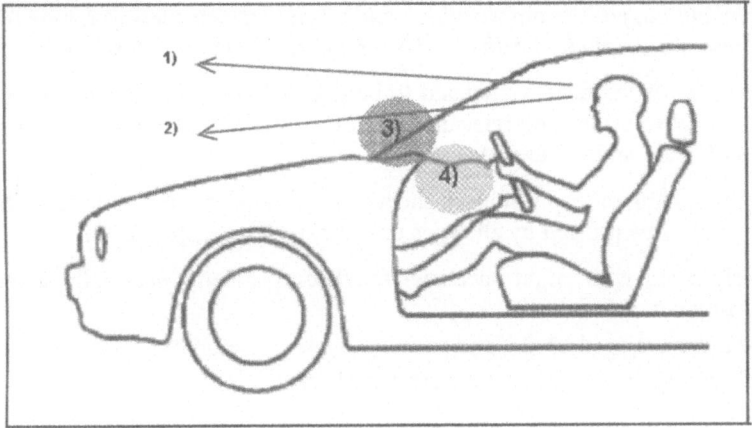

Figure 5.6 Virtual image and real display in head-down and head-up modes.

project. A general plan of scenarios to be tested in laboratory and field studies include:

- highway/rural roads;
- night, dusk/dawn situations;
- light or heavy fog; and
- light or heavy rain conditions (wet road).

All possible combinations of these conditions are of interest. The usability/ usefulness of such a system during good weather conditions and daytime may be controversial and needed specific testing.

The user needs analysis showed clearly that a simplified and cheaper version of the interface could be used. Users did not seem to need the system during daytime and in high-density traffic scenarios. Moreover, they seemed to need the system mostly in fog condition and less during rain. They did not like haptic warnings. All these findings simplified the whole system architecture and allowed the designers to focus on the issues of major interest for the users.

From the technological point of view a HUD was found to be the best *design option* for such a system, consequently:

- *The virtual image display, head-up mode (VID-HU) was selected for the RCWS.*

However, the driver can be also supported by a cheaper and less complex system, for example a VID-HD, or even by images displayed on the instrument panel, with the support of only audio and not haptic messages.

As a result of these studies, the warning channel based on haptic warnings and interventions was removed from the design of the RCWS. This further simplified the architecture interface and consequently the following *design option* was made:

- *Visual output, with sound warnings, and possibly spoken messages, were selected as presentation modes for the RCWS (no haptic mode was considered).*

The position of the displays needed to be decided after considering all possible technological solutions and relevant limitations, and taking into account the preferences of users and HF experts.

Finally, according to the findings of the user needs analysis, it was decided that the system should be utilised by all types of users, with an emphasis on:

- elderly users, who might have specific vision problems, especially at reduced visibility conditions;
- young drivers aged 18–25; and
- professional drivers, in particular truck drivers.

Issues of Concern

From the analysis of the questionnaires and relevant bibliography a number of issues arose in reference to the user needs and preferences regarding advanced warning systems, such as the RCWS under development.

Most issues concern cognitive aspects of HMS, and are of paramount importance to the development of a useful and adaptive human–machine Interface. In brief these issues concern the following:

- *Training.* To be effective the system must be comprehensible and immediately clear to the driver. The need of training is a questionable issue. For example, Italian drivers seem to disagree with this point, while Swedish drivers accept it. An open issue consists on the organisations involved in this kind of training: government authorities, driving schools, car manufacturers, or car dealers.
- *Risk homeostasis theory.* This theory predicts that drivers would compensate the risk reduction by engaging in more risky behaviours, thereby negating the benefits of such a safety system (Wilde, 1994).
- *Complacency.* The complacency is the problem arising when the user trusts the system so much that he/she ceases to monitor it sufficiently (Wiener, 1981). Some of the main findings on trust between human and machine are the following:
  —Trust grows as a positive function of reliability of the automation and efficiency of the system.
  —Trust declines if faults occur (according to the magnitude of the faults).
  —Drivers often assign differential trust to differential parts of the system.

Selection of Model of User

The selection of the model of user was of particular interest to HF specialists, as it focused on the definition of the cognitive functions and processes involved in the interaction of the driver and the anticollision warning system.

In performing this selection another essential aspect of the HMI needed specific consideration, i.e., the integration of the RCWS in the overall interaction process of the driver, vehicle, and environment. The selection of the cognitive model of the driver and the integration in the overall interaction realm and processes led to the definition of variables and thresholds implemented in the design, and to the definition of the experimental quantities to be studied in the successive steps and phases of the work.

From the definition of the characteristics of the tool and warning strategies and from the analysis of the user needs, it resulted that the cognitive functions of particular interest in this case are the *perception* of the warning signal, followed by a "rapid" *interpretation* of the rule associated to a certain sign of imminent collision, and the necessary and prompt *execution* of corrective actions that may be derived. This is a typical process of information processing.

From the analysis of the literature of cognitive models associated with the "information processing system" paradigm, performed in Chapter 3 (§3.1.1), it is clear that models such as the SRK (Rasmussen, 1986), the RMC (Wickens, 1984; Cacciabue, 1998) and COCOM (Hollnagel, 1993) are all well suited to represent these specific cognitive functions of the driver. In particular, all three models consider the possibility that the cognitive function *planning* may be cut short during certain situations which demand "skill-based behaviour" and immediate response to certain stimuli of the context.

The models RCM and COCOM include in their architectures a specific representation of the cognitive functions of *perception*, *interpretation*, and *execution*. Therefore, they have been considered more suited than the SRK model for the case at hand.

The RMC focuses on the four cognitive functions (*perception, interpretation, planning*, and *execution*) and defines the allocation of resources and the knowledge base as the fundamental cognitive processes that govern the interaction between cognitive functions. The correlation between cognitive functions and processes is quite simple, as the cognitive processes define the availability and activation of the cognitive functions leading to the actual performance of actions.

COCOM considers the four cognitive functions, in its part called VSMoC, as the basic functions that allow description in a simple but structured way of how information is perceived and eventually generates actions. In addition, COCOM considers certain cognitive process in the control part of the model.

In COCOM, four control processes affect the transfer of information between the cognitive functions. The lower the control process (scramble and opportunistic) the less amount of information is transferred and managed between cognitive functions, and action/reactions are only relatively correlated to stimuli. On the other hand, higher control processes (strategic and tactic) allow complete and structured information management between all cognitive functions, including planning.

The control model governs the competence part of COCOM, and it is in turn governed by contextual and psychological factors, such as the outcome of a previous task or action and the subjective estimation of available time (Hollnagel, 1993).

Table 5.1 Basic characteristics of cognitive model of driver for HMI studies

Characteristics of Cognitive Model of Driver	
Cognitive functions of RMC	• *Perception*
	• *Interpretation*
	• *Execution*
Contextual conditions affecting the cognitive process of *allocation of resources* of RMC	• *Time available*
	• *Physical environment*
	• *Workload*
	• *Type of obstacle*

Both models seem to frame well the interaction of the driver with the RCWS system. However, the RMC is simpler at a level of cognitive processes and can be applied more straightforwardly. Therefore, for this study, the following *design option* was made:

• *The reference model of cognition was selected as the paradigm for describing cognitive and behavioural performance of the driver and was applied through-out the whole design process.*

At this stage of development of the design process, three aspects of the previous activities contributed to further defining the specifications of the RMC, namely:

• the analyses on user needs;
• the technological considerations about performance of the infrared and radar sensors; and
• the preliminary design choices on the type of warning system, i.e., non-intervening system, and main aspects of warning strategies, i.e., information content, timing, levels, display modes, and channels.

It was clear that a system like the RCWS must act primarily on the cognitive functions of *perception* and *interpretation*, favouring the *execution* of an adequate evasive manoeuvre by the driver.

Moreover, the time available for intervening, the physical environment, and workload were identified as the most important contextual conditions affecting driver behaviour. This implies that such contextual conditions, in addition to the type of encountered obstacle, are the factors that affect the cognitive process of *allocation resources*, which governs the interplay and effectiveness of cognitive functions.

Table 5.1 summarises the cognitive functions that are mostly operational in describing driver behaviour and the contextual conditions that affect the cognitive process of *allocation of resources*.

The selection of the model of cognition utilised for reference during the design process concluded, ideally, phase 1 of the work and led to the development of more specific activities related to the RCWS system under development. These were carried out in phases 2 and 3.

In parallel to this activity of human factors analysis the technical development of the system continued according to the process already discussed earlier (Figure 5.2).

5.2.2 Phase 2: System Characteristics and Design Development

As already mentioned, there was continuity and complete integration between the three phases of application of the HERMES methodology and the technical design. The development of the anticollision warning system evolved in parallel and in synergy with the HF studies.

The human factors method mostly applied during this phase of work were:

- discussion and brainstorming between HF specialists and designers of the technology;
- consideration for the principles of the user-centred design approach at design level; and
- involvement of end-users in iterative evaluation and revision process.

The previously developed analysis of user needs and the paradigm selected for describing driver behaviour were of guidance. In this respect, the three cognitive functions of *perception*, *interpretation*, and *execution* were constantly accounted for when making choices between possible alternatives.

In particular, phase 2 of the work was important in defining and formalising the plan for experiments and testing in a more structured and integrated way than what was preliminary done in phase 1. Indeed, the design of experiments in different work settings (laboratory, simulator, and real road environments) and the associated scenarios could be more focused and developed in detail once the characteristics of the specific warning system under development were selected and defined.

Some of these experiments were carried out in the early stages of the technical design development, e.g., for the selection of the most appropriate symbols and icons and for the positioning of the head-up display. Other experiments were, instead, only planned in phase 2 and actually performed in the final stages of testing the RCWS in real road and realistic driving scenarios (phase 3).

In order to finalise the scenarios for designing the set of experiments and tests, the analysis of past experience derived from accidents investigations and reporting was particularly useful.

The fact that a preliminary general plan of experiments was developed in the early design stages ensured coherence and continuity of work between system and human factors engineering. The final plan of experiments concerning road tests and realistic driving scenarios was completed during the last step of phase 2 of work.

In detail, the activities of phase 2 of work were centred on four steps, namely:

1. *Phase 2, Step 1.* Selection of characteristics of the RCWS.

   Step 2. Selection of type of tests and plan of experiments.

   Step 3. Study of accidents and past experience.

   Step 4. Definition of accident scenarios for testing and evaluation.

Selection of Characteristics of the Road–Collision–Warning System

The choice of developing a nonintervening warning system requires the selection of adequate means of supporting the cognitive functions associated with driver reaction to the indication of a possible hazard.

From the results of studies (questionnaires and interviews) on user needs carried out in phase 1 of the work, these means have been limited to visual and aural warnings. The RCWS indications must ensure that *perception* of signals is granted and *interpretation* of the associated signs lead the driver to *execute* the appropriate evasive manoeuvre by braking or steering to avoid the obstacle.

At the same time, the environmental conditions in which a system like the RCWS operates are of poor visibility due to darkness, fog, rain, and difficult roads and traffic situations. Therefore, the timing for intervention becomes a critical variable to analyse.

Type of display and position are also essential aspects to be defined in order to ensure effectiveness of the system and efficiency with respect to driver tasks, performance, and safety.

All these factors were considered in parallel while making the design choices for the interface and technical properties of the tool, and the outcome of such harmonisation activities resulted in the definition of the most relevant characteristics of the RCWS in terms of:

• timing and levels of warning strategies, and
• types and modes of presentation of the human–machine interface.

These two results will be discussed in the next subsections.

RCWS: Timing and Levels of Warning Strategies

The HI of a collision warning system must attract the attention (*perception*) of the driver to a dangerous object or road user obstacle, without generating distraction, while triggering the awareness of a hazard (*interpretation*) and activating the most effective reaction (*execution*).

The more time a driver has to react to the alarm by changing the driving parameters (speed and direction) the less dangerous the situation is going to be, and vice versa. Therefore, the warning to the driver by the collision avoidance system should be activated in different levels.

The design decision to operate the RCWS on two levels of warning strategies, i.e., *cautionary warning* and the *imminent warning*, was already made during the previous phase of work. The choice to be made during this step of work concerned the selection of the actual warning distances and timings based on the usual measures of safety, i.e., Time To Collision (*TTC*) and Headway (*HW*), utilised in collision hazards and described earlier (Eq. 5.1 and Eq. 5.2) (Polychronopoulos et al., 2003).

Specific thresholds of *TTC* and *HW* for these dangerous situations can be developed from the National Highway Traffic Safety Administration (NHTSA, 1996):

- *Cautionary warning*:
  TTC: $5 \leq TTC < 10\text{--}15\,sec$
  HW: $1.5 \leq HW < 2\text{--}3\,sec$
- *Imminent warning*:
  TTC: $3 < TTC < 5\,sec$ or less
  HW: $1 < HW < 1.5\,sec$ or less

According to the above discussion, the predicted minimum distance is the minimum distance between the owner's car system and a potential obstacle. Therefore, the *PMD* represents an extra distance to be added to the Δx, separating the leading vehicle's rear bumper and the owner's car, in order to grant more time to react thanks to the sophisticated combination of the radar acquisition and real-time processing of the infrared images. The *PMD* is defined in such a way that the Time to Predicted Minimum Distance (*TPMD*) and the *TTC* coincide when *PMD* = 0. Therefore the *TPMD* and corresponding *HW* for the RCWS can be defined as:

$$TPMD(RCWS) = TTC + \Delta t = \left| \frac{\Delta x + PMD}{v_{system} - v_{leading}} \right| \qquad \text{Eq.5.3}$$

$$HW(RCWS) = \frac{\Delta x + PMD}{v_{system}} \qquad \text{Eq.5.4}$$

where:

Δx = distance between the leading vehicle's rear bumper, or obstacle, and the owner's car;

v_{system} = velocity of the owner's car; and

$v_{leading}$ = velocity of the leading vehicle or obstacle.

A definition of the *TPMD* that combines physical and human factors data can be considered as:

$$TPMD(RCWS) = t_s + t_d + t_r + t_v$$

where:

t_s = time for the system to react and to enter the warning mode;

t_d = time for the driver to realize there is a potential danger;

t_r = time for the driver to act on the brake and/or on the steering wheel;

t_v = time for the vehicle to stop, depending on physical variables such as vehicle speed, brakes, tyres, and road surface.

The specific *design option* selected for the RCWS was to set the following values of thresholds for *TPMD* and *HW* in the two cases of *cautionary* and *imminent* dangerous situations:

- *Cautionary warning* for the RCWS:
  TPMD (RCWS): $4 \leq TPMD\ (RCWS) \leq 8\ sec$
  HW (RCWS): $1.5 \leq HW\ (RCWS) \leq 3\ sec$
- *Imminent warning* for the RCWS:
  TPMD (RCWS): TPMD (RCWS) < 4 sec
  HW (RCWS):   HW (RCWS) < 1.5 sec

The calculation of the warning distances shows that the activation distance of nearly 290 m in the *cautionary warning* at a working speed of $v_{maximum} = 130\,km/h$ exceeds the reliable object classification distances of the sensors. In the *cautionary warning* the driver gets information about the detected object. This information shows an exposure to danger, but the system is not able to reliably classify the object at a higher distance than 150 m.

RCWS: Types and Modes of Presentation

In order to enhance perception and interpretation of the warning signals, and to reduce to a minimum the time of eyes off the road, the information must be presented in the central field of view and should bear certain ergonomic characteristics of sensory perception (colour and sound) that favour efficient perception, interpretation, and execution of adequate evasive manoeuvres.

The specific characteristics of the Virtual Image Display in Head-Up (VID-HU), chosen as the most effective visual information presented to the driver, are to:

- allow the superimposition of warning symbols;
- have a sufficient image brightness in fog;
- enhance the image size;
- avoid the use of a special windscreen for image projection;
- obtain a wide eye box; and
- allow suitable vehicle integration.

A number of symbols have been defined to represent different categories of obstacles, i.e., humans, cars and trucks (including cyclists or motorbikes), or not defined objects/obstacles. In addition, generic warning symbols associated with an imminent hazard have also been considered.

The symbols presented to the driver in HU mode for specific categories of obstacles are shown in Figure 5.7 and the symbols chosen for generic hazards are presented in Figure 5.8. An example of presentation in a vehicle that combines the symbols selected for warning and the concept of VID-HU mode, is shown in Figure 5.9. The following *design options* were selected:

- *A sound of varying intensity and frequency, depending on the level of warning, has been associated to the VID-HU to further attracting attention.*

Warning Symbols per category of Obstacles

Humans   Car & Truck   Undefined Obstacle

Figure 5.7 Warning symbols for category of objects of the RCWS.

Generic Warning Symbols

Figure 5.8 Generic warning symbols of the RCWS.

Figure 5.9 In-vehicle presentation of image and warning symbols for the RCWS.

- *In a situation of <u>cautionary warning</u>, the visual warnings given to the driver is differentiated according to object classification and the icons overlap in colours on the far infrared image that is projected on a combiner in the driver's peripheral viewing area. Brightness of images can vary according to the approaching limit for <u>imminent warning</u>.*

When an *imminent warning* occurs (*TPMD (RCWS) < 4sec*) and the detected obstacle has a high probability to be hit, a unified warning is given to the driver with a sound, and an icon of generic warning is projected in the direct field of view (Figure 5.8). The image blinks to enhance *perception*.

The use an unified warning in this case comes from the need to reduce as much as possible the driver's cognitive workload that would derive from the *interpretation* of different types of warning with a consequent longer time to execute an evasive maneuver. Consequently, one of the three alternatives shown in Figure 5.8 had to be selected as generic symbol for a sign of imminent collision. This process required further interaction with users by means of interviews that led to the selection of the following *design option*:

- *In a situation of <u>imminent warning</u>, the hazard triangle with an exclamation mark in the middle was selected as the most significant icon to alert users that an immediate action is required.*

In most cases, an *imminent warning* follows a *cautionary warning*, which is released first. Therefore, obstacles are displayed in the head-up display first, classified by the system and tagged with a warning symbol according to the *cautionary warning* specifications. When the brightness increases and the threshold of *imminent warning* is reached, the generic warning icon is displayed in the windshield and the warning sound is played. The generic warning icon blinks for as long as it is shown. In this case both modes, *cautionary* and *imminent*, are active and the content in the head-up-display does not change.

Obstacles occurring suddenly and releasing the *imminent warning* immediately are also releasing the cautionary mode at the same time. In this case the warnings of *cautionary* and *imminent* mode are appearing at the same time.

The selection of icon typology, including size, colour, levels of brightness and blinking mode, the location of images and icons, the types of sound (voice, acoustics, etc.) have all been developed according to the user-centred design methodological approach and in conformity with the existing norms and standards. In particular, different standardization issues have been considered in respect to (a) different European Directives, such as 590/630/EEC on field of vision, 78/316/EEC on identification of controls, and 78/632/EEC on interior layout; (b) ISO standards, such as ISO 4513 on driver's eye location, ISO 15005 on traffic information and control systems, ISO 15008 on ergonomic aspects of in-vehicle information presentation, ISO 9241-11 on usability (ISO, 1993), and ISO/FDIS 13407 on human-centred design (ISO, 1999); and (c) the European statement of principles on human–machine interface for in-vehicle information and communication systems (EC, 1999).

Selection of Type of Test and Plan of Experiments

According to the human-centred design principles, all design solutions are iteratively reached by involving end-users through experiments and tests (Figure 5.2). The development of a plan of experiments that guide the whole design and testing process for the RCWS system was formalised as part of phase 2 of the work, even if it covered the whole design process, as it involved context of use and user needs, i.e., typical aspects dealt with in phase 1 of the work, and affected design solutions, typical of phase 3 of the work.

The experimental plans and the relative experiments were progressively refined and executed as the design process evolved.

Two aspects were particularly important in the definition of the variables and boundary conditions that affect drivers and play a role in the interaction with the warning system: the selection of the cognitive model of the driver, and the peculiarities of the automotive environment.

Both aspects have been studied in detail in phase 1 of the work. The model of cognition that was adopted focuses on the cognitive functions *perception, interpretation*, and *execution* and on the contextual and temporal conditions that affect the control of such functions. This implies that tests and experiments must examine the effectiveness and efficiency of the warning system towards these three cognitive functions and that the contexts in which the system is operating plays a critical role. Therefore, the experimental conditions, i.e., the scenarios, defining boundaries for the tests are equally important.

The peculiarities of the automotive environment imply that the ethnographic studies and, in particular, the analysis of the socio-technical aspects affecting drivers must account for the enormous diffusion of this means of transportation and for the variety of user needs that such popularity implies.

On the basis of these conditions, and in agreement with the user-centred design approach, the generic plan of experiments focused on three major characteristics: *system usability, system technical functionality*, and *driving performance safety*.

As we are concerned with human factors analysis, and the description of the whole design development of the RCWS is outside the scope of this book, we will not discuss the *system technical functionality* tests. These tests concerned the evaluation of the effectiveness and efficiency of the warning system merely from the technical viewpoint.

In order to study *usability* and *driving performance safety* from the HF perspective, three types of environments and three different set-ups of the system were considered, namely:

- Driving environments
    Static driving simulator
    Dynamic driving simulator
    Real traffic

Table 5.2 Test environments and set-ups

	Static driving simulator	Dynamic driving simulator	Real traffic scenario
HMI virtual prototypes	X	X	
HMI mock-up	X		
HMI integrated on demonstrator vehicles			X

Figure 5.10 Static driving simulator.

- Driving set-ups
    Virtual prototypes
    Muck-ups
    Demonstrator vehicles

Table 5.2 gives an overview of the combination of the tests that could be carried out according to set-ups and environments.

Figure 5.10 and Figure 5.11 show the simulator environments where the tests with virtual prototypes and HMI mock-up were performed.

More in detail, testing *system usability* implies measuring and evaluating that a driver is able to:

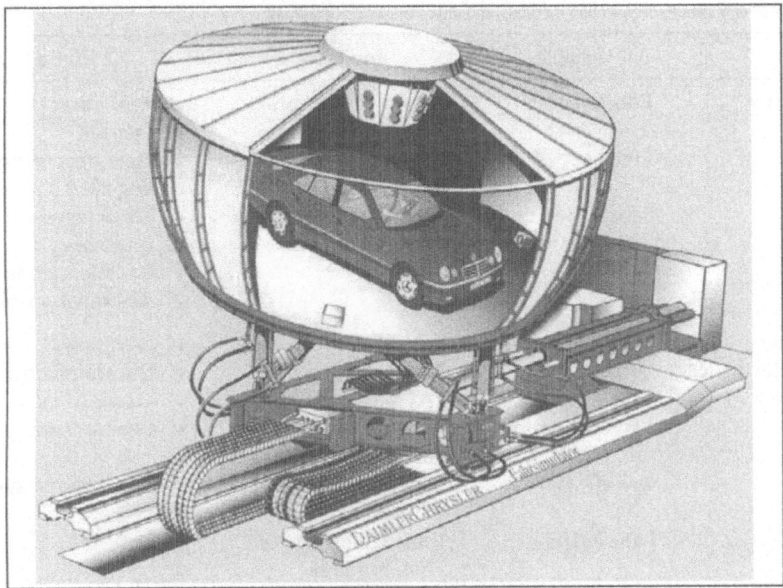

Figure 5.11 Dynamic driving simulator.

- understand information given by the system and execute corrective manoeuvres in order to avoid collisions (*effectiveness*);
- understand information given by the system and execute corrective manoeuvres with short time to reaction and without increasing workload (*efficiency*);
- consider the system comfortable including willingness to install it on board and use it (*satisfaction*).

Testing *driving performance safety* requires evaluating the improvement in safety level obtained by the use of such type of instrument, limiting the task demanded to the driver. Consequently, a number of variables associated with driver behaviour and performance were identified, such as:

- workload;
- visual attention and eye-off-road;
- line keeping, lateral, and longitudinal control;
- missed detection;
- missed interpretation;
- uncertain and/or inefficient behaviour, i.e., errors of execution.

For the evaluation of these variables in all set-ups and test environments, two types of measures were considered, namely:

- *Subjective measures*, by typical ethnographic study techniques, as questionnaires, interviews, thinking aloud processes, and direct observations; and
- *Objective measures*, by utilising standardised tools and techniques for measuring overt behaviour, such as eye-tracking systems, and video recording of behaviour combined with driving data recorders.

Table 5.3 Variables, data collection methods, and indicators for test planning

	Variable	Indicator	Method
Usability	Effectiveness	• Errors of perception • Errors of execution	✓ Field observations ✓ Log-file
	Efficiency	• Errors of interpretation • Reaction time	✓ Field observations ✓ Log-file
	Satisfaction and acceptance	• Subjective preferences • Comments	✓ Thinking-aloud ✓ Interviews ✓ Questionnaires ✓ Satisfaction scale
Driving performance safety	Workload	• Subjective measures • Errors of perception • Errors of execution	✓ Interviews ✓ Questionnaires ✓ Standard methods
	Visual attention	• Glance frequency and duration	✓ Eye-tracking system
	Eye-off-road	• Time of keeping eye-off-road	✓ Eye-tracking system
	Lane keeping	• Vehicle mean speed • Time headway • TPMD	✓ Log-file
	Lateral control performance	• Time-to-line crossing	✓ Log-file
	Inefficient strategies	• Errors of interpretation • Errors of execution	✓ Field observations
	Signs of uncertainty	• Errors of interpretation • Errors of execution	✓ Field observations

Table 5.3 summarises the general set of variables, indicators and methods that could be applied in performing the tests and experiments for all three types of settings, i.e., Static and Dynamic Driving Simulators, and Real Traffic situations.

Another important set of variables that harmonised the three test environments was the spectrum of users or subjects. In general, at least 20 drivers took part in each series of experiments. The sample of subjects were selected according to the following characteristics:

• range of age between 23 and 55 years;
• mixed gender;
• driving exposure of at least 10.000 km/year;
• different attitudes towards new technologies;
• if glance behaviour measures were collected, people wearing glasses were also identified.

A combined approach of *"within and between subject"* was used for the Experimental Designs that were developed for all three set of experiments. In each experiment, subjects were required to meet a series of tasks/situations leading them through all aspects to evaluate. The sequence of tasks/situations was randomised in order to avoid sequential effects.

The three types of tests environments with these coherent and harmonised set of variables were different for their specific goals. In particular:

1. *Static driving simulator.* Tests were performed first and almost in parallel with the work of phase 1. They were dedicated to the definition of the most suitable location (head-down versus Head-up Display HUD), dimension, and presentation mode (static versus dynamic image) of the warning signal.

2. *Dynamic driving simulator.* Tests were performed for studying usability and driving performance safety, comparing three different settings of the RCWS, namely:

   (a) a system based on HUD, plus Infrared (IR) monitor, where the IR-image on a separate monitor was mainly meant to support lane keeping, whereas the symbols in the HUD provided a warning for special obstacles;

   (b) a system based only on HUD, with the lane keeping support integrated in the HUD in a symbolic way (no additional monitor was necessary); and

   (c) no system at all.

3. *Real traffic situations.* Tests were firstly designed during phase 2 of the work and then carried out in phase 3 by studying real drivers' behaviour in different traffic scenarios, using special vehicles equipped with the RCWS. Therefore, the system could be tested in all aspects of usability and driving performance safety.

The critical aspect, common to all these tests, was the definition of the experimental scenarios that were designed combining all the independent variables and, in particular, the contextual conditions. The analysis of past experience and accident reports is particularly useful to define the scenarios for testing. Therefore, a specific activity of phase 2 of the work was dedicated to the development of this aspect of definition of accident scenarios, which affected the whole plan of experiments.

Study of Accidents and Past Experience

In the domain of road safety a universal definition of "accident scenarios" and a general methodology for scenario identification are not available. For this reason, considerable effort was spent in defining the most critical scenarios, in relation to the analysis of past events and accidents.

The development of road accident scenarios started from the investigation of the main European road accident databases (EUROSTAT, CEMT, OECD). This first analysis allowed the identification of general accident trends and causes all over Europe. Problems in international data comparability were encountered. In spite of the effort spent so far to standardise European data, many discrepancies still remain in conceptual definitions, taxonomies and methodologies for data collection. Sources analysed definitely contained a large amount of data about accidents, but the information provided was not sufficient to accurately identify scenarios.

Data collected are reported only in the aggregated level, so that it is impossible to discriminate among single accidents and define factors contributing to accident scenarios.

Alternative sources have been identified in insurance companies, car manufacturers and police databases. Unfortunately these sources turned out to be inaccessible, because of the high level of confidentiality of the information contained therein.

Finally, newspaper articles offered the most useful information for reconstructing accident scenarios. A total number of 300 articles were selected among those published in the year 2000 by the same major European daily newspaper. The selection focused on the richness and detail of the information provided: particular attention was dedicated to the presence of data on weather conditions, traffic conditions, seriousness of accident consequences, and accuracy of accident dynamic description. The use of newspaper articles, even if with some inconveniences (e.g., almost only road accidents with the most serious consequences are reported), proved to be powerful for longitudinal studies. The reference to a whole year and to all subsequent articles on a specific road accident allowed the updating of data on causes and seriousness of consequences.

This analysis process led to the final identification of a set of accident scenarios that supported the definition of specific requirements for testing the new anticollision system. A number of 32 types of accident scenarios were sorted out accordingly to the frequency of the accident dynamic, the presence of bad weather conditions, of multiple causes for the accident and of a recognised contribution of human factors. Finally, each road accident scenario could described by three sets of variables (Cacciabue et al., 2003):

1. *Actors*. This category includes driver-related variables, i.e., age, sex, and psycho-physiological condition immediately before the accident, and vehicle-related variables, i.e., number of vehicles involved and vehicle power.
2. *Setting*. The accident scene includes a number of environmental variables, i.e., type of road, route, paving type and maintenance condition, and infrastructures.
3. *Accident dynamic*. It contains events sequence and outcomes of the accident. Actions performed by the actors and external events are considered within this category.

An example of the schedule produced for a specific accident according to this classification is shown in Table 5.4.

This generic structure of scenario variables was retained for the finalisation of the experimental plans to be performed on the road with the RCWS.

Definition of Accident Scenarios for Testing and Evaluation

The results of phase 1 on the analysis of the needs of users of an anticollision warning device and the definition of the characteristics of the tool that derived

Table 5.4 Example of the scenario structure

Scenario	Variables	Consequences
• *Actors*: Young male driver (high-power engine) coming back from a party, overtaking a car near to a bend, is involved in a frontal crash.	⇒ Drunkenness ⇒ Familiar road ⇒ Passengers on-board	Frontal collision with two deaths and one seriously injured person.
• *Setting*: Collision occurs at the end of a straight road, continually lined, lightly uphill, tree-lined, and with a damaged surface.	⇒ Main state road ⇒ Night	
• *Accident Dynamic*: Collision between two vehicles, one with medium and one with high-power engines.	⇒ Forbidden manoeuvre ⇒ High speed	

Table 5.5 RCWS system characteristics tested during experiments

System characteristics	Cognitive functions	Features
Warning signal	Capture driver's attention and ensure *detection* of warning	• Acoustic • Visual (HUD) • Driver to control the volume and light levels without disabling option
Warning sign	Avoid cognitive planning effort: only *interpretation* followed by action	• This information should explain what happens • No time spent to scan environment in order to find hazard nature
Evasive manoeuvre suggestion	Reduce driver response time to *execute* corrective action	• Visual information reinforced by oral message on evasive manoeuvre • Five seconds before collision is a safe rate to alert the driver in time

from the progressive selection of warning strategies and display modes, were the precursors for the definition of the scenarios for tests and experiments.

The system characteristics that fit the user needs analysis and consider the most relevant cognitive functions affected by the warning are associated with (a) the relevance of the signal for enhancing detection; (b) the information content of the sign for enhancing the correct interpretation of the *imminent* or *cautionary warning*; and (c) the execution of an adequate evasive manoeuvre. These characteristics are summarised in Table 5.5.

These characteristics were further combined with the findings of step 2 of the work on the selection of types of test for the evaluation of usability and driving

Table 5.6 General scenario conditions for tests and experiments

Scenario characteristics		Features (static/dynamic–environmental–individual)
Accident dynamic	• Traffic situation	✓ The preceding vehicle has lower speed ✓ The preceding vehicle brakes hard ✓ Stationary vehicle ahead ✓ Persons or animals crossing the road ✓ Steady/slowly moving obstacle on the road ✓ Oncoming traffic in the same lane ✓ Vehicle in or coming from a right-turn road ✓ Queue situation in stop-and-go traffic ✓ Overtaking preceding vehicle
Setting	• Road type	✓ Urban ✓ National, mixed single and dual carriage ✓ Rural ✓ Highway/motorway ✓ Tunnel
	• Weather conditions and daytime	✓ Rain ✓ Fog ✓ Snow ✓ Daytime ✓ Night time
Actors	• Target type	✓ Vehicles (trucks, motorbikes, bicycles) ✓ Pedestrians ✓ Animals ✓ Undefined obstacles
	• User characteristics	✓ Young drivers ✓ Elderly drivers ✓ Professional drivers

performance safety (Table 5.3) and with the study of accidents for the selection of the most suitable scenarios for experiments in different environmental contexts.

The definition of a common basis of conditions and variables could then be developed. These are summarised in Table 5.6.

The combination of the above generic scenario conditions with the contexts and environments that could be reproduced in static and dynamic simulators and in demonstrator driving vehicles, led to the definition of series of specific tests that were carried out throughout the whole design process, as shown in Table 5.7.

5.2.3 Phase 3: Safety Matrix and Final Experiments and Tests

The final phase of the work coincided with the design and implementation in a vehicle of the RCWS system and the performance of tests and experiments in real driving contexts.

Table 5.7 Plan of experiments and tests for the RCWS

Plan of experiments	Variables	Scenarios
HMI laboratory tests	• Different HMI concepts • Importance of secondary info channels (i.e. audio / haptic warnings) • Warning strategies. • Three groups of drivers: elderly, middle age, young.	**Test 1.** Night, rural road, low traffic density, bad lighting **Test 2.** Night, highway, heavy rain, medium lighting, medium traffic density **Test 3.** Dawn, fog, rural road, low traffic density in the driver's direction, mean density in the opposite direction
Driving simulator tests	• Functionality test in a complex environment. • Three groups of drivers: elderly (above 60) middle age (25 ≤ years ≤ 60) and young (below 25).	**Test 1.** Daytime with light rain; heavy rain; sunshine; vehicle emerging from rural road. **Test 2.** Night, good visibility then very poor; rain; pedestrian attempting to cross road unexpectedly; bike badly visible **Test 3.** Dawn with mist. Suddenly a wild animal enters the road.
Real traffic contexts	• System usability, user acceptance and impact on driver behaviour. • Three groups of drivers: elderly, middle age, young.	**Test 1.** Night, urban context, good lighting conditions and mean traffic; then highway, mean lighting and traffic density; then rural road with bad lighting and low traffic density. **Test 2.** As test 1, but under rain. **Test 3.** As test 1, but in daytime with fog or heavy rain.

These tests were the final iteration between the HF analysts with the designers of the system technology.

An essential contributor to the development of the final testing programme was the correlation of these experiments with the reference model of driver behaviour, selected in the early phases of work (phase 1) and already applied during the laboratory and simulator tests (phase 2). This granted harmonisation and coherency throughout the whole design process, as resulting from the application of the HERMES methodology.

From the human factors perspective, the major activity carried out in phase 3 consisted in two main steps, namely:

1. *Phase 3, Step 1*. Development of a *safety matrix* to be utilised by the evaluators during the experiments in vehicle.

2. *Phase 3, Step 2.* Performance of road tests and feedback to the final design iteration.

Development of a Safety Matrix

The *safety matrix* contains the set of most relevant and essential variables to be evaluated during the different tests and experiments for evaluating efficiency, effectiveness, and user friendliness of the designed system from a safety point of view.

In order to be as complete as possible, a *safety matrix* should give indication also on the means by which it is possible to measure the variables, the type of analysis that should be carried out, and a series of comments that are intended to support the analysts in performing the evaluation and review of the variables.

The RCWS *safety matrix* is oriented to the evaluation of driver response and behaviour with a specific warning system and to define the improvement of safety obtained by the implementation on-board vehicles. Therefore, two aspects are particularly critical, namely:

• the reference model of driver selected to support the design process; and
• the safety-relevant aspects of the HMI in terms of system responses and human performances.

The integration of these two aspects is of crucial relevance for the design process. Therefore, in essence, the *safety matrix* is a tool for supporting the analyst in evaluating, on the one hand, the cognitive resources necessary to adequately react to the warning, and, on the other hand, the support offered by interface to maximise effectiveness and efficiency of intervention.

The *safety matrix* was developed on the basis of subjective ratings of drivers and evaluators concerning the actual utilisation of the warning system, and on the numerical analysis of objective data derived from system performance in a variety of road situations.

The complete set of variables that, in principle, could be evaluated is shown in Table 5.8a–d, which contain many different types of measure and methods for data collection. The overall *safety matrix* has been split up in four tables corresponding to the three cognitive functions of *perception, interpretation,* and *execution* and the cognitive *resources* available for an effective use of the system. These four tables are now discussed individually, in more detail.

Safety Matrix: Variables Affecting Perception of Signal

The variables observed with reference to the cognitive function *perception* of signals focus on the effectiveness of the auditory and visual indicators to reach the driver's sensorial perception (Table 5.8a). The following variables can be evaluated:

1. *Warning symbols visibility and clearness,* which evaluates the easiness of visual perception and visibility of symbols, based on subjective rating by driver and evaluator.

Table 5.8.a Safety matrix variables affecting perception of signal

#	Perception	Measure	Method	Comments
1	Warning symbols visibility and clearness	Questionnaire	Subjective rating by driver and evaluator	Easiness of symbols, visual and visibility perception
2	Warning sounds audibility	Questionnaire	Subjective rating by driver and evaluator	Easiness of hearing sounds
3	Warning symbols distraction	Questionnaire	Subjective rating by driver and evaluator	Distraction from driving tasks due to RCWS symbol
4	Warning sounds distraction	Questionnaire	Subjective rating by driver and evaluator	Distraction from driving tasks due to RCWS sounds
5	Not perceived warnings	Video camera	Numerical analysis of data	Rate of warnings not perceived

2. *Warning sounds audibility,* which evaluates the easiness of hearing the particular sounds and is based on subjective rating by driver and evaluator.
3. *Warning symbols distraction,* which evaluates the degree of distraction from driving tasks due to warning symbols, from the perception point of view. This measure is based on subjective rating by driver and evaluator.
4. *Warning sounds distraction,* which measures the degree of distraction from driving tasks due to warning sounds and it is based on subjective rating by driver and evaluator.
5. Number of *not-perceived warnings,* which evaluates the rate of warnings not perceived by the driver. This variable can be measured from recorded performances by a video camera installed onboard of test vehicles.

Safety Matrix: Variables Affecting Interpretation of Sign

The cognitive function *interpretation* concerns the content and salience of the signs associated with perceived signals.

Obviously, no interpretation is possible if there is no perception. However, once a signals has been perceived thanks to its auditory and visual qualities, then the meaning associated with it, i.e., the sign, must be effective in supporting the correct diagnosis or interpretation of the perceived information. Consequently, in order to measure the effectiveness of the warning system from the viewpoint of the interpretation of the "perceived" signs, the following variables are selected (Table 5.8b):

6. *Warning sign clearness,* which measures the easiness of comprehension and interpretation of signs associated with the visual aspects of the warning indicators. The interpretation of the sign is obviously strictly correlated to the

Table 5.8.b Safety matrix variables affecting interpretation of sign

#	Interpretation	Measure	Method	Comments
6	Warning sign clearness	Questionnaire	Subjective rating by driver and evaluator	Easiness of comprehension, interpretation of signs
7	Warning sounds unambiguousness	Questionnaire	Subjective rating by driver and evaluator	Easiness of sound comprehension, interpretation
8	Warning sign distraction	Questionnaire	Subjective rating by driver and evaluator	Degree of distraction from driving tasks due to warnings
9	Warnings discrimination	Questionnaire	Subjective rating by driver and evaluator	Degree of confusion of RCWS with other warnings
10	System annoyance	Questionnaire	Subjective rating by driver and evaluator	Degree of annoyance induced by the use of RCWS system
11	Errors in system monitoring	Video camera	Numerical analysis of data	Degree of discrimination of missing / improper warnings (compliance with the system)

perception of the signal. However the two aspects (*perception* and *interpretation*) of driver cognitive behaviour have to be evaluated separately. This variable is based on subjective rating by driver and evaluator.

7. *Warning sound unambiguousness*, which measures the easiness of comprehension and interpretation of warning sound, and is based on a subjective rating by driver and evaluator. The same comment as above applies concerning perception and interpretation of the warning sound.

8. *Warning sign distraction*, which measures the degree of distraction from driving tasks due to warnings signs, and is based on a subjective rating by driver and evaluator. The same comment as above applies concerning perception and interpretation of the signal and sign.

9. *Warnings discrimination*, which defines the degree of confusion of the specific warning sign with respect to other warnings. This variable is based on a subjective rating by driver and evaluator.

10. *System annoyance*, which evaluates the degree of annoyance induced by the use of the warning system, and it is based on subjective rating by driver and evaluator.

11. Number of *errors in system monitoring*, which evaluates the rate of missed or improper interpretations of signs and may be combined also with the degree of compliance with the system. This variable can be measured from recorded performances by a video camera installed onboard test vehicles.

Table 5.8.c Safety matrix variables affecting execution of corrective manoeuvre

#	*Execution*	Measure	Method	Comments
12	Incorrect reactions to warnings	Video camera	Numerical analysis of data	Rate of incorrect actions taken to avoid collision
13	Reaction time	Video camera	Numerical analysis of data	Mean reaction time value for starting the recovery action
14	Efficiency	Video camera	Numerical analysis of data	Rate of correct actions taken with respect to the mean reaction time
15	Standard deviation of steering wheel angle (SDST)	DDR log file	Numerical analysis of data	Variability of the steering wheel angle
16	Mean speed	DDR log file	Numerical analysis of data	Mean of vehicle speed
17	Standard deviation of speed (SDSP)	DDR log file	Numerical analysis of data	Variability of vehicle speed
18	N° of reactions to *cautionary warning* ($4 \leq TPMD \leq 8$ sec.)	DDR log file	Numerical analysis of data	Number of reactions (corrective manoeuvre) following first-level warning
19	N° of reactions to *imminent warning* ($TPMD < 4$ sec.)	DDR log file	Numerical analysis of data	Number of reactions (corrective manoeuvre) following second-level warning

Safety Matrix: Variables Affecting Execution of Corrective Manoeuvre

The cognitive function *execution* fconcentrates on the performance of corrective manoeuvres.

Therefore, most variables can be quantified from data collection during the experiments. In addition to measuring directly the driver performance, it is possible to measure the response of the vehicle by collecting information from the Drive Data Recorder (DDR) system, which is fitted on board of most experimental vehicles. This gives further information about the manifestation of driver behaviour. The variables that can be measured with respect to driver performance are (Table 5.8c):

12. Number of *incorrect reactions to warnings,* which can be evaluated from the numerical analysis of the data collected by video recording the experiments.

13. *Reaction time,* which can be evaluated by measuring the mean time taken to start the recovery action. This variable can be evaluated from the numerical analysis of the data collected by video recording the experiments.

14. *Efficiency*, which can be measured by the rate of correct actions taken with respect to the mean reaction time. Also this variable can be evaluated from the numerical analysis of the data collected by video recording the experiments.

15. *Standard deviation of steering wheel angle (SDST)*, which is evaluated from the log-file of the DDR system.

16. *Mean speed*, which also can be derived from the log-file of the DDR.

17. *Standard deviation of speed (SDSP)*, which also can be derived from the log-file of the DDR.

18. Number of *cautionary warnings* (4 ≤ TPMD ≤ 8 sec.), measured from the log-files of the DDR.

19. Number of *imminent warnings* (TPMD < 4 sec.), measured from the log-files of the DDR.

Safety Matrix: Variables Affecting Allocation of Cognitive Resources

The *resources* available for governing the cognitive functions are effected by the contextual conditions and personal characteristics of drivers.

According to the reference model of driver behaviour, the cognitive resources are related to aspects such as the available time, or perceived available time, for performing a task, and the associated complexity. The possibility to evaluate cognitive recourses is related to the possibility to measure and evaluate a number of physical variables, as well as certain subjective indicators of driver workload, and other measures of satisfaction with the interaction with the system.

In particular, the following variables are considered (Table 5.8d):

20. *Driver subjective workload*, which is based on a subjective rating by driver, and data are colleted by means of a questionnaire.

21. *Heart rate variability*, which can be evaluated by means of special instruments installed on-board of experimental vehicles.

22. *Glance frequency*, which can be evaluated by means of an eye-tracking system installed on-board of experimental vehicles that records the number of glances to the human–machine interface to acquire information.

23. *Glance duration*, which can be evaluated from the log-file of an eye-tracking system that records the average duration of glances to acquire information.

24. *Eye-off-road*, which can be evaluated from the log-file of an eye-tracking system that records the rate of time of keeping eye-off-road.

25. *Driving comfort*, which measures, by means of a questionnaire, the subjective rating of driver and evaluator in terms of degree of comfort while using the system.

26. *Safety benefits*, which measures, by means of a questionnaire, the subjective rating of driver and evaluator in terms of expected contribution of RCWS in increasing safety.

27. *Driving quality*, which measures, by means of a questionnaire, the subjective rating of driver in terms of quality of driving performance with respect to usual, with the RCWS system installed on board.

Table 5.8.d Safety matrix variables affecting allocation of cognitive resources

#	Resources	Measure	Method	Comments
20	Driver subjective workload	Questionnaire	Subjective rating by driver	Driver evaluation of the workload while using the system
21	Heart rate variability	Heart rate monitor	Numerical analysis of data	Heart rate variability
22	Glance frequency	Eye-tracking system	Numerical analysis of data	Mean number of glances to the human–machine interface to acquire information
23	Glance duration	Eye-tracking system	Numerical analysis of data	Average duration of glances to acquire information
24	Eye-off-road	Eye-tracking system	Numerical analysis of data	Rate of time of keeping eye-off-road
25	Driving comfort	Questionnaire	Subjective rating by driver and evaluator	Degree of comfort while using the system
26	Safety benefits	Questionnaire	Subjective rating by driver and evaluator	Expected contribution of RCWS in increasing safety
27	Driving quality	Driving quality scale	Subjective rating by the driver	Quality of driving performance with respect to usual standard

Performance of Road Tests and Feedback: the Final Design Iteration

The last step of the design process consisted in the performance of the tests and experiments in real traffic contexts and in feeding the results back to the design process for the final assessment of the warning system before implementation in commercial vehicles.

The plan for the experimental tests generated during the previous phases of work (Table 5.6 and Table 5.7), was combined with the *safety matrix*.

The experimental vehicles utilised for the experiments were fitted with the RCWS and the data recording system on-board allowed the collection of information about most parameters defined in the *safety matrix*. However, certain variables could not be measured because the specific systems were not available on-board the experimental vehicles at the time of the experiments. In particular, the heart rate monitor was not available for the tests and consequently the relative variable could not be measured.

The questionnaires associated with the *safety matrix* were also developed in such a way that drivers undergoing the tests had to answer questions before and after the tests. However, the objective was to spend as little time as possible on the questionnaires in favour of the actual driving experiments.

For brevity, the description of how questions were designed and presented is not reported here. However, in order to give some guidance also in this area, the reader is referred to Chapter 7, which discusses the development of a questionnaire for subjective ratings of systems and organisations, which follows the same general principles utilised in this case, even though it is implemented in a different domain of application.

The tests carried out on system usability, user acceptance and behaviour followed the specifications of the plan of experiments shown in Table 5.7. In particular:

- Three populations of users were examined, with different age characteristics (young drivers, <25; middle age drivers, 25–60; and elderly drivers, 60+) and in the presence of a variety of potential obstacles (*target types*).
- Different settings (*road types* and *weather conditions and daytime*) were mixed.
- Different traffic situations were examined covering, as much as possible, low, mean, and high traffic in all possible settings of road types, weather conditions, and time of day.

All together, 60 subjects were tested, 20 for each population of drivers.

The tests were performed over a considerable period of time, during the winter, and the number of tests that could be carried out during a day depended on the weather conditions.

Data were collected and analysed by standard statistical means and the results were fed back to the final design stage.

In principle, these tests confirmed the initial assumptions and demonstrated that the design solutions adopted during the final stages of the design were adequate for the warning strategies selected for the RCWS.

Drivers showed interest and willingness to utilise the system, especially elderly drivers. No tests were performed on long-term behavioural effects, leading to possible aberration of behaviour, due to complacency and risk homeostasis. These studies could be performed for further refining the instrument in a second stage of development and following some time of inclusion in the market. However, this type of assessment is outside the scope of application of the HERMES methodology and will not be dealt with here.

The design process was completed with minor modifications and changes resulting from the road tests. The experiments served as validation exercise for the design and demonstrated also the power of the approach based on user-centred methodology for including user-dependent considerations in a formal fashion since the early phases of design. Moreover, it demonstrated that the application of consistent HF method in all step of the design process is very helpful in improving the effectiveness, efficiency, and user satisfaction as well as in reducing design modifications after development of the system as result of field experiments and validation tests.

5.3 Summary: Application of HERMES to the Design of an Anticollision Warning System

5.3.1 Application of HERMES

This chapter has presented a practical application of the HERMES methodology to the design of an HI.

The five standpoints identified as the basic conditions from performing a sound HF approach (Chapter 2) have been respected:

1. The HERMES methodology was applied to the area "design" (standpoint 4).
2. The specific goals and objectives of the tool to be developed were established early in the design process from the viewpoint of HEAM and UCD principles (standpoint 1).
3. A user model of reference was selected and maintained throughout the whole design and testing process (standpoint 2).
4. A logical interplay was conducted between the data collected from experiments, field assessments, and review of past experience (retrospective analysis) and the design choices (prospective analysis) (standpoint 3).
5. The measure of safety improvement resulting from the implementation of the new HEAM system was carried out through the development of a *safety matrix* applied during the evaluation of the system in real traffic scenarios (standpoint 5).

The application of HERMES to this case study shows that the methodology is applicable in practice and gives valuable and significant results for developing and ascertaining safety standards.

The methodology was applied in three phases, which respected the UCD principles. Phase 1 of the work focused on (a) the definition of the characteristics of the tool and warning strategies from a safety perspective; (b) the definition of user needs; and (c) the selection of model of user and identification of cognitive functions and processes associated to warning strategies.

Phase 2 was centred on: (a) the selection of characteristics of the (RCWS); (b) the definition of the type of tests and plan of experiments; and (c) the accident scenarios for testing and evaluation.

Phase 3 consisted of two main steps, namely (a) the development of a *safety matrix* to be utilised during the experiments by the evaluators, and (b) the performance of road tests and feedback the final design iteration.

The HF characteristics and features of the road collision warning system that have been obtained form the study described above and from the *design options* selected during the whole process are summarised in Table 5.9.

Table 5.9 Human factors features of the road collision warning system

HF features of the road collision warning system	Description
Warning strategy Interface characteristics	• Anticollision "warning" non-intervening system • Visual and aural, but no haptic warnings; • No voice suggestion about evasive manoeuvre; • Head-up display; display on windscreen

Levels of warning indication	• Two levels of crash avoidance warning:
	  ✓ *cautionary*; and
	  ✓ *imminent.*

Driver behaviour reference model	• Reference model of cognition
	  ✓ Cognitive functions of relevance:
	    Perception of warning signals;
	    Interpretation of the
	      associated warning signs; and
	    Execution of corrective manoeuvre;
	  ✓ Cognitive processes critical for decision making:
	    Allocation of resources

Tests for usability and driving performance safety	• Driving environments:
	  ✓ Static driving simulator
	  ✓ Dynamic driving simulator
	  ✓ Real traffic
	• Driving set-ups:
	  ✓ Virtual prototypes
	  ✓ Muck-ups
	  ✓ Demonstrator vehicles

Assessment of safety levels	• Safety matrix

5.3.2 Conclusions

This chapter has dealt with the major methodological issues concerning the application of HF in automotive environments and on the impact that these may have on a whole design process of a HMS and HI. The objective was to demonstrate the importance of considering a general framework when dealing with HMS that adequately account for human models at an equivalent level to machine models.

Theoretical and numerical means exist for including human behaviour in processes aiming at developing "human-oriented" design, safety assessment, training, as well as accident investigation. Similarly, data and test beds are available for supporting models with validated parameters and constants, necessary for performing predictive analyses.

The only basic requirement is that the analysts must preserve a coherent approach throughout the process, i.e., the prospective and prospective methods must be logically correlated during the entire process, independently of whether this is design, safety assessment, training, or accident investigation. This ensures that the models applied for predictive studies are based on data and parameters drawn from sources and experiments related to taxonomies coherent with the same models and paradigms.

The HERMES methodology, developed for the precise objective of respecting these essential conditions, has shown to be a very useful roadmap for coherent and constructive implementation of exiting methods, models, and paradigms of HMI in the area of *design* of HMSs.

6

Application of HERMES in Training: A Human Factors Course in the Aviation Domain

This chapter focuses on the application of the HERMES methodology for the development of a (non-technical) Human Factors training course in a domain with strong characteristics of Human–Machine (HMI) and Human–Human (HHI) Interactions.

Contrary to the case discussed in the previous chapter, where the design of a HMS required the contribution of a number of experts in different technological areas, the Human Factors (HF) training course was completely designed and developed by HF specialists, with a certain knowledge and experience in the domain of application.

However, the active collaboration of the organisation for which the course was developed, at all its levels, from the top management to the primary actors involved in managing technical systems, was essential to ensure quality and validity of the final product. In practice, the attitude of the organisation allowed the HF analysts to acquire further experience in the domain, and gain specific knowledge about habits and everyday practices, as well as information on past events and accidents. Not only these factors contribute to building the corporate memory and culture of an organisation which strongly affect behaviour and human relations, but they are essential material for an HF training course.

The development of training courses is a typical application of the HERMES methodology, as all steps can be performed in sequence. The following sections will commence with defining the problem relative to the specific HF training course. Subsequently, the way in which the HERMES methodology was applied in generating a consistent set of models and criteria will be described. These models were applied in the analysis of data about the organisation, derived from the study of past experience and field observations. All these elements affected the development of the specific training course, which was derived from standard HF training courses and adapted to the needs and characteristics of the organisation.

The course was then put into operation and the attainment of its expected goals was monitored for a period of time. Finally, the effectiveness was evaluated, according to the requirements of the DSME (Define, Select and Implement, Monitor, and Evaluate) process for developing Human Error and Accident Management (HEAM) measures, described in Chapter 4 (§4.3).

6.1 Problem Statement

In the domain of aviation, Human Factors (HF) training has become compulsory in the last several years, and nowadays all major airlines and professional schools carry out HF training based on well-developed and documented methods and approaches (FAA, 1978; ICAO, 1991; Taggart, 1987; Wiener et al., 1993; Bilton, 1997; Masson and Koning, 2001).

This type of training was originally devoted to pilots, but it has progressively developed to cover all other actors in the aviation realm, starting from air traffic controllers, to cabin assistants, airline managers, and maintenance personnel. These types of courses are generically identified as Crew Resource Management (CRM) training courses. However, their application to different areas of aviation has generated specific acronyms, such as, for example, TRM (Team Resource Management) or MRM (Maintenance Resource Management) (Orasanu and Salas, 1993; Hopkin, 1995; Baranzini et al., 2001).

With the application to different aviation systems and environments, trainers and researchers have become increasingly more aware of the critical influence and impact of national and organisational culture on the effectiveness of CRM training. While certain aspects of CRM are universally endorsed as basic subjects to be trained, e.g., personality factors, leadership, situation awareness, management of fatigue and stress, and crew coordination, other CRM subjects, such as command styles, acknowledgement of stress, and attitudes toward the use of automation are specifically affected by national and organisational cultures. Therefore, for CRM training to be successful, it must be tailored to the organisational context in which it is delivered (Merritt et al., 1996).

In practice, the development of a training course requires the accurate analysis of the socio-technical context and working environment to which is dedicated. Moreover, the familiarisation with habits and attitudes typical of a certain organisation can be further improved by the study of past events and accidents that affected the history of the organisation.

This process is typical of the HERMES methodology, and its application guided the development of a training course for the pilots of a major European airline, as well as the generation of an associated training for the "facilitators," i.e., the pilots and co-pilots, who were appointed the role of trainers of the CRM for the whole company. Therefore, the goal of the project was dual:

1. On the one hand, to develop a dedicated CRM, able to capture the critical safety factors and issues existing within the organisation.
2. On the other hand, to select a group of pilots and to provide them with further knowledge and understanding of HF principles as well as training skills to generate a specific ability in facilitating or lecturing their colleagues.

In this chapter, the findings of the ethnographic studies and the analysis of past events within the airline are discussed in view of their effect on the CRM training course. Then, the criteria and methods that were considered for developing the CRM and for preparing the facilitators to manage the classrooms will be presented.

6.2 The CRM Course and Training of Facilitators

Given the confidential nature of the information contained in this work, this chapter will not enter into detailed evaluation of the data, field observations, and answers to the questionnaires. The airline will be generically identified with the acronym FCA. Also the content of the CRM described here varies slightly from the actual one developed for the airline. However, the substance of the work remains unchanged.

According to the HERMES methodology, the goals of the activity were defined from the start of the work. Then, the methods most suited for the development of the course were selected and progressively implemented. During this process, an adequate monitoring of the attainment of the goals led to some important improvements. Finally, once the CRM was put into practice, the results were evaluated over a period of time, to measure effectiveness and quality of the product.

The airline for which the CRM was developed consisted of almost 2000 pilots and covered routes from short regional to long-haul flights. Therefore, a wide variety of aeroplanes and piloting tasks were to be considered.

Moreover, the problem of dealing with automation was reckoned of primary importance. Consequently, special attention was dedicated to all problems and issues associated with the use and exploitation of automation.

Given the complexity of the problem and the dimension of the airline, the Human Factors Team (HF team) was composed of:

- a group of five pilots, three captains and two first officers, with different flying experience and expertise in Human Factors;
- a group of psychologists and aeronautical engineers, with experience in Human Factors and some experience in flying; and
- a team of experts in Information Technology (IT), to support the work in computer technology and simulation. This team joined the group in a second stage of development, as it will be discussed in the following.

The development and testing of the CRM demanded several months of work, and was performed in three correlated and timely distributed phases (Figure 6.1):

Phase 1. Evaluation of the socio-technical context of the work, definition of the theoretical stand, and familiarisation with the airline flying practices, habits, and tasks.

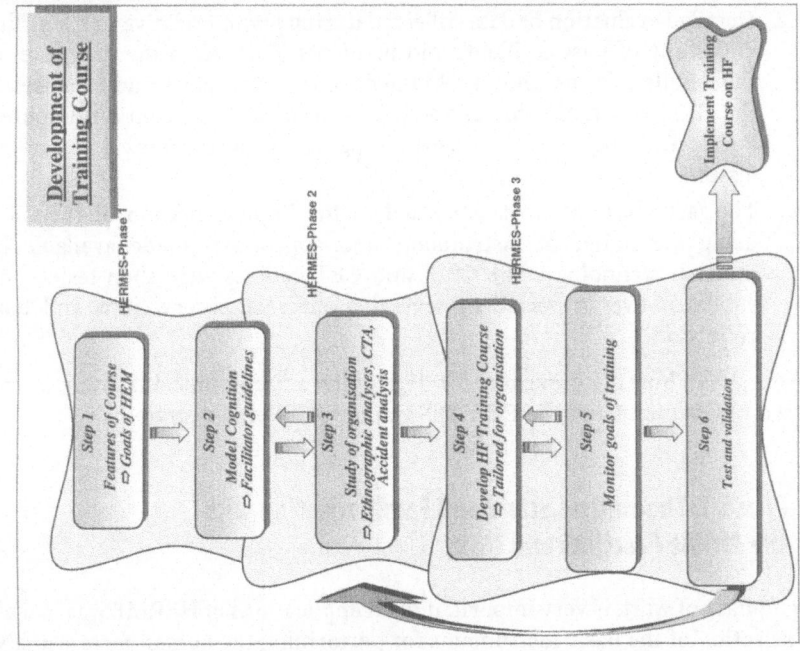

Figure 6.2 Stepwise procedure for developing nontechnical training course and HERMES phases.

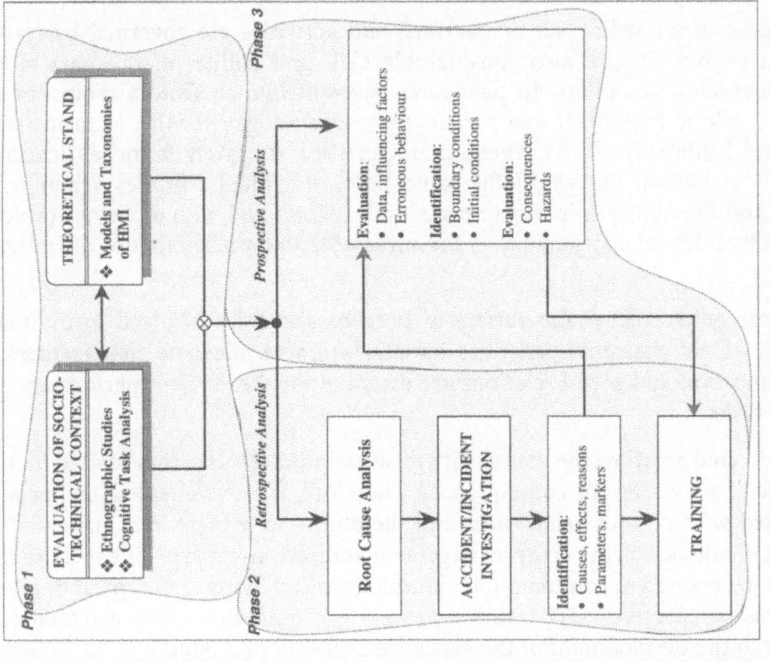

Figure 6.1 Application of HERMES for the development of a CRM course.

Phase 2. Detailed evaluation of data collected during *phase 1* relative to the airline and study of past accidents and incidents. This led to the definition of the detailed design and development of the core issues and exercises of the CRM course. In this phase, it was decided to develop a Computer-Based Training (CBT) course in support to the theoretical classroom lessons.

Phase 3. The last phase of work consisted in the implementation of the CRM using a number of instruments and approaches made available by modern technology. The CRM and CBT courses were then tested and validated over a period of several months of observations and data collection.

Figure 6.2 shows how the procedure for developing a nontechnical training course discussed in Chapter 4 and the HERMES framework were integrated.

6.2.1 Phase 1: Theoretical Stand and Familiarisation with the Airline Practices and Tasks

The first phase of work is very important in all applications of HERMES, as it leads to the selection of the basic instruments for performing the human factors analysis and implies the concrete and practical evaluation of the organisation as a whole and of the procedures and practices applied in everyday life.

In the case of an airline, all interactions and activities are governed by formal procedures, but require also considerable skill and ability on the part of the main operators, i.e., pilots. In particular, relevant human factors requisites for a pilot combine individual and personality aspects together with team management and leadership ability. These characteristics are rarely found as "natural" attitudes in human beings. At the same time, individual attitudes versus technology and the ability to perform tasks under stress and with a heavy workload are equally relevant components in the successful accomplishment of the job of a pilot.

At a more general level, the variety of persons and roles involved in the management of air transport generates situations where adequate management of communication and social relations are essential for the efficient performance of various tasks.

Finally, in civil aviation, the complexity of situations and tasks may lead to human errors with very relevant consequences. Therefore, the awareness of the hazards associated with certain behaviours, and the ability to prevent and manage dangerous situations, i.e., the error management skill, are also a type of expertise that needs to be trained in combination with the technical flying skill typical of pilots. From this perspective, a HF training process is a quite normal as essential contributor to the development of the expertise and skill of a pilot.

Moreover, the procedures and practices that exist within an airline are related to national and organisational cultures. This generates conditions that must be considered in Human Factors training, which may vary according to the type of organisation for which the HF training is developed.

At the same time, human factors expertise must not be associated with common sense. Trainers must have a sound and consolidated knowledge in HF that allows them to deal in a formalised and structured way with all typical issues derived from common sense, which is usually well developed in pilots as result of their experience and age. Moreover, piloting is a profession that requires high technical skill and ability. Therefore, it is essential that HF trainers are equally knowledgeable and experts in aviation, in particular, in piloting; otherwise it becomes very difficult to establish a valuable interaction with the trainees.

These are the reasons why the first phase of application of HERMES was entirely dedicated to:

- the definition of goals and objectives of the specific CRM (CRM-FCA) course to be developed;
- the familiarisation with the philosophises and polices and, especially, procedures and practices existing within the airlines; and
- the development of sound knowledge and identification of reference models of human behaviour for the "facilitators" of the CRM.

In practice, phase 1 of the work was structured into three steps distributed among the members of the HF team, in order to generate familiarisation and experience with the airline:

1. *Phase 1, Step 1.* Definition of the goals and features of the CRM-FCA.

    Step 2. Evaluation of the sociotechnical contexts of the airline FCA.

    Step 3. Selection of reference model(s) of cognition and development of trainers ("facilitators") guidelines.

Definition of the Goals and Features of the CRM-FCA Training Course

The introduction and extensive use of automation in aeroplane cockpits has increased the efficiency, precision, and safety of aviation operations. It is nowadays widely accepted that present and future aeroplane environment and glass cockpit aeroplanes (A-320, 330, 380 . . . , MD-11, . . .) are easier to operate than their predecessors (e.g., DC-9, B727, etc.).

The human-centred design principles that are utilised by manufacturers when developing a flight deck design should aim at maintaining the central role of the pilot in supervising flight management and control. This requires that pilots are constantly ahead of the aeroplane's behaviour and are able to master perfectly the control principles and procedures applied by the automatic systems. Unfortunately, this does not always occur, and, in non-normal circumstances and failure scenarios, pilots may become very confused, because of the mismatch between their expectations and the aeroplane's performance.

According to recent studies on glass cockpit design and human factors–related safety a number of problems are encountered, such as, for example, automation design principles are often not defined or documented for certification (Sarter and Woods, 1994; FAA, 1996a,b); cognitive aspects of task performance are not sufficiently considered in design (Moricot, 1996; Gras et al., 1994); crew feedbacks are not given relevance in making design trade-offs and in developing safety measures or training practices (Degani and Wiener, 1994a; Tennery et al., 1995).

Nowadays, even more than in the past, it is essential that pilots exploit all resources available in terms of "information, equipment, and people" (Lauber, 1984). A CRM course aims precisely at optimising (a) the human–machine interaction; (b) the acquisition of timely and appropriate information; and especially (c) the interpersonal activities, including leadership, teamwork, problem solving, decision making, situation awareness, stress management, etc.

In aviation there are two types of formal training: simulator and classroom. Simulator training is a hands-on process where pilots are performing according to their technical and nontechnical skill. Classroom training focuses on the acquisition of notions on human factors affecting behaviour in everyday work and in managing emergencies. CRM is mainly carried out in classrooms. Both types of training are nowadays mandatory for pilots in order to obtain and maintain their flying licences. The requirement to apply human factors training during simulator sessions (Line-Oriented Simulations, LOS) was originally proposed by the U.S. Federal Aviation Administration in 1978 (FAA, 1978). Since then, many innovations have been made in order to enable a clear differentiation between circumstances when pilots are trained or are performing recurrent training sessions, or are "checked" for their flying skill. In particular, a number of different types of training/evaluation sessions are foreseen, namely: Line-Oriented Evaluation (LOE), Line-Oriented Flight Training (LOFT) and Line-Oriented Simulations-Recurrent Training (LOS-RT).

In this scenario, the goals of the airline FCA were:

• The development of a consolidated programme of training that integrated and harmonised simulator and classroom activities, especially for what concerned the HF issues.

• The formulation of a new CRM, adapted to the needs and peculiarities of the airline, and representing the formal and logical connection between simulator and classroom training.

For pilots of modern aircrafts, the sessions in the simulator are, in the vast majority of the cases, the only circumstances in which serious system failures are experienced. In these cases, the knowledge of automation principles and design features are directly tested, together with the ability of the crew to act as a team and to interact with an autonomous, intelligent, strong system that performs according to design criteria, but may not be fully understood by the pilots.

While the technical skill of pilots is relatively easily evaluated in simulator sessions, the analysis and assessment of human factors is much more difficult to perform,

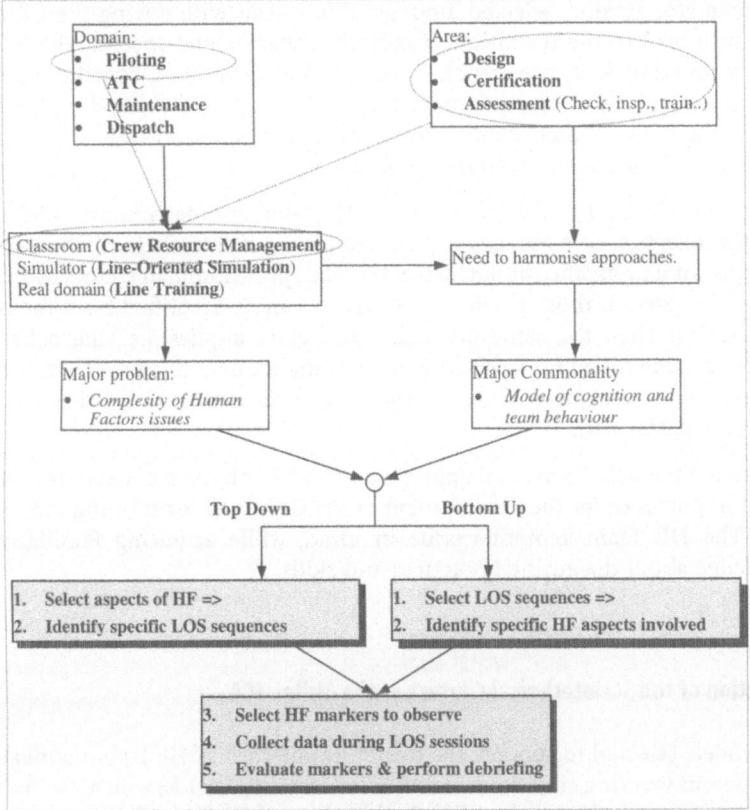

Figure 6.3 Integration of CRM in the training practices of the airline FCA.

mainly because of the cognitive and socio-technical aspects involved. These require specific knowledge of psychology and familiarity with the organisational and cultural factors affecting the airline.

The consideration of a reference model of cognition is very useful to overcome these difficulties, as it represents a formal structure of human behaviour, which can be used as reference in all cases. Moreover, starting from a simple model that contains only the fundamental functions of cognition, it is possible to build a much more complex paradigm by which to consider factors like cultural, organisational, and contextual dependent behaviours and errors.

The process selected to harmonise the consideration of human factors issues is simulator and classroom training, aimed at merging the use of a reference model of cognition with the practical aspects of flying practices and skills (Figure 6.3). The CRM and associated models serve as a basis for developing simulator sessions. Two alternative ways may be followed. The first one, also referred as a "top-down" approach, starts by evaluating a number of human factors aspects to be observed in a simulator session, selected amongst those dealt with during the CRM. This could also lead to the definition of certain scenarios and specific phases of the flight to observe with preference. These choices generate a specific number of behavioural indices or markers to be observed. Finally, following the collection of data during the simulator sessions, the markers are evaluated and a debriefing can be performed on the observed behaviours.

The second alternative, also called the "bottom-up" approach, consists in starting from the selection of a number of LOS sequences of interest for technical needs, or simply for a preference of the instructor. The specific human factors aspects that can be evaluated in those particular scenarios can be identified from the content of the CRM. Then the same procedure as before applies, i.e., the behavioural markers are identified for selected phases of flights, and, finally, after the simulator sessions and collection of data, markers are evaluated and a debriefing with the trainees is performed.

In any case, the definition of an appropriate model of human behaviour is of paramount importance for the development of the CRM and for training the "facilitators." The HF Team kept this issue in mind, while acquiring familiarity and knowledge about the airline's practices and skills.

Evaluation of the Sociotechnical Context of the Airline FCA

The models selected to support the overall framework of HF training must cover the different working situations encountered in everyday life, which can be appreciated by Ethnographic Studies (ES). In this phase of work, the ES that were carried out consisted in "jump seats," interviews, and questionnaire.

Jump Seats

In order to become familiar with the procedures and practices of the airline and to acquire further knowledge of the tasks carried out in a cockpit, all members of the HF Team performed a number of "jump seats."

During a "jump seat" experiment, a crew was observed for a whole shift or period of work. In the case of short flight, i.e., national or European routes, this consisted in three to four flights per day. While, in the case of long-haul flights, the interaction with the crew covered two or three days, i.e., the flight from home base to the destination, the rest period, and the flight back home.

During these periods, the objective of each member of the HF Team was to focus on few specific tasks, such as, for example, the use of the Flight Management System, or the communication between members, i.e., amongst the pilots, and with the other actors contributing to the flight, including air traffic controllers, ground personnel, and cabin assistants.

In addition, during these flights an intensive process of exchange of opinions and discussion took place, as all pilots were informed of the role of the HF Team member participating in the jump seat. These informal discussions, with no structuring or predefined goal, were very useful in developing the individual expertise of FT Team members.

Initially, the HF Team had to overcome scepticism and gain the acceptance and trust of the pilots. Then, the different subjects discussed during the jump seats were left totally to the pilots. In general two subjects were discussed: technical aspects concerning the tasks and duties of captains and first officers, and social and management issues specific to the airline.

This type of discussions were reported at the end of each jump seat mission by the HF Team member and were considered as further information to be added to the outcome of the more formal results of interviews and questionnaires performed within this step of the work. In particular, they were analysed together with the "open questions" of the questionnaire, as reported below.

Interviews

The interviews aimed primarily at identifying the philosophies and policies of the airline. For this reason, the managers of the airline were interviewed, including "head-pilots," flight operation managers, safety and administration top managers.

The procedure that was applied was based on a semi-structured interview, focused to identify management styles and attitudes towards safety and training.

The results of these interviews were added and analysed together with the outcome of the informal discussions during the "jump-seats" and the data collected from a more specific questionnaire distributed to the entire category of pilots.

Flight Management Attitude Questionnaire

To be successful, CRM training must be tailored to the organisational context. A new strategy to harmonise CRM with organisational and national culture has been developed within the aerospace crew research project, of the NASA-UT (NASA-University of Texas) and FAA. This strategy is based on the Flight Management Attitude Questionnaire (FMAQ), which aims at distinguishing cross-cultural attitudes toward performance in commercial flight operations (Helmreich et al., 1993; Helmreich and Merrit, 1998).

The theoretical foundations of FMAQ lie in the research work on the relevance of cultures in working contexts and the impact that these issues have in aviation as well as in other domains (Berry, 1980; Hofstede, 1980; Helmreich, 1984).

The questionnaire was based on the FAMQ 2.0 (international), originally developed by the NASA team (Merritt et al., 1996) and adapted to the specific aims of the airline FCA.

The goal of the FMAQ is to depict the attitudes of crewmembers toward leadership and command, communication and coordination, automation usage, reactions to stress, organizational climate, quality and type of training, manuals and checklists, safety-reporting procedures, management issues, and teamwork with other employee groups. Differences due to position, fleet, base, and/or level of experience are also investigated. The outcome of the analysis of data focuses on recommendations for training based on strengths and weaknesses existing within the organisation.

Used as a baseline measure, the result of the first FAMQ can be compared with attitudes at a later date, for example, after training or some organizational change, such as a merger or fleet change.

The final version of the questionnaire (FMAQ-FCA), which derived from the FAMQ 2.0 and was adapted to the specific characteristics of the airline FCA, can be found in Annex 6.1.

In the questionnaire, a part from the translation into the national language of the airline, a number of items were added or modified with respect to the original version. Hence the FMAQ-FCA version was finally structured in six sections, namely:

Part I   Pilots' and flight engineers' view of the company
          A Satisfaction with training and flight operations (14 items).
          B Quality of teamwork and cooperation (7 items).
          C Perception of organisational climate (24 items).
          D Recommendations for improving training and flight operations (3 *free text opinions* for each topics).
Part II   Flight management attitudes (34 items).
Part III Leadership style (perceived and preferred; 2 choices).
Part IV   Work values and goals (13 items).
Part V   Cockpit automation (17 + 11 new items).
Part VI   Background information.

Respondents could answer the items by using a 5-point scale. As anonymity and confidentiality are essential to the project, airlines and pilots data were always deidentified.

Models of Cognition and Facilitators Guidelines

One of the major concerns of the HF Team was to generate and consolidate a basic knowledge of Human Factors for the "facilitators" of the CRM, as the selected persons within the airline were pilots and co-pilots with valuable experience in flying, but with little background in HF.

Consequently, a relevant activity within the first phase of work was dedicated to the development of a "CRM facilitator manual" which could serve as reference text (Amat et al., 1997).

The manual concentrated on two main aspects: the basic modules of the CRM and the reference model of cognition:

1. Three main modules were considered, as this is quite standard for CRM courses and is required by most civil aviation authorities (Wiener et al., 1993; ICAO, 1991). These modules were individual behaviour, group dynamics, and operational factors.
2. A model of cognition was selected to serve as a reference paradigm to represent human behaviour in all circumstances and to generate a deeper understanding of the reasons and fundamental principles associated with human behaviour.

The Three Basic Modules of the CRM Facilitator Manual

The aim of any CRM training course is to foster a proficient utilization of all available resources, in terms of information, equipment, and people, to achieve safe and efficient flight operation.

The contents and design of the CRM course developed for FCA are based on the available literature and on the field study carried out in order to identify the specific characteristics of the airline.

The classroom part of the CRM-FCA was based on a three-day-seminar, made of three modules:

1. *Individual behaviour*
   - Personal attitudes
   - Situation awareness
   - Decision making
2. *Group dynamics*
   - Crew as a team
   - Team work
   - Communication
   - Leadership
3. *Operational factors*
   - Advanced automation
   - Operational fatigue
   - Stress

The facilitator manual contains technical information and basic notions of Human Factors, which are presented and discussed during the training sessions and lectures of the three-day seminar. In particular, the theoretical background and content of three modules are described in detail together with an exhaustive literature review.

Reference Model for Describing Pilot Behaviour

In the CRM-FCA a generic model of human behaviour sustains all different aspects relative to the activity of a pilot and correlates the three main modules of the course. This represents a way to integrate and correlate all theoretical aspects of the course.

Generic behaviour of a human being has been categorised and represented formally for many years. The use of computers and information technology in the cockpit has strongly affected the work and the behaviour of pilots. Consequently, also the formal models of the human behaviour have been focused on the mental activity of the pilot, rather than on manual performances.

Moreover, communication, interaction, and group work intracockpit and between the cockpit and ground operators have become fundamental issues to consider in flight safety. Consequently, the model adopted for representing pilot behaviour must be able to account for individual as well as crew work, i.e., for describing the generic human activity within the cockpit (Figure 6.4).

A pilot is asked to simultaneously perform three basic activities, namely: to *aviate*, i.e., maintain control of the aeroplane; to *navigate*, i.e., remain informed of its state and position; and to *communicate*, i.e., exchange information with the ATC and other crew members.

At the level of individual behaviour, two fundamental aspects need consideration: Decision Making (DM) and Situation Awareness (SA). These are strongly affected by individual characteristics, which are governed by personality features.

At the level of crew-work, the generic activity involves teamwork, communication, leadership and consideration of the crew as a team. These aspects involve concrete behavioural features like efficiency and explicitness in communication, attitudes within the team, and sharing of models between pilots and coordination.

At the level of individual behaviour, it becomes essential to consider a simplified model that covers all activities of a pilot, without becoming too specialised or cumbersome. A model of this nature was considered with reference to the paradigm of the information processing system (Neisser, 1967) and was based on four basic cognitive functions, as shown in Figure 6.5. This model was called the Generic Individual Behaviour (GIB) model.

GIB and Existing Models and Paradigms

On the basis of GIB, all concepts and activities were modelled. In some cases, specific functions of cognition were used as reference, while in other cases, GIB was applied as a generic envelope of well-known models that could better describe the specific human activity under discussion.

As an example of this approach, the consideration of Situation Awareness (SA) of a pilot was defined with reference to the two upper cognitive levels of interpreta-

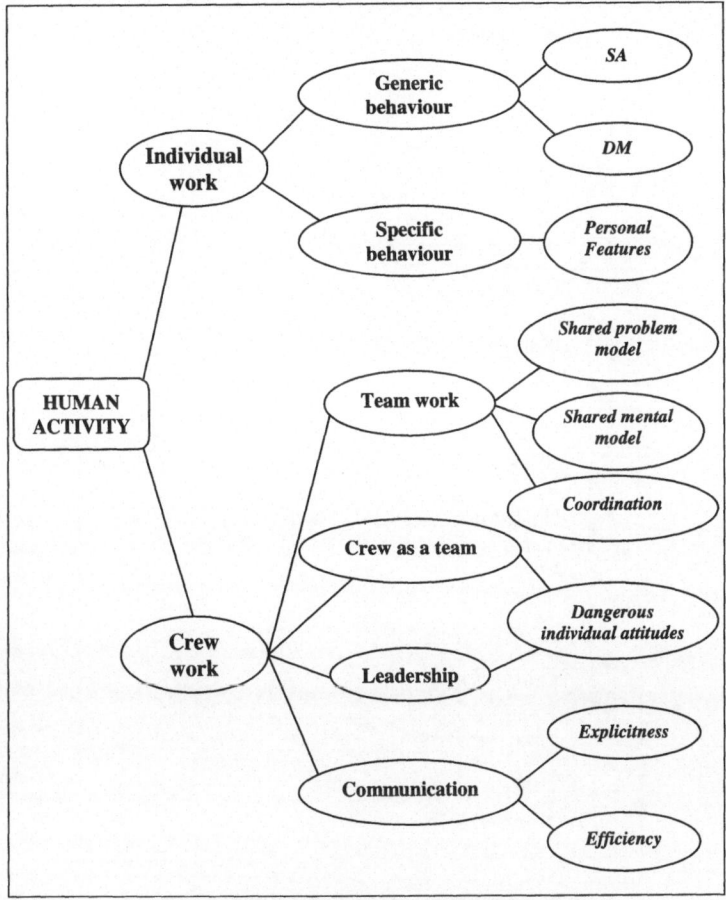

Figure 6.4 Generic model of human activity within a cockpit.

tion and planning (Figure 6.6). According to the accepted literature, SA can be defined as "the *perception* of the elements in the environment, within a volume of time and space, the *comprehension* of their meaning and the *projection* of their status in the near future" (Endsley, 1994).

Another example of correlation between the GIB model and well-known paradigms was defined for representing human error. The classification developed by Norman (1981) and further modified by Reason (1990) that identifies four types of human errors as *slips, lapses, mistakes,* and *violations* can be correlated to the cognitive functions of the GIB model (Figure 6.7).

The GIB model was also correlated to the well-known and established Skill–Rule–Knowledge (SRK) model (Rasmussen, 1983, 1986) in order to offer the CRM facilitators another paradigm of reference for discussing human behaviour. In particular, the comparison of GIB and SRK models was applied in the case of decision-making processes.

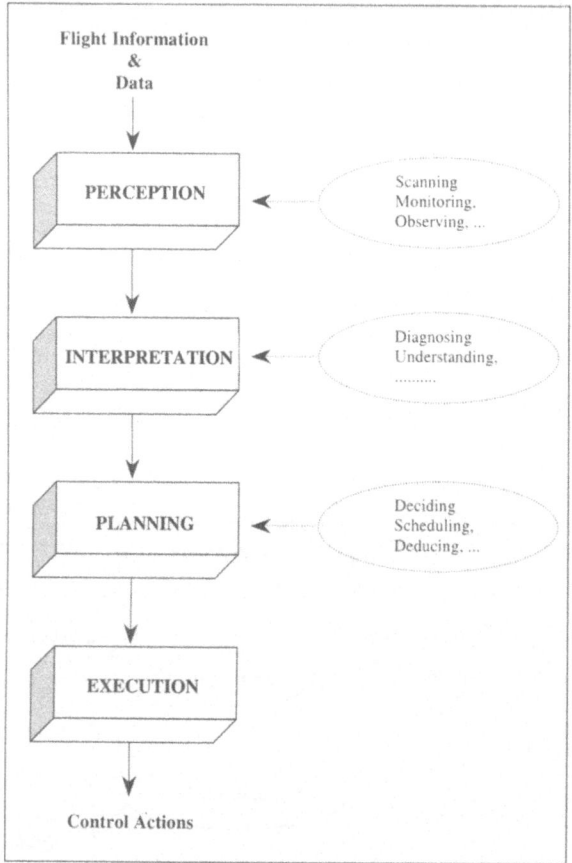

Figure 6.5 Representation of the generic individual behaviour model (GIB).

If one takes for granted that every action has been decided (consciously or not) by a person, then also determining when to press a button or when to perform a check list or, how to perform it, implies making a decision, even if it might be seen as a "small" or an "obvious" one. The application of a procedure is also the result of a decision. In this case, the pilot follows the *rule* "if the current situation is recognised as situation A, then procedure A1 has to be performed."

Moreover, the automatism demonstrated when flying manually, by reactions such as pushing on the pedal for changing the rudder position when losing one engine, is the product of a decision. In these cases, the behaviour is based on the reflexes of the pilot, i.e., according to SRK model, a *skill-based* behaviour.

When the type of decision is more complex, such as in the case of a diversion, or a choice between two different types of approach to the landing runway, e.g., visual or instrumented, then the pilot must plan using experience and expertise and needs to account for several parameters. This is a typical *knowledge-based behaviour*.

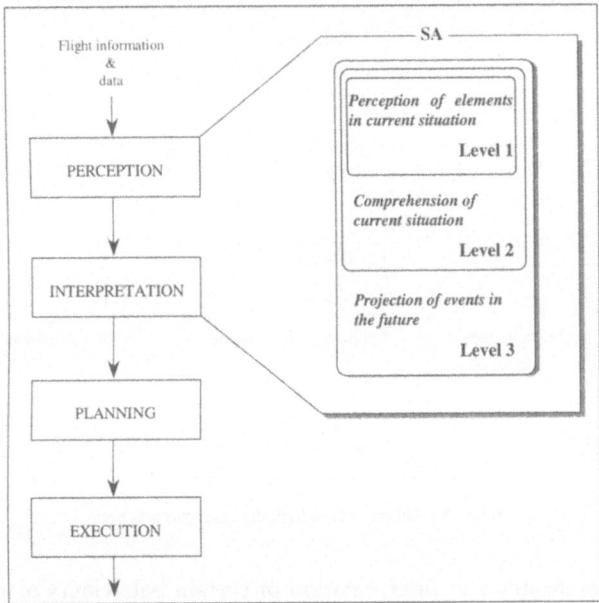

Figure 6.6 Situation awareness and GIB model.

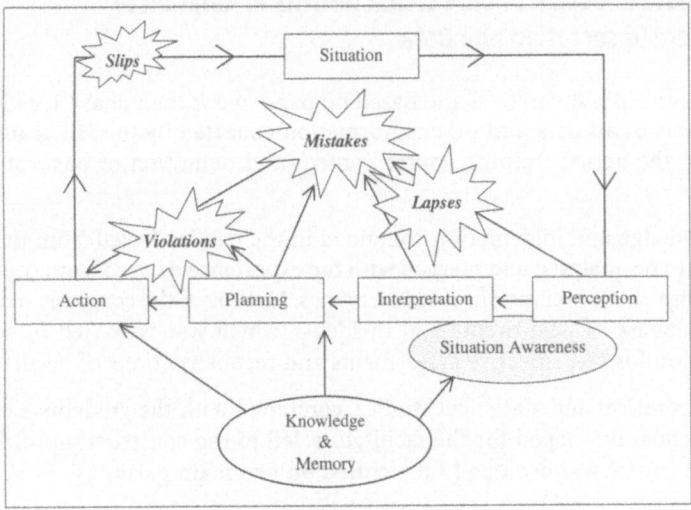

Figure 6.7 Mistakes, lapses, slips, and violations integrated in the GIB model.

These three processes can easily be accommodated within the GIB model and this gives the correlation that exists between GIB and SRK models (Figure 6.8).

Situation awareness and decision making can be interpreted with reference to the GIB and SRK models in a general theoretical framework that can support the dis-

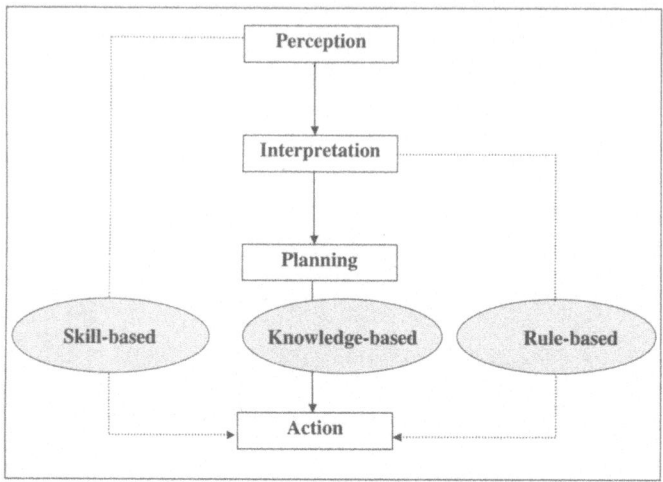

Figure 6.8 GIB and SRK models for describing pilot behaviour.

cussion in classrooms and interpretation of certain behaviours of pilots (Figure 6.9).

6.2.2 Phase 2: Design of CRM Characteristics in Relation to Field Observation and Data

Phase 2 was more theoretical and based on paper work than phase 1, as it covered the analysis of all data and other information collected by the HF Team and the design of the actual training course content and definition of basic supporting material.

The knowledge and information contained in the data collected from the FMAQ-FCA had to be analysed and merged with the experience derived from the outcome of the jump-seat observations and interviews. Moreover, the company made available a database of past events and incidents, which was very rich in additional information for retrospective assessments and reconstructions of occurrences.

These theoretical and statistical studies, combined with the guidelines contained in the manual developed for the facilitators, led to the construction of the CRM-FCA. The course was eventually structured on two main parts:

1. an introductory session that made use of a Computer-Based Training (CBT), for presenting in a simple and interactive way the basic HF concepts.
2. a main CRM-FCA set of sessions containing much more extended and detailed discussions, examples, and practical tests on HF issues of relevance for the pilots of the airline.

In practice, phase 2 of work was structured into three steps performed in collaboration with the pilots that were selected by the airline for training as "facilitators" of the CRM-FCA. These three steps were:

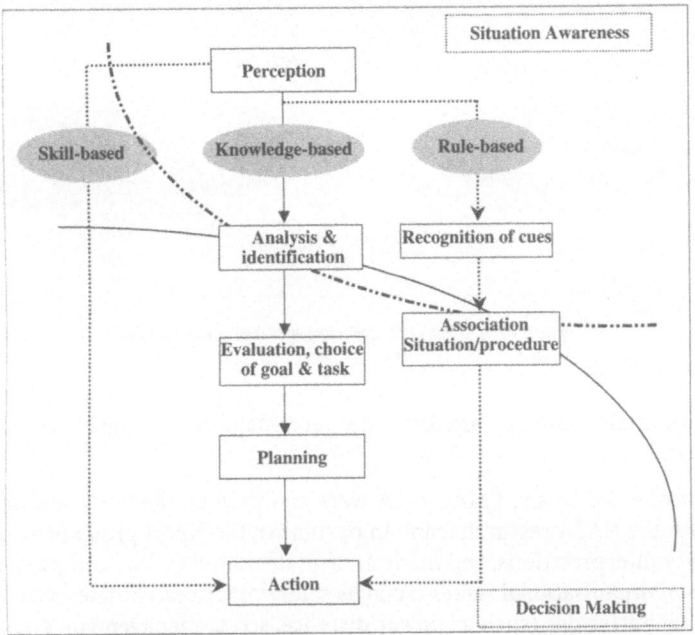

Figure 6.9 Situation awareness and decision making in the frame of SRK and GIB models.

1. *Phase 2, Step 1.* Analysis of data collected on pilots attitudes by means of the FMAQ-FCA questionnaire and study of incidents from existing databases;

   Step 2. Design and development of the CBT, as introductory part of the CRM-FCA;

   Step 3. Detailed design and development of the CRM-FCA classroom course.

Analysis of Data Collected on Pilots Attitudes and Incidents

Results of the FMAQ-FCA

The Flight Management Attitude Questionnaire (FMAQ) study was originally intended for harmonising and tailoring the CRM-FCA to the organisational and national culture of the airline. However, thanks to some crucial changes to the original version of the questionnaire, the results reached a much wider scope, permitting to outline the airline profile perceived by pilots in terms of safety culture, company climate, quality of training, and line operations, and to support the design of the new elements of the CRM-FCA course.

The questionnaire was completed and returned by 517 respondents, over a total of 1800 pilots. Figure 6.10 shows the pilot population subdivided according to their category and fleet. In bold characters we show the percentage with respect to

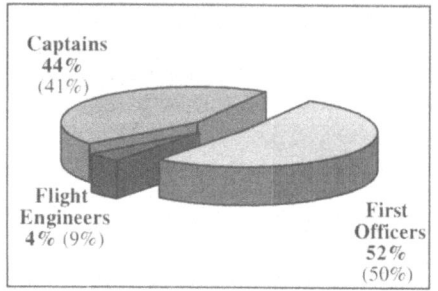 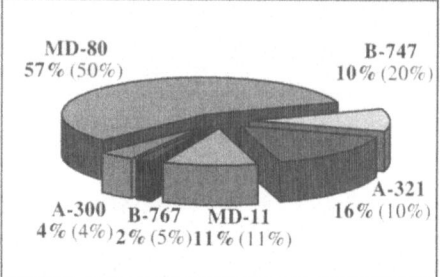

Figure 6.10 FCA pilots subdivided according to their category and fleet.

517 questionnaires and in brackets the percentage with respect to FCA whole population.

The data collected by the FMAQ-FCA were analysed by the HF Team in collaboration with the NASA research team. In particular, the NASA group provided complementary interpretations, and made an overall analysis of the company, specially focusing on organisational issues (such as safety practices, climate, etc.) and flight management attitudes (such as power distance, stress management, etc.).

Respondents answered to the items of the FMAQ-FCA by using a 5-point scale: very low (1), low (2), adequate (3), high (4), very high (5), or disagree strongly (1), disagree slightly (2), neutral (3), agree (4), agree strongly (5).

The data were analysed from many different perspectives, aiming at understanding general attitudes and peculiarities of the airline as a whole, and differentiating amongst different categories, e.g., captains and first officers, pilots of automated and nonautomated aeroplanes, pilots of different age and generation, etc. In order to maximise differences and simplify the already complex task of statistical analysis, the data were grouped into three categories, by grouping point values 1–2 and 4–5 into either "low" and "high" or "disagree" and "agree," respectively.

Special attention was dedicated to the free text comments and suggestions that almost all pilots provided.

Though the FMAQ is composed of different sections, each one addressing a particular topic (e.g., flight management, automation), the answers within and across two or more sections were analysed for investigating links between these topics. In practice, by grouping different items of the questionnaire, the overall results could be structured around a number of "headings" such as, for example, organisational issues, flight management attitudes, cockpit automation, training needs, etc. Four examples of the graphical representation and grouping of some items contained in the questionnaire are shown in Figure 6.11, Figure 6.12, Figure 6.13, and Figure 6.14, for evaluating, respectively, safety practices and culture, task sharing, crew coordination, and automation management.

The overall results showed that the areas of most concern were the organisational and safety culture and flight management attitudes. Even if the main reason for this result could be largely connected with the particular moment in the history of

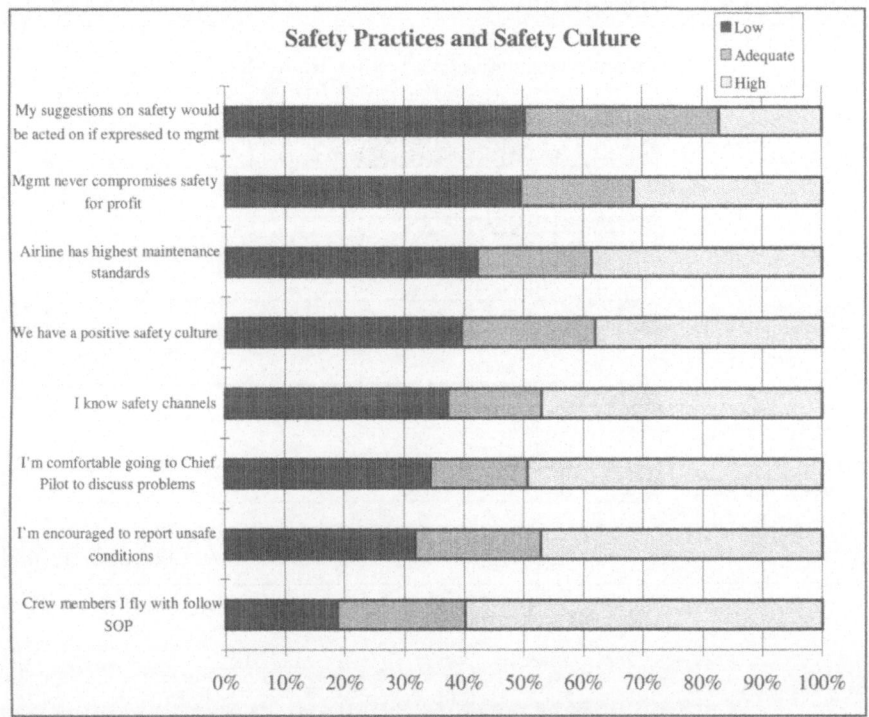

Figure 6.11 Grouping of items for measuring safety practices and safety culture.

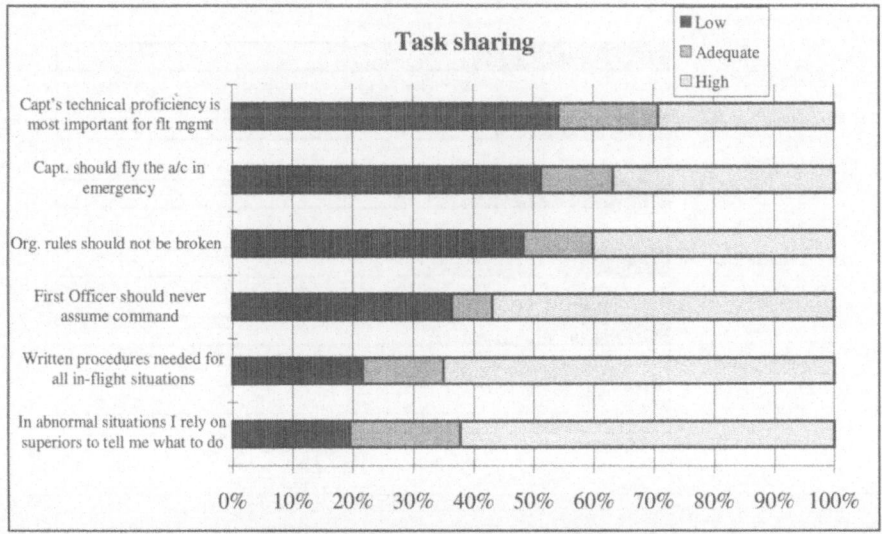

Figure 6.12 Grouping of items for measuring task sharing.

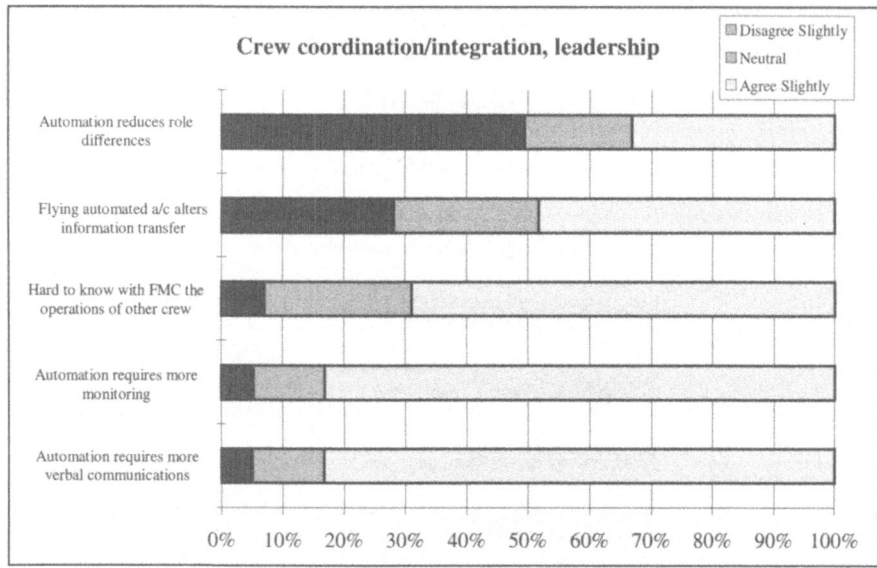

Figure 6.13 Grouping of items for measuring crew coordination.

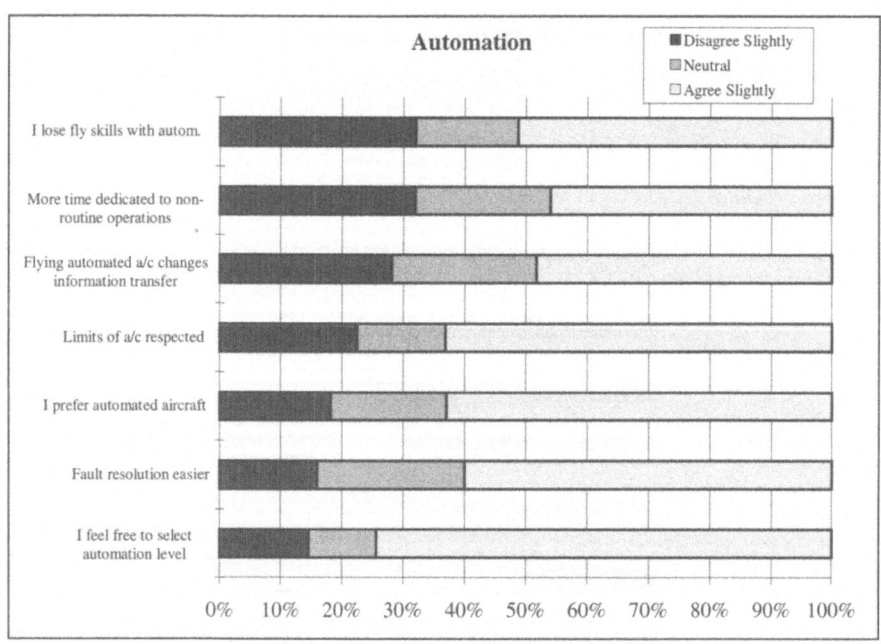

Figure 6.14 Grouping of items for measuring automation issues.

the airline, it was clearly demonstrated that the major goal of the CRM-FCA would be to tackle these factors, while a strong recommendation for the airline was to operate at social level in order to ascertain that these problems were gradually resolved and removed.

The flight management attitude and the command style exhibited in general by FCA pilots endorsed a hierarchical command style. However, the generation change in captains and the new attitudes required to fly highly automated aeroplanes have shown differences in command styles of the younger generation of pilots. Attitudes toward stress and performance were generally unrealistic and reflected the individual attitude of "machismo" and invulnerability.

Pilots revealed a positive tendency toward automation, both within automated and nonautomated fleets. Pilots reached a good compromise with automation and were aware of the need to improve verbal communication and crosscheck between crew members and changes on information flow in automated aircraft. Consequently, it could be argued that the pilots did not show a particular attitude of complacency and loss of situation awareness.

Focusing on the findings that could have a spin-off for the training programme, it could be concluded that the CRM-FCA course should mainly concentrate on *staff motivation, interpersonal and professional cockpit relation, assertiveness.*

The analysis of free text opinions gave a measure of the professionalism of FCA pilots and flight engineers, despite the general attitude of remarkable detachment from the organisation and the company policies, and showed a number of constructive statements with respect to training needs and development. In particular, FCA pilots demanded training contents closer to line operations and to realism and integration with personnel from other operational areas as well as better and more training, especially in simulator sessions.

Extended Outcome of FMAQ Data Analysis

The analysis of the questionnaires completed by the pilots allowed the HF Team to develop other safety indicators and recommendations, in addition to the insight that was gained with respect to the CRM course.

These can be interpreted as an additional result of the ethnographic analyses carried out within this phase of work and further input for the airline to improve safety levels and standard within the company.

The following safety issues and indicators were identified:

- Constitute a "confidential safety reporting system" within a safety office with strong credibility with line personnel, that encourages and does not punish disclosure and feedback of results.
- Develop and/or make more evident the responsiveness of flight operations management group.

- Improve the quality and extend training programmes, especially by ameliorating the realism of simulator sessions and focusing more on training content rather than on formal checks.
- Ameliorate the selection of trainers, check pilots, and instructors, by identifying the best captains and granting more responsibility to the more experienced first officers.
- Improve further the "nontechnical" training (CRM), focusing more on team work, interpersonal relations within the cockpit, assertiveness, always in strict contact with flight operations.
- Improve the airline communication and exchange of expertise by developing new CRM courses for nonpiloting staff, including cabin crew, maintenance personnel, and company corporate management.
- Verify improvement of the safety level within the organisation as a result of the new training exercise, by a further assessment of the "attitudes of pilots," approximately one year after the implementation of the CRM-FCA course.

Study of Incidents Within FCA

The HF Team completed the evaluation of the airline by studying a number of incidents and accidents that occurred in the history of the airline. Access to confidential data and reports was granted.

The retrospective approach that was utilised for evaluating root causes is typical of the HERMES methodological structure and will not be discussed in detail here, as a whole chapter of this book (Chapter 8) is dedicated to the way to carry out an accident investigation.

In this case, the characteristics of the airline were sufficiently highlighted by the ethnographic studies and the analysis of the FMAQ-FCA. Therefore, the study of past incidents was mainly utilised to confirm and strengthen the findings of field studies and statistical analyses.

The main conclusion of the process of this step of the project consisted in the definition of the training needs and the adoption of a complementary approach to classroom training, typical of CRM courses that consisted of CBT.

The objective of the CBT was to support the theoretical training and collective discussion by an individual and interactive session where the pilots could experience and identify, in a multimedia environment, real working situations and concepts or factors that were then extensively discussed in the classroom sessions.

Design and Development of the Computer-Based Training

Main purposes of the Computer-Based Training (CBT) for the airline FCA (CBT-FCA) were to introduce human factors to the participants, to prepare deeper discussions to be performed in later classroom sessions, and to generate an interest in the HF subject matter by making the participants ponder about a series of general problems in the aviation domain, and specifically in the context of flight management.

The CBT-FCA was planned as the first step of the course and was designed for individual sessions to last approximately 45–60 minutes. Each participant in the CRM sat in a CBT session before starting the formal classroom lectures.

As the CBT was planned at the initial stage of the training programme, an important step was to get the participants used to thinking about human factors problems and to express a personal opinion about them.

A secondary purpose was to introduce the most basic theoretical framework on human factors, giving a brief description of notions like communication, decision making, situation awareness, etc.

The major methodological choice was to combine the CBT with the presentation of real accident cases where human factors have played a decisive role in the development of the accident. This would automatically provide the CBT session with an authentic atmosphere, further underlining the relevance to the participant.

The format of a CBT was based on an interactive multimedia software that provided a level of interaction where each participant could choose the pace of presentation, i.e., to stop and study details of situations or pass by quickly, and returning on issues of interest at any time during the interactive session. This format reduces the feeling of being forced through a training course of no interest.

To guide the participants through the subject, and to support the mental processes of thinking over the problems, the participant was asked a series of open questions relating to the accident cases shown in the CBT. These questions, given in a broad form, allow personal interpretation of the situation, in accordance with the philosophy that the mental processes themselves are more important than the conclusion they lead to.

The CBT-FCA contained two accident cases, presented following the chronological order of the events leading to the accident. The free form of interaction allowed the participant to go back and see any part of the accident desired. The presentations were divided into events based on the cognitive processes involved, and the questions or issues for pondering were related to an event as a whole.

The result of the interactive presentation of real accident cases and the use of open but guiding questions was to lead participants to develop their own analysis of the accident. Given the limited material provided, such conclusions could not necessarily be compared with the official analysis of the accident, but it provided a good basis for an introduction to human factors.

Structure and Content of CBT-FCA

The CBT-FCA follows a simple structure divided into three modules for each accident: (a) an introduction; (b) a central part with the presentation of the accident case and questions; and, (c) the conclusions.

The introduction was mainly based on video shots from cockpits, control rooms, and control towers with the purpose of creating confidence in the CBT and a sensation of familiarity with the subject.

The central part of the CBT presents two accident cases through an interactive screen with a time line representing a series of events, each described by a video clip, and textual excerpts from the official accident report. The relevant questions are presented and further theoretical explanations are available for the participant who might desire it. The two accident cases are Kegworth (Case 1) and Tenerife (Case 2).

- *Case 1, Kegworth*

  The Kegworth accident happened on January 8, 1989, involving a Boeing 737–400 G-OBME. In all, 47 passengers died, 74 others suffered serious injury. The cause of the accident was that the operating crew erroneously shut down the No. 2 engine after a fan blade had fractured in the No. 1 engine. Different factors contributed to the incorrect response of the flight crew such as loss of situation awareness, stress, lack of training and experience and poor communication exchange between the two pilots and amongst the pilots and Air Traffic Controllers (ATC), as well as cabin crew.

  Communication, training, and stress are the main issues on human factors that have been illustrated throughout this accident from the pilots' view and experience. These issues well match and complete the air traffic controllers' experience and task view.

  The Kegworth case is presented in a video that contains parts of the communication between the flight deck and ATC and shows the task management on the flight. The accident report excerpts and some data from the "flight data recorders" are used to give information not included in the video, like changes of radio-frequency. Last, information about altitude and heading is provided in textual form.

- *Case 2, Tenerife*

  The Tenerife accident happened on March 27, 1977, when KLM flight 4805 and Pan Am flight 1736 collided on the runway of the Tenerife airport with a loss of 583 lives. From the accident analysis it is concluded that the combination of interruption of important routines, a loss of cognitive efficiency, and a loss of communication accuracy created an environment for the rapid diffusion of multiple small errors.

  This case is particularly interesting for the training of pilots and also air traffic controllers, as the development of the accident and the events leading to it are heavily correlated with the interaction between the ATC and the two aircrews, between the two aircrews, and within each crew. Radio communication is the single most important means of communication and information interchange between ATC and aircrews, and the communication problems seen in the accident at Tenerife are of great relevance to a human factors discussion.

The concluding module is brief and has the main purpose of giving an integrated view of the problems presented and preparing the participant for the discussion with other participants that follows the CBT in classroom sessions.

Development of the CBT-FCA

The actual development of the CBT-FCA has been performed in three steps: the study of the accidents, the definition of the questions and comments associated with the pages of the presentation, and the development of the interface.

As a first step, the two accidents of Kegworth and Tenerife were studied in detail using the official reports (AAIB, 1990; ICAO, 1978). From the reports, some important excerpts were transcribed and considered in the CBT. For each accident a storyboard was defined, dividing the accident into phases and each phase into events. Each event corresponds to a screen presentation in the CBT. All event-screens have the same basic layout and interaction format.

The second step in the development of the CBT was the association of questions and comments to each event. Questions and comments have been introduced in order to activate the processes of thinking about the root causes and associated HF issues. In certain cases, comments have been introduced to clarify and specify the events in more detail. The questions cover the basic human factors concepts and, in particular, concentrate on the phraseology and management of communication, and on standard and emergency operating procedures.

The third step in the development of the CBT was the design and development of the user interface. It was important to ensure that that the CBT did not give the impression of being either rigid or boring, as this could easily be counterproductive in respect of the goal of raising awareness towards human factor issues. The introduction, which has the purpose of familiarising the user with the CBT and giving an idea about human factors training is very short and does not require any interaction with the user.

The interface of the central part was designed to leave participants free to explore the events contributing to the accident, and provides, at the same time, a simple and chronologically accurate presentation of the accident. Experience from similar projects has shown that participants initially navigate through the events in a time sequence order, and then go back to examine parts of particular interest, dealing with questions and answers (Bellorini et al., 1997).

As an example in the case of the accident of Tenerife, the main window (Figure 6.15) contains the most important elements for the presentation. The timeline is seen to the left, with an excerpt from the accident report to its right. The timeline is itself a navigation instrument, as it allows the participant to go back to earlier events to revisit details or to go forward to the next event. The right of the screen shows a video with the reconstruction of the event and, below that, a map of the airport with the positions of the involved aircraft.

Two buttons for opening a new window with further *comments* and *questions* are located below the excerpt from the accident report with communications between pilots and air traffic controllers. The comments-window explains certain aspects of the event in the light of the theoretical parts of the main CRM course. This could, for instance, focus on a Situation Awareness (SA) problem, with the possibility of opening another window with further comments and discussion on SA. Both comments and further reading are optional, with the choice left to the participant.

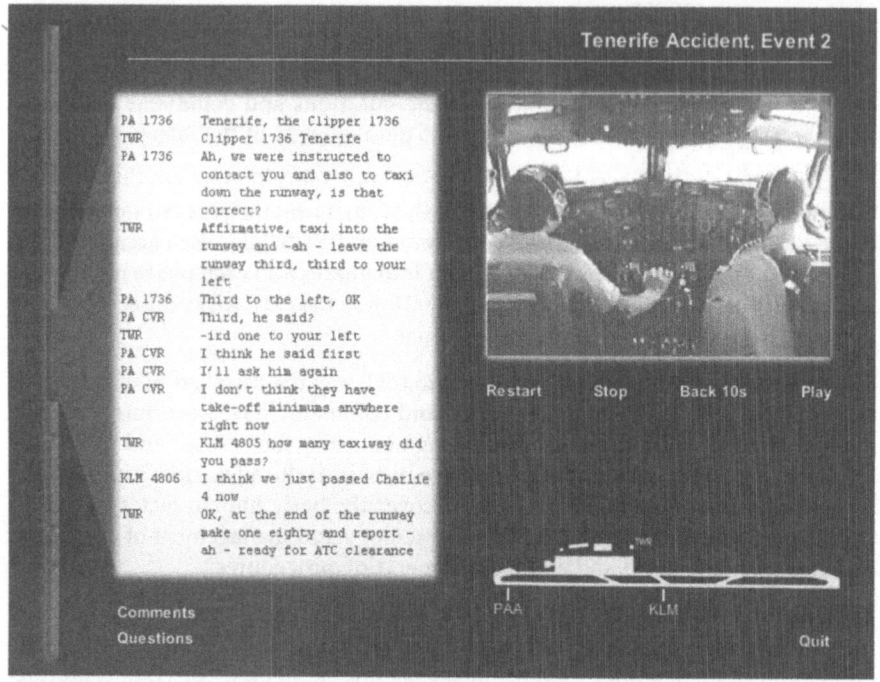

Figure 6.15 Central page of CBT-FCA presenting the accident of Tenerife.

The question-window guides the participant through the analysis by asking specific questions and offering possible answers. All answers are logical and reasonable, but only one is correct, as it takes into consideration all aspects of the human factors involved.

Detailed Design and Development of the CRM-FCA Classroom Course

The major component of any CRM is the human factors seminar. The final CRM-FCA was designed in the form of a 17-hour-long intensive seminar for a group of pilots, facilitated by the airline's own instructors, and distributed over three consecutive days of lectures.

The seminar involves participants as individuals and as a group through guided discussions, exercises, and lectures using video material, questionnaires, and games. According to the basic design options, defined during phase 1 of the work, the seminar is structured around three major modules (individual activity, group dynamics, and operational factors) and an introductory module. These modules are distributed according to the diagram shown in Figure 6.16.

Introduction to Human Factors Module

The first module (*Introduction to Human Factors*) introduces human factors in the flight deck. This module makes active use of the results derived from the

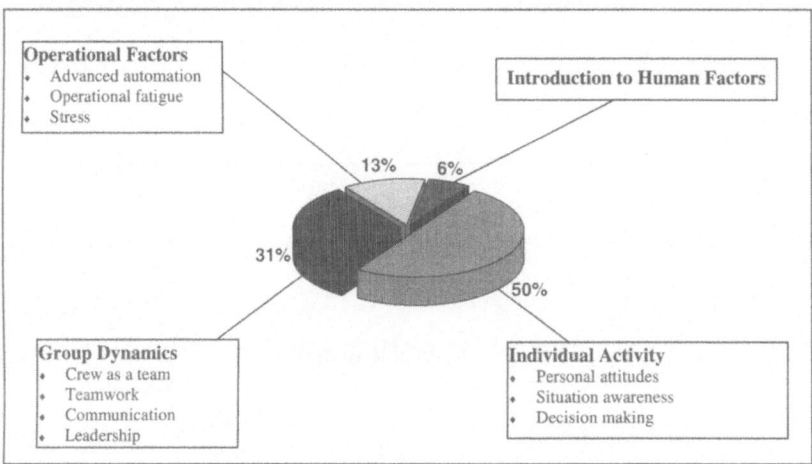

Figure 6.16 CRM-FCA/human factors seminar.

computer-based preparatory course. The basic concepts, definitions, and expectations of participants are discussed.

The supporting material consists of slides, while participants discuss their expectations from the CRM during a roundtable session. A video is presented on one of the accidents analysed during the CBT.

This introductory module was planned to last 30 minutes.

Individual Activity Module

The second module (*Individual Activity*) focuses on the individual pilot behaviour and activity and is made of three parts: personal attitudes, situation awareness, and decision making.

The notion of cognition is demystified through a better knowledge of the pilot's own personal features and on the factors interfering with cognitive activity. Personal features are addressed including aspects that are inherent to the individual, and the individual's specific history and experience. The focus lies on the danger derived from hazardous attitudes, such as antiauthority, machismo, invulnerability, resignation, and impulsiveness.

Staying aware of the situation (situation awareness) is crucial for being fully in control of the aircraft. Distractions and heavy workload often trigger phenomena that lead to a wrong picture of what is going on. Based on case studies, and by sharing experiences, solutions for preventing and detecting loss of situation awareness are elaborated (Figure 6.17).

Decision making on the flight deck encompasses the process that starts with diagnosis of what is happening from the perceived situation, up to the implementation of a course of action. The context in which decisions are taken makes the pilot particularly vulnerable to certain biases.

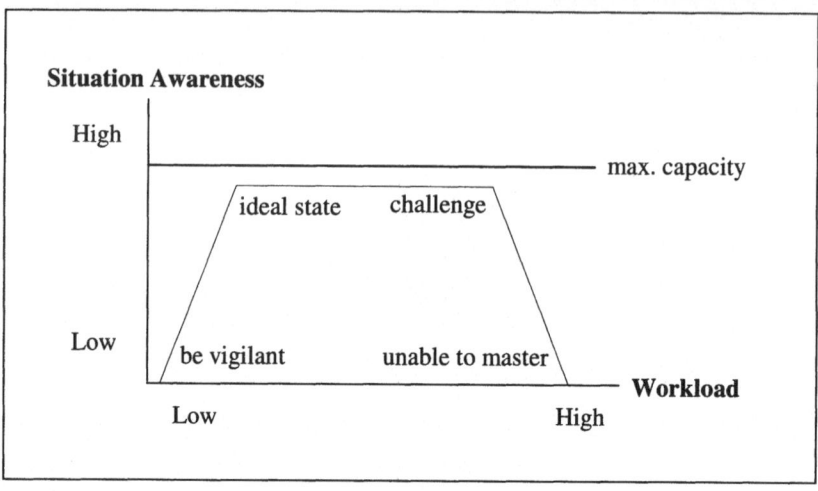

Figure 6.17 Relationship between workload and situation awareness.

The individual activity module is made of facilitator presentations, based on trans-parencies and figures, guided discussions, small questionnaires, and exercises ("games") to be performed in groups and followed by general discussions. The durations of the three parts of this module are as follows:

- *personal attitudes*, one hour and 50 minutes,
- *situation awareness*, three hours and 50 minutes;
- *decision making*, one hour and 40 minutes.

Group Dynamics Module

The third module (*Group Dynamics*) concerns the performance of a team of pilots and focuses on the coordination of activity, communication, and interpersonal relationship.

The cockpit crew forms a team in the cockpit, sharing goals, and coordinating activity. The aim of this training module is to illustrate how the performance of a team can exceed the performance of an individual.

The characteristics of a team are outlined and examples of problems in the estab-lishment of teamwork are discussed, such as coordination, group decision making, and achieving a shared understanding of the situation.

This training module focuses on the function of communication in the flight cockpit, on the factors that might influence the communication, and on the achievement of effective communication.

Leadership, expressed through the ability of facilitating the crew activity towards the accomplishment of its task, is the key to sustain teamwork. This training module identifies different leadership styles and their impact on the crew performance.

The Group Dynamics Module starts with a well-known "game" and test on team performance and introduces the teamwork section. As for the previous module, the sustaining materials and formats for this module are presentations by view-graphs, short videos, and roundtable discussions about real events. The duration of the four sections of this module are as follows:

- *crew as a team*, 50 minutes;
- *teamwork*, two hours and 30 minutes;
- *communication*, one hour and 45 minutes;
- *leadership*, two hours and 10 minutes.

Operational Factors Module

The forth module (*Operational Factors*) concerns the management of factors influencing operative effectiveness. The specific elements of this training module depend on the characteristics of the airline.

The impact of fatigue due to, for example, long-haul flights, and stress that is a natural part of flight operations, affects the performance of the individual pilot as well as that of the flight crew. This module discusses the background for understanding fatigue and stress, provides solutions on how to avoid factors that may enhance them, and suggests how to cope with them.

A major issue discussed during this module concerns the pitfalls of automation. Automation is usually considered as supporting flight deck operations, leaving the pilot with more cognitive resources to cope with other tasks. However, problems such as compliance, overreliance, self-confidence, etc., may endanger the interaction of pilots with automation. This module aims to provide the background for understanding when it is desirable and necessary to be aware of these issues and exploit the advantages of automation.

As in the case of the two previous modules, module 3 is also based on viewgraphs, figures, videos, and games to be presented in the classroom by the facilitators and on roundtable discussions and exercises involving all participants either as a whole group or subdivided into specific subgroups.

The three sections of this module are planned for the following durations:

- *stress*, 50 minutes;
- *operational fatigue*, two hours and 5 minutes;
- *advanced Automation*, one hour and 15 minutes.

6.2.3 Phase 3: CRM Implementation and Quality Testing

The final phase of work was divided into two distinct steps: the actual development of the training material and delivery of the course, and then the evaluation of the effectiveness of the course. These steps are very different from the method-ological and practical viewpoints.

The first step does not present any particular complexity as it consists in the imple-mentation of previously planned activities. The second step is, in essence, the rep-etition of the FMAQ-FCA questionnaire, already carried out in phase 1. In this case, however, the questionnaire has been adapted to the aim of assessing the improve-ments and changes that the CRM generated both at the level of crew behaviour and safety attitudes of the airline.

From a methodological point of view there were no specific aspects that could be considered innovative or new with respect to the methods that had been applied during the previous phases of work. Consequently, the details of the questionnaire and analysis carried out in this phase will not be reported here, as they are the replicas of human factors approaches already discussed.

However, this process of assessing quality and effectiveness represents a critical step in improving safety performances, and it is well formalised in the stepwise procedure for developing training (Figure 6.2). It represents the last crucial stage in the DSME process of development of HEAM measures (§4.1).

Moreover, the sets of indicators and markers that can be developed from this analy-sis become essential elements for the evaluation of the safety levels. These safety levels should be revised by means of an audit at regular intervals and every time there is major change within the company/airline. Examples of such changes are: technology or organisational restructurings, due to mergers between companies; diversification in the fleet, due to the acquisitions of new routes or new types of aeroplanes; expansion of the crew population by hiring of new pilots of different nationalities and cultures (Helmreich and Merrit, 1998; Smallwood and Fraser, 1995).

In essence, the major activity carried out in phase 3 consisted of two main steps, namely:

Phase 3, Step 1. Development and implementation of the CRM-FCA (classroom sessions).

Step 2. Evaluation of quality and effectiveness of the CRM-FCA.

Development and Implementation of the CRM-FCA

The development of the training material consisted in a very productive activity that involved the HF Team mainly for quality testing and evaluation.

In particular, the "master" copies of all documentation and supporting elements (viewgraphs, videos, games, photos, and questionnaires) were developed by professionals of graphic and multimedia tools and were preserved for future improvements and repetition of the course.

Evaluation of Quality and Effectiveness of the CRM-FCA

In association with the airline, a specific approach for evaluating the effectiveness of the CRM-FCA training programme was developed.

This programme involved an evaluation of the attitudes of pilots pre- and posttraining. A questionnaire was developed on the basis of factors that were identified and tested in previous assessments of CRM or defined by regulatory authorities (Helmreich and Wilhem, 1991; Helmreich and Merrit, 1998; Bilton, 1997; FAA, 1997). In particular, the line/LOS checklist developed at the University of Texas (Helmreich and Merrit, 1998) was considered and further elaborated, as was done in the case of the FMAQ. The resulting questionnaire was utilised for data collection over a period of 8 months, during the courses based on the CRM-FCA that were carried out within the airline for normal nontechnical training.

This approach is the most consistent way to evaluate safety improvements and organisational changes.

The analysis of the data collected showed positive trends in performance changes that could be credited to specific sections or modules of the CRM. At the same time, it could be observed that the general climate in the airline was improving as the result of a series of actions promoted by the management, the CRM-FCA being one of them.

6.3 Summary: Application of HERMES to the Development of a CRM Course

6.3.1 Application of HERMES

This chapter has presented a practical application of the HERMES methodology (Human Error Risk Management or Engineering Systems) to the area of HF training. In particular, the development of a CRM course in the aviation domain was described.

The five standpoints to be retained in order to perform a sound HF approach (Chapter 2) have been respected:

1. The HERMES methodology was applied to the area of application "Training" (standpoint 4: area of application).

2. The goals and objectives of the course were set from the viewpoint of HEAM (standpoint 1), focused on training pilots to prevent and manage risky situations and emergencies.

3. A cognitive model of reference was selected and applied throughout the course (standpoint 2).

4. The logical interplay between retrospective and prospective approaches was maintained by ensuring that information and data derived from field assessments ("jump seat" flights), interviews, questionnaires, and studies of past events were

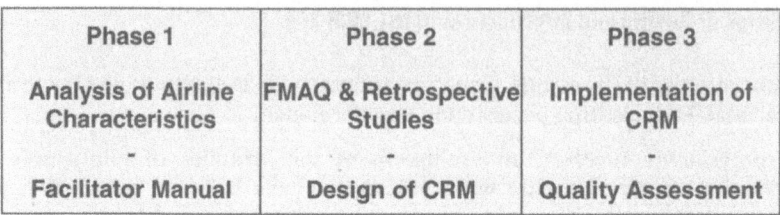

Figure 6.18 Phases of development of the CRM-FCA course.

taken into consideration in specifying the detail content of the CRM (standpoint 3).

5. The outcome of the analysis of the safety level and climate existing within the airline was utilised to define markers of safety levels, as well as indicators of areas of concern. These markers offered also a valuable reference point for measuring changes that may occur within the airline in terms of safety and operating standards (standpoint 5).

The methodology was applied in three phases (Figure 6.18). Phase 1 of the work focused initially on the evaluation of the socio-technical context existing within the airline and on the familiarisation with flying practices and habits. These ethnographic studies were complemented by the development of a questionnaire for the evaluation of pilot attitudes and safety climate. Finally, a theoretical model of pilot behaviour was selected to guide the development of the CRM and a facilitator manual was devised.

Phase 2 focused on the detailed evaluation of all data collected by field studies and retrieved from past events. This led to the detailed design of the CRM-FCA course. In this phase, it was also decided to implement a Computer-Based Training (CBT) in support to the more theoretical classroom sessions of the CRM.

Phase 3 consisted in the development of the actual training material and software for the CRM and CBT. After the full development of the CRM and CBT, a period of testing and evaluation took place to measure the quality and effectiveness of training.

6.3.2 Conclusions

The application of HERMES showed that the methodology is applicable in practice and can give valuable and significant results for developing training courses.

The stepwise procedure for developing a training process and the HERMES methodology have been followed in detail (Figures 6.1 and 6.2). Moreover, the general process for Definition, Selection and Implementation, Monitoring and Evaluation HEAM measures has been fully implemented, including the final step of the evaluation of the product.

Appendix 6.1: The FMAQ-FCA Questionnaire
Modified FAMQ-2.0 International for the Airline FCA

The success of the survey depends on your contribution, so it is important that you answer questions as honestly as you can. There are no right or wrong answers, and often the first answer that comes to mind is best. **Individual responses are absolutely confidential.**

Part I: FCA Pilot Views

This portion of the questionnaire is designed for you to express your view of your company. Please answer by writing beside each item a letter from the corresponding scale.

A	B	C	D	E
Very Low	Low	Adequate	High	Very High

<u>A.</u> Please evaluate your level of *satisfaction* with these different aspects of flight operations.

___ 1. Ground School

___ 2. Simulator-based training

___ 3. Line relevance of training material

___ 4. Quality of new-hire training

___ 5. Instructor skills

___ 6. Flight training overall

___ 7. Fairness of checking

___ 8. Flight manager/chief pilot availability

___ 9. Flight ops management

___10. Disciplinary policy and enforcement

___11. Pilot utilization and productivity

___12. Operations manuals

___13. Checklists

___14. Procedures manuals

<u>B.</u> Please describe your personal perception of the *quality of teamwork and cooperation* you have experienced with:

___1. Other cockpit crew members

___2. Gate agents

___3. Ramp personnel

___4. Flight attendants

___5. Dispatch

___6. Maintenance

___7. Crew scheduling

C. Please answer the following by writing your response beside each item using the following scale.

A	B	C	D	E
Disagree Strongly	Disagree Slightly	Neutral	Agree Slightly	Agree Strongly

____ 1. The managers in flight operations listen to us and care about our concerns.

____ 2. I know the proper channels to direct questions regarding safety practices.

____ 3. Check that airmen are respected role models in our airline.

____ 4. I feel comfortable going to the chief pilot's office to discuss problems or operational issues.

____ 5. Our training has prepared pilots and flight attendants to work as a well-coordinated team.

____ 6. My last line check was a positive learning experience.

____ 7. I am encouraged by my supervisors and coworkers to report any unsafe conditions I may observe.

____ 8. Working here is like being part of a large family.

____ 9. Flight management staff constructively deal with problem pilots.

____10. I am proud to work for this organization.

____11. Management will never compromise safety concerns for profitability.

____12. My last simulator instructor effectively debriefed the flight's human factors issues.

____13. My suggestions about safety would be acted upon if I expressed them to management.

____14. Check that airmen across all fleets hold pilots to the same standard.

____15. Our Instructors have a good understanding of line operations.

____16. Senior management at FCA is doing a good job.

____17. Cabin personnel should be included in the crew briefing at the start of a duty day.

____18. Crew members that I fly with comply with FCA SOPs.

____19. Pilot morale is high.

____20. Flight operations management fully supports my daily efforts on the line.

____21. I like my job.

____22. Pilots trust senior management at FCA.

____23. This airline has a positive safety culture.

____24. FCA practices the highest maintenance standards.

<u>D.</u> Please name your top three recommendations to improve FCA flight operations and training.

_____

_____

_____

Part II: Flight Management Attitudes

Please answer the following items by writing your response beside each item using the following scale

A	B	C	D	E
Disagree Strongly	Disagree Slightly	Neutral	Agree Slightly	Agree Strongly

____ 1. The captain should take physical control and fly the aircraft in emergency and nonstandard situations.

____ 2. Captains should encourage crew member questions during normal flight operations and in emergencies.

____ 3. Even when fatigued, I perform effectively during critical times in a flight.

____ 4. The airline's rules should not be broken – even when the employee thinks it is in the airline's best interests.

____ 5. I expect to be consulted on matters that affect the performance of my duties.

____ 6. Senior staff deserve extra benefits and privileges.

____ 7. I let other crew members know when my workload is becoming (or about to become) excessive.

____ 8. Captains who encourage suggestions from crew members are weak leaders.

____ 9. My decision-making ability is as good in emergencies as in routine flying conditions.

____10. Junior crew members should not question the captain's or senior crew members' decisions.

____11. It is better to agree with other crew members than to voice a different opinion.

____12. The captain's responsibilities include coordination between the cockpit and cabin crew.

____13. I am more likely to make judgment errors in an emergency.

____14. Successful flight deck management is primarily a function of the captain's flying proficiency.

____15. If I perceive a problem with the flight, I will speak up, regardless of who might be affected.

____16. I am ashamed when I make a mistake in front of the other crew members.

___17. In abnormal situations, I rely on my superiors to tell me what to do.

___18. Crew members should not question actions of the captain except when they threaten the safety of the flight.

___19. I am less effective when stressed or fatigued.

___20. My performance is not adversely affected by working with an inexperienced or less capable crewmember.

___21. To resolve conflicts, crew members should openly discuss their differences with each other.

___22. Crew members should monitor each other for signs of stress or fatigue.

___23. Personal problems can adversely affect my performance.

___24. A truly professional crew member can leave personal problems behind when flying.

___25. Except for total incapacitation of the captain, the first officer should never assume command of the aircraft.

___26. Written procedures are necessary for all in-flight situations.

___27. Crew members should mention their stress or physical problems to other crew before or during a flight.

___28. Good communication and crew coordination are as important as technical proficiency for flight safety.

___29. Effective crew coordination requires crew members to consider the personal work styles of other crew members.

___30. During periods of low work activity, I would rather relax than keep busy with small tasks.

___31. A true professional does not make mistakes.

___32. An essential captain duty is training first officers.

___33. How frequently, in your work environment, are subordinates afraid to express disagreement with their superiors?

1	2	3	4	5
Very frequently	Frequently	Sometimes	Seldom	Very seldom

___34. How often do you feel nervous or tense at work?

1	2	3	4	5
Always	Usually	Sometimes	Seldom	Never

Part III: Leadership Styles

Please read the following descriptions of four different leadership styles, and answer the questions that follow.

Style A Leaders usually makes decisions promptly and communicate them to subordinates clearly and firmly. Expect subordinates to carry out the decisions loyally and without raising difficulties.

Style B Leaders usually makes decisions promptly, but, before going ahead, try to explain them fully to subordinates. Gives reasons for the decisions and answers whatever questions there are.

Style C Leaders usually consult with subordinates before reaching decisions. Listen to their advice, consider it, and then announce decisions. Expect all to work loyally to implement the decision, whether or not it is in accordance with the advice given by the subordinates.

Style D Leaders usually call a meeting of subordinates when there is an important decision to be made. Puts the problem before the group and invites discussion. Accepts the majority viewpoint as the decision.

___1. Which one of the above styles of leadership would you *most prefer* to work under?

___2. In your organization, which style do you find yourself most often working under?

Part IV: Work Values and Goals

Please answer the items below by writing beside each item a letter from the scale below.

A	B	C	D	E
Of Very Little or No Importance	Of Little Importance	Of Moderate Importance	Very Important	Of Utmost Importance

Please think of your *ideal* job – disregarding your present job. In choosing an *ideal* job, how important would it be to you to:

___ 1. Maintain good interpersonal relationships with fellow workers or crew members?

___ 2. Have an opportunity for advancement to higher-level jobs?

___ 3. Have security of employment?

___ 4. Live in an area desirable to you and your family?

___ 5. Have a changing work routine with new, unfamiliar tasks?

___ 6. Have a warm relationship with your direct superior?

___ 7. Have an opportunity for high earnings?

___ 8. Have challenging tasks to do, from which you get a personal sense of accomplishment?

___ 9. Know everything about the job, to have no surprises?

___10. Have sufficient time left for your personal or family life?

___11. Work with people who cooperate well with one another?

___12. Find the truth, the correct answer, the one solution?

___13. Observe strict time limits for work projects?

Part V: Cockpit Automation

The following items deal with attitudes regarding flight deck automation. For the purposes of this survey, automated aircraft are defined as those with a programmable Flight Management Computer (FMC).

If you are currently flying an automated aircraft, base your answers on your experience with this aeroplane.

If you have not flown such an aeroplane, base your answers on your expectations regarding such type of aircraft.

Please answer by writing beside each item a letter from the scale below.

A	B	C	D	E
Disagree Strongly	Disagree Slightly	Neutral	Agree Slightly	Agree Strongly

____ 1. I prefer flying automated aircraft.

____ 2. Under abnormal conditions, I can rapidly access the information I need in the FMC.

____ 3. The effective crew member always uses the automation tools provided.

____ 4. I am concerned that the use of automation will cause me to lose flying skills.

____ 5. It's easy to forget how to do FMC operations that are not performed often.

____ 6. I look forward to more automation – the more the better.

____ 7. Pilots should avoid disengaging automated systems.

____ 8. There are modes and features of the FMC that I do not fully understand.

____ 9. Automated cockpits require more verbal communication between crew members.

____10. I regularly maintain flying proficiency by disengaging automation.

____11. Automated cockpits require more cross-checking of crew member actions.

____12. My company expects me to always use automation.

____13. I feel free to select the level of automation at any given time.

____14. Automated systems should be used at the crew's discretion.

____15. Flying highly automated aircraft alters the way crew members transfer information.

____16. I try to use automation as much as possible during flight operations.

____17. It is difficult to know with FMC what operations the other crew members are performing.

____18. In automated aircraft, sometimes, things happen that surprise pilots.

____19. A day of work, in similar operating conditions, is less tiring in automated aircraft.

____20. In automated aircraft, nonroutine operations require more time than in conventional aeroplanes.

___21. In automated aircraft, workload is not reduced as more monitoring and supervisory activities are required.

___22. In automated aircraft, more time is necessary to insert data in the FMC than what is needed to pilot manually or simply use the auto-pilot.

___23. Older pilots need more time than younger pilots to insert data in the FMC.

___24. I prefer the automated presentation of faults and the electronic checklist, as they facilitate solution of problems and improve safety in decision making.

___25. The computerised presentation of systems minimises the work of memorisation by the pilot.

___26. The increase of automation reduces the differences between captain and first officer.

___27. In automated aircraft the solution of failures and faults is easier.

___28. The automation ensures that working limits of aircrafts are never passed.

Part VI: Background Information

- Gender (M or F) _____
- Commuter _____
- Years at FCA _____
- Years in aviation _____
- Fleet (aircraft type and series) _____
- Experience in this aircraft (Y) _____

- Flying background (tick one) _____ Military _____ Civilian _____ Both
- Crew position: _____ Captain _____ First Officer _____ Second Officer
- Status: _____ Line Pilot _____ Instructor _____ Check Pilot _____ Management

Nationality _____

Experience of using a computer:

1	2	3	4	5
None	Small	Sufficient	High	Professional

7

Application of HERMES for Safety Assessment: The Safety Audit of a Large Railway Organisation

This chapter will consider the application of the HERMES (Human Error Risk Management for Engineering Systems) methodology for Safety Assessment studies.

The HERMES methodology, as already stated earlier, represents the framework for applying consistently and fruitfully a set of techniques and data so as to integrate retrospective and prospective type analyses for safety purposes, with particular reference to human factors. The quality and soundness of the results depend on the accuracy with which methods are applied and on the reliability of the data utilised. Lacking these two conditions, the outcome may be very misleading and unrealistic.

As in the previous cases, the application that will be discussed concerns a real case and the reader must consider that the overall process, presented in this book in a condensed form, required in practice a consistent period of study and a number of thorough and dedicated analyses in order to achieve its objectives.

The stepwise and methodological guidelines for developing a safety assessment within the HERMES framework, discussed in Chapter 4, will be followed in detail.

The application of HERMES for safety assessment considers three possible forms of analysis of human factors–related issues:

1. Design Basis Accident (DBA), for deterministic assessment of maximum credible accidents;
2. Quantitative Risk Assessment (QRA), for the definition of frequencies of occurrence of unwanted events, in reliability and risk assessment terms; and
3. Recurrent Safety Audits (RSA), for the evaluation of safety levels of an organisation and identification of possible areas of concern.

This chapter focuses on the application of HERMES for the development of a RSA approach aimed at studying the safety level of a large organisation in the domain of railway transportation systems.

As for the HERMES applications discussed in previous chapters, the data reported here and a number of methods and models have been slightly modified from the real case in order to preserve confidentiality. However, the essence of the study has not been changed and the propedeutic characteristics of the exercise remain unchanged and valid.

7.1 Problem Statement

The top management of a large European Railway Company, which will be called ERC for convenience, felt the need to carry out a study of the human factors and safety-related issues existing within the organisation, with particular attention to the population of Train Drivers (TD). The objective of such study was the identification of most relevant areas of intervention for improving safety and reliability of the service, and in general to ascertain the current state and possible needs for human error and accident management (HEAM) measures.

The existence of a distributed HF problem at different levels within the organisation and the need to carry out an audit on the safety state of ERC was also enhanced by the occurrence of a major accident that resulted in several injuries and the death of a number of passengers and ERC staff.

As discussed in Chapter 2 of this book, the need to develop preventive actions of safety is very often the reaction to an accident and the resulting "proactive" measures, if well developed and designed, enable the consideration of a wider variety of issues and offer real improvement of safety that goes beyond the development of countermeasures solely focused on the causes and factors associated to the specific accident.

In this case, the ERC top management realised the need to expand the scope of the reaction measures resulted from the accident investigation, and demanded the support of external analysts and specialists in human factors for the overall assessment of the safety level of the organisation and for the identification of human factors–related problems affecting this level of safety.

This is a typical case for the development of an audit concerning a large organisation such as a railway company that offers an important and reliable service to the community and is perceived and expected to be as absolutely safe by the population at large. When an accident occurs, then the entire system is put under scrutiny and, if the safety levels have not been adequately monitored and kept, the audit identifies where problems lie and can, in certain cases, give hints on how to solve them.

The railway company presented a variety of problems and issues that had been developing over the years and were creeping through the organisation even before the accident. Moreover, the evolution and enormous growth of the service that occurred over the last years and the massive utilisation of automation in the design of modern train cabins further contributed to aggravating the complexity of problems.

In such a scenario, i.e., the enormous dimension of the organisation to be assessed and the complexity of the problems existing within the organisation, the HERMES methodology turned out to be the most appropriate approach to apply for a Safety Audit, as it offers a wide variety of methods and models amongst which to choose the most appropriate ones for solving the variety of problems presented during the audit, while paying attention to the integration and balance between the different approaches.

7.2 Application of HERMES for the RSA of the Railway Company ERC

The HERMES methodology was applied for auditing the organisation, as well as offering the ERC a set of safety indicators and markers that would provide the organisation with a structured format for regularly checking its safety state and identifying areas of concern. According to the RSA approach, this is a typical route for performing preventive safety assessment.

The dimension of the ERC organisation consisted of more than 100,000 employees, engaged every day in the management of railway traffic, with a population of more than 70% train drivers. The technology and means available at the ERC consist in a high number of "trains" presenting a wide variety of technological solutions on board, from the most modern fully automatic controlled, high-speed machines, to very conventional locomotives, based practically on full manual control. At the time of the study, the railway network covered almost 20,000 kilometres over a vast territory and a variety of terrains. The ERC also managed the underground systems of some metropolitan areas. Approximately 10,000 trains per day ensured the movement of several hundreds of thousands of passengers.

As a consequence of the complexity and dimension of the ERC organisation, the application of HERMES for the Safety Audit alone covered an exercise of several months of work by a team of Human Factors experts (HF Team).

The problem of ERC was mostly that of identifying areas of concern for the safety performance of the rail transport service. Therefore, the application of HERMES focused on the identification of safety critical factors, or Indicators of Safety (*IoS*), and Recurrent Safety Audit Matrices (*RSA-Matrices*) that serve the purpose of ascertaining the existing level of safety within the organisation and identifying the reference measures for future audits (Figure 7.1).

The study ended with a preliminary Safety Audit of the organisation, and no further analysis, such as the identification of *training* needs, was carried out. The performance of a formal *Recurrent Safety Audit*, to be performed some time after the preliminary Safety Audit, did not take place within the boundary of the study presented here.

The work was performed in three correlated phases, which aimed at transferring the maximum possible experience and insight about the company to the HF

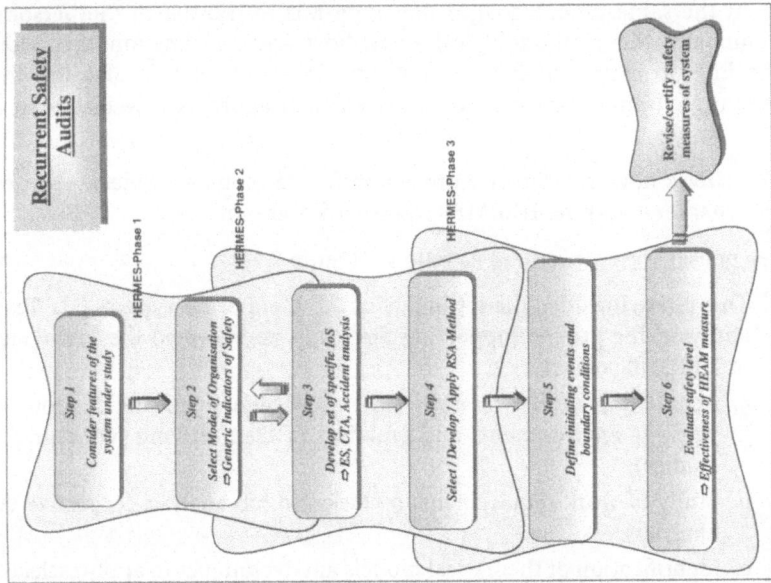

Figure 7.2 Stepwise Procedure for developing Recurrent Safety Audit and HERMES phases.

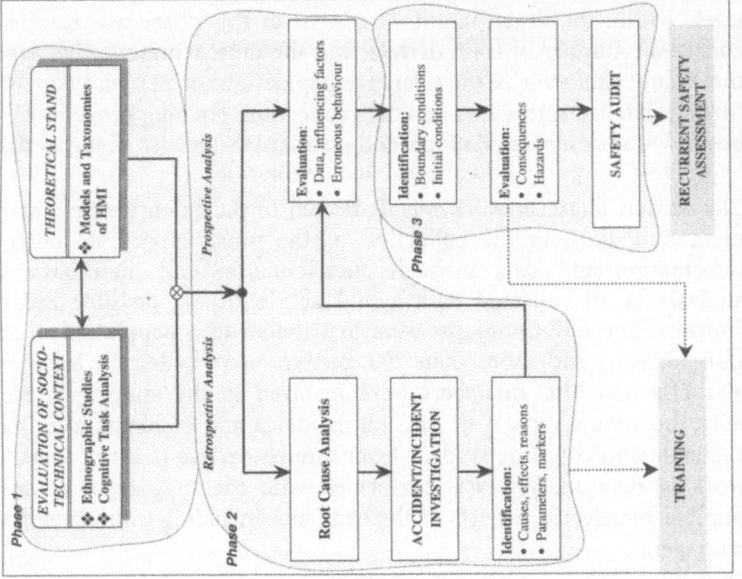

Figure 7.1 HERMES application for the development of the ERC Safety Audit.

analysts. At the same time, a goal of the study was to transfer to the personnel of the company the methodological know-how and information that would allow the ERC to carry out future audit exercises and safety evaluation from within the organisation, which is the most efficient approach (Cacciabue et al., 1999).

Figure 7.2 shows how the stepwise procedure for developing a Safety Audit discussed in Chapter 4 and the HERMES framework were integrated.

The three phases were structured as follows (Figure 7.1):

Phase 1. This phase included the setting up of the Team of HF experts (HF Team) and working groups supporting the study, and covered the initial steps of HERMES, namely:

a. Acquisition of information and knowledge in the field about the working environments and practices of train driving (ethnographic studies).

b. Study of work requirements by task and job analysis (cognitive task analysis).

c. Identification of theoretical models and techniques to apply (selection of models and taxonomies of HMI).

This activity involved some staff members and managers at different levels within the organisation. The focus of this phase was to select a consistent number of train drivers, and the most representative depots, that would characterize the reference populations of TDs and depots for further detailed interviews and data collection. Training procedures and books with norms, regulations, and standards were also made available by the ERC, as part of theoretical documentation.

Phase 2. The second phase of work was dedicated to the extensive field assessment and, thus, to the collection of the most important source of information and data through questionnaires and interviews. The analysis of all collected data aimed at identifying possible areas of concern. The questionnaires were distributed to a population of 2500 train drivers, and more than 700 answers were collected. More than 300 TDs and ERC managers were involved in the interviews. Moreover, the annual reports of ERC on incidents and accidents were made available and could be studied from a retrospective point of view. The work of data analysis was performed with the support of a limited number of selected experts of the company, including train drivers and managers.

Phase 3. The third phase was totally dedicated to the core development of the safety audit and preparation of recommendations. The identification of safety indicators and matrices, as well as the recommendations on safety improvements, were performed with reference to the results of phase 2 and by exploiting the experience and creativity of the analysts.

7.2.1 Phase 1: Evaluation of Sociotechnical Context of the ERC and Theoretical Stand

Phase 1 covered the setting up of working teams, the collection and analysis of theoretical and background material, and the initial field work of acquisition of experience with a real working environment. This process led to the definition and selection of an appropriate theoretical stand for the whole study.

Phase 1 extended for a period of a few months of work and consisted in five main steps, namely:

Phase 1, Step 1. Creation of a human factors team and steering committee.

Step 2. Preliminary selection of model and taxonomy.

Step 3. Preparation of ethnographic studies.

Step 4. Task analysis and performance of initial ethnographic studies.

Step 5. Adjustment and validation of model and taxonomy.

Creation of a Human Factors Team and Steering Committee

The initial step in *phase 1* was the definition and identification of a team of experts (HF Team) that would contribute to the study and of a steering committee made up of members of the management of the ERC, as well as of staff and train drivers.

The HF Team of experts was made of two engineers, with experience in various areas of transport, two psychologists and one sociologist, all with experience in field observation and data collection and analysis, and a computer specialist, with experience in databases and data analysis. All members of the HF Team were experts in Human Factors and familiar with the HERMES approach or were adequately trained in the application of the HERMES methodology.

The creation of a steering committee was of critical importance in establishing reciprocal appreciation and trust between the HF Team and the ERC management and train drivers. In this way, it was possible to ensure that analysts of the HF Team could access and examine real working conditions, everyday problems, and critical information about safety issues. On the other hand, the company felt confident that the confidentiality required in this type of analyses was granted and preserved.

The steering committee was initially asked to develop for the Human Factors Team a set of very general descriptions and indications on the work of a TD and on the major issues associated to the existing working contexts.

In addition, the steering committee made available all supporting material, such as training manuals, books on rules and regulations, and reports on accidents and statistics that had occurred over a number of years within the company.

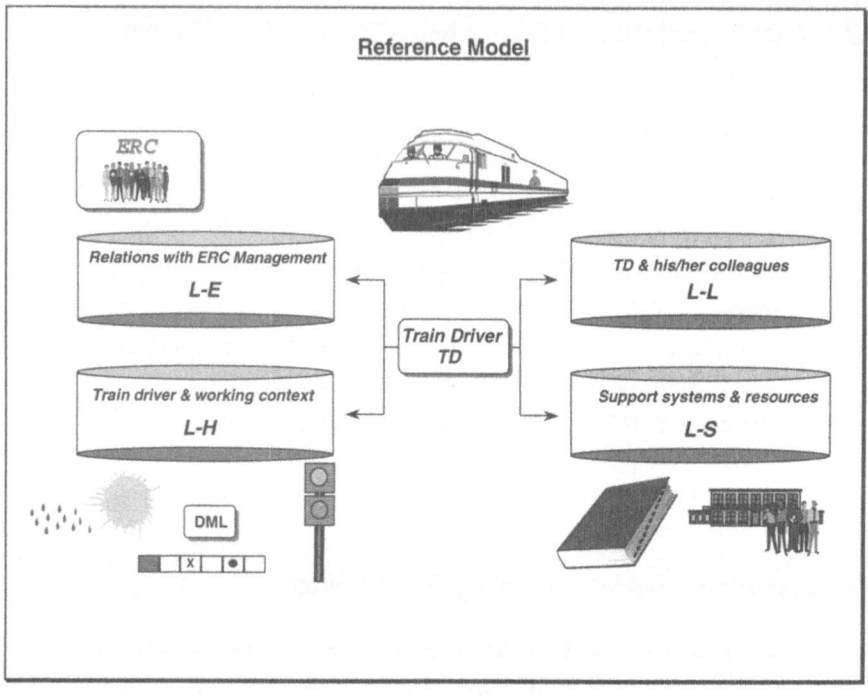

Figure 7.3 Modified SHELL model utilised for the interviews.

Preliminary Selection of Model and Taxonomy

Having established mutual confidence and a collaboration spirit between members of the steering committee and the HF Team of analysts, the second step aimed at selecting an appropriate model and taxonomy for performing the study.

The group of safety analysts started by examining the guides that were delivered by the steering committee and by performing some preliminary interviews with a number of train drivers, as well as with managers of the ERC. In addition to the steering committee, a "test group" of TDs was also established and started cooperating with the HF Team. The "test group" consisted of 15 Train Drivers selected by the management of the company.

The interaction with the steering committee members, with managers of the ERC and the unstructured interviews carried out with the 15 TDs of the test group, led the analysts to select a reference model for the analysis.

The model SHELL (§3.1) was selected as a reference for the analysis of the behaviour of train drivers and their working context and environment because of three major considerations (Figure 7.3):

1. SHELL is the reference model adopted in other domains strongly affected by human factor issues, such as aviation, for the classification of accidents and

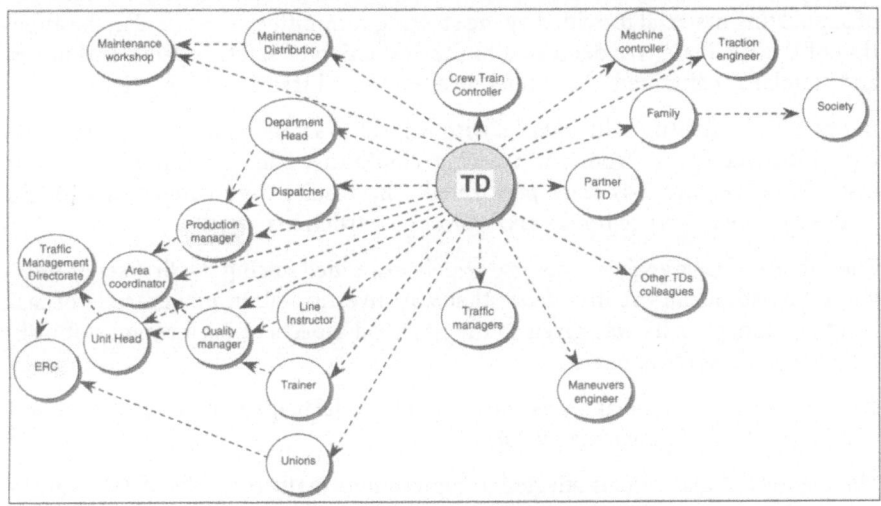

Figure 7.4 TDs socio-technical relations chart.

incidents. For example, the data classification ADREP 2000 of the International Civil Aviation Organisations (ICAO) is based on SHELL.

2. The SHELL model is a consolidated framework of reference, developed in the early seventies, which has been validated and widely applied in other real working context studies.

3. The SHELL model was known to the experts of the ERC and had already been used in the past within the ERC for similar studies.

Preparation of Ethnographic Studies

The relationship of TDs with other ERC personnel and staff was further elaborated in a chart of relationships, which was widely applied during the interviews (Figure 7.4). Moreover, the results of the first interviews and workshop helped in:

- Identifying an initial set of important factors that affect performance and behaviour of TDs. These were called *Performance Influencing Factors* (PIFs).
- Preparing the guidelines for the interviews based on the SHELL model. It was also decided to terminate each interview with a short questionnaire, aimed at collecting written feedback on the PIFs and on a number of special issues, identified by the management of ERC as important points to be analysed. The questionnaire is included in Appendix 1 of this chapter.

Task Analysis and Performance of Initial Ethnographic Studies

The following step of development focused on the definition of the *tasks* of a Train Driver. This activity required the combination of filed observation with the study of supporting material provided by the steering committee. In particular, the attention of the HF Team was dedicated to the evaluation of safety aspects, and norms and standards contained in the reference books and training instructions.

In practice, to identify critical tasks with respect to safety, a number of interviews with ERC managers (from the steering committee and test group) and "line instructors" (Figure 7.4) were performed. The results were compared with the analysis of the reference books and training instructions.

The subject of task analysis was also widely examined during the first set of interviews carried out in the first depot that was investigated. In this exercise of task identification, priority was given to the practical aspects of performance of tasks rather than theoretical ones.

The combination of task analysis and a SHELL model helped in refining the important *Performance Influencing Factors.*

The theoretical instrument selected to represent formally the tasks of TDs was the so-called *Hierarchical Task Analysis* (HTA) (Kirwan and Ainsworth, 1992). HTA describes in a simple treelike format the sequence of tasks to be carried out by TDs. It starts by defining the high-level tasks (such as "preparation of the train," "train driving") and arrives at detailed specific action level (such as "opening of doors on arrival at a station").

A number of interconnected HTAs were developed for the crucial and most important tasks of TDs, from top-level HTA to detail HTA for specific tasks, namely:

- Top-level HTA "main TD mission," consisting of collecting a train from a depot, performing the whole mission, and delivering the train for a subsequent mission (Figure 7.5).
- HTA for "main TD mission," with specific attention to detailed levels 1 and 2. Tasks: "Register" and "prepare locomotive/train" (Figure 7.6).
- HTA for "main TD mission," focusing on the critical part of the job, namely, level 3. Task: "Drive train," subtasks 3.1–3.5 (Figure 7.7).
- HTA for "main TD mission," specific for the detailed level 3. Tasks: "Drive train," subtasks 3.6–3.9 (Figure 7.8).

Adjustment and Validation of Model and Taxonomy

The friendliness of the SHELL model and its applicability to the TDs of the ERC was initially tested in a workshop with the "test group," i.e., the group of 15 Train Drivers selected by the Management.

From this first workshop, the definitions of the four areas of the SHELL model were organised in a slightly different form than the modelling structure. In this way, it was expected that the interaction with the TDs, planned in the following steps, would have been more efficient and the interviews would have been more fruitful (Figure 7.3).

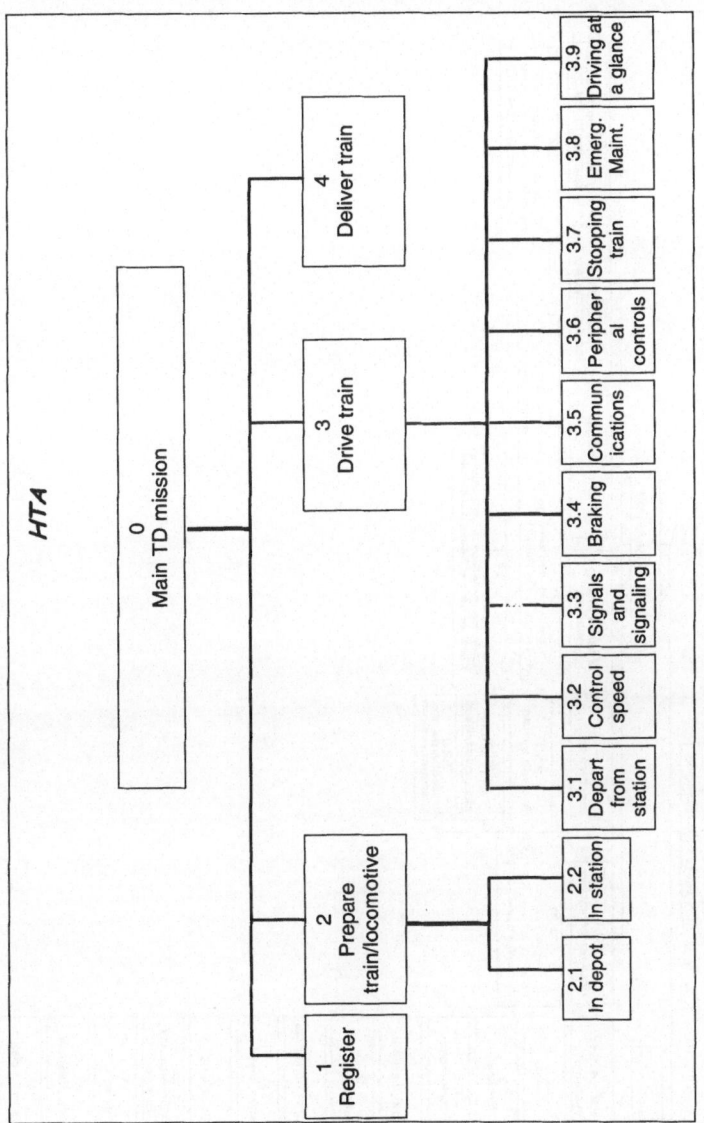

Figure 7.5 Hierarchical task analysis – Top level. Task: Drive train safely and efficiently from depot to station.

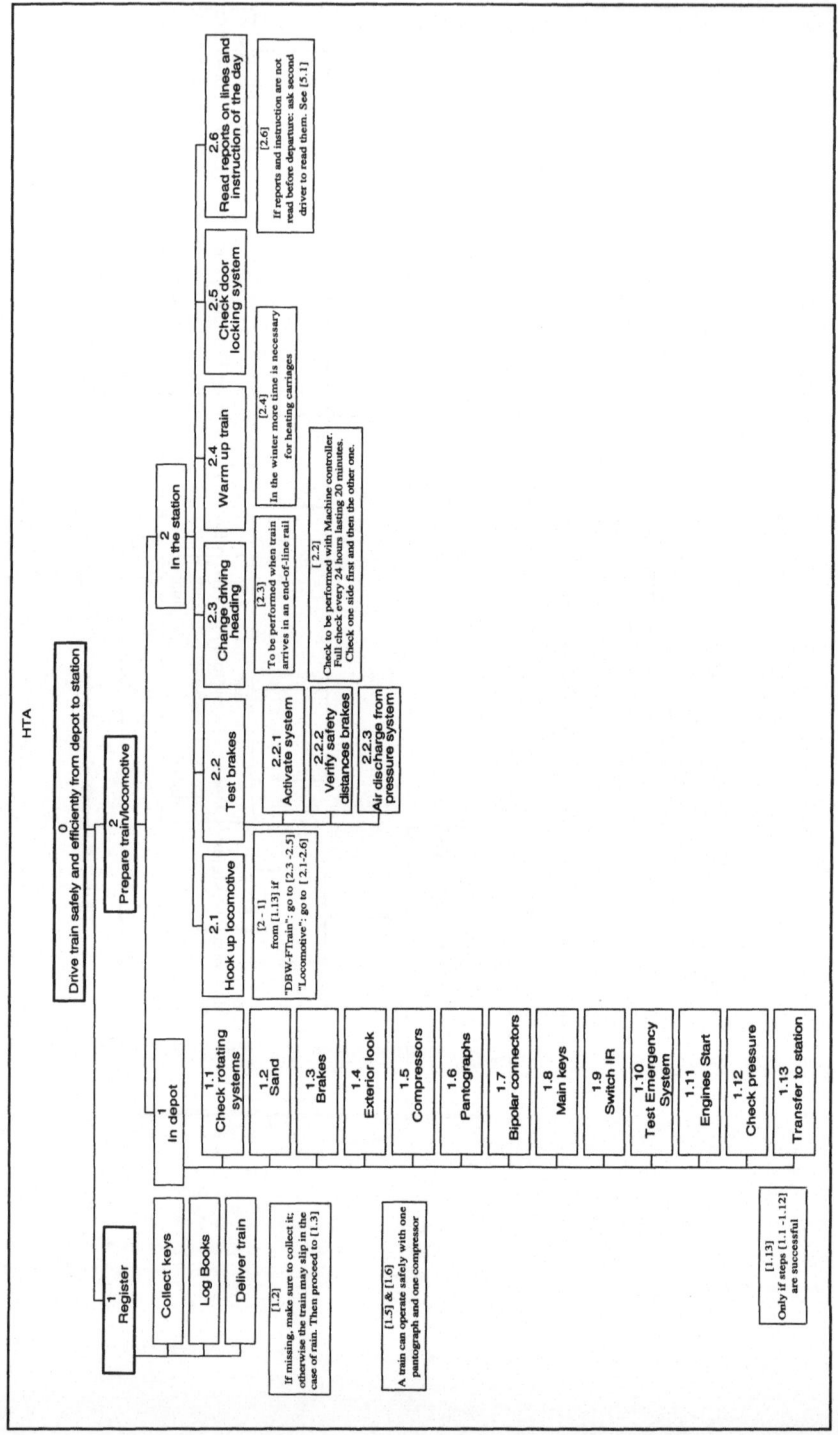

Figure 7.6 HTA – Detailed levels 1 and 2. Tasks: "Register" and "prepare locomotive/train."

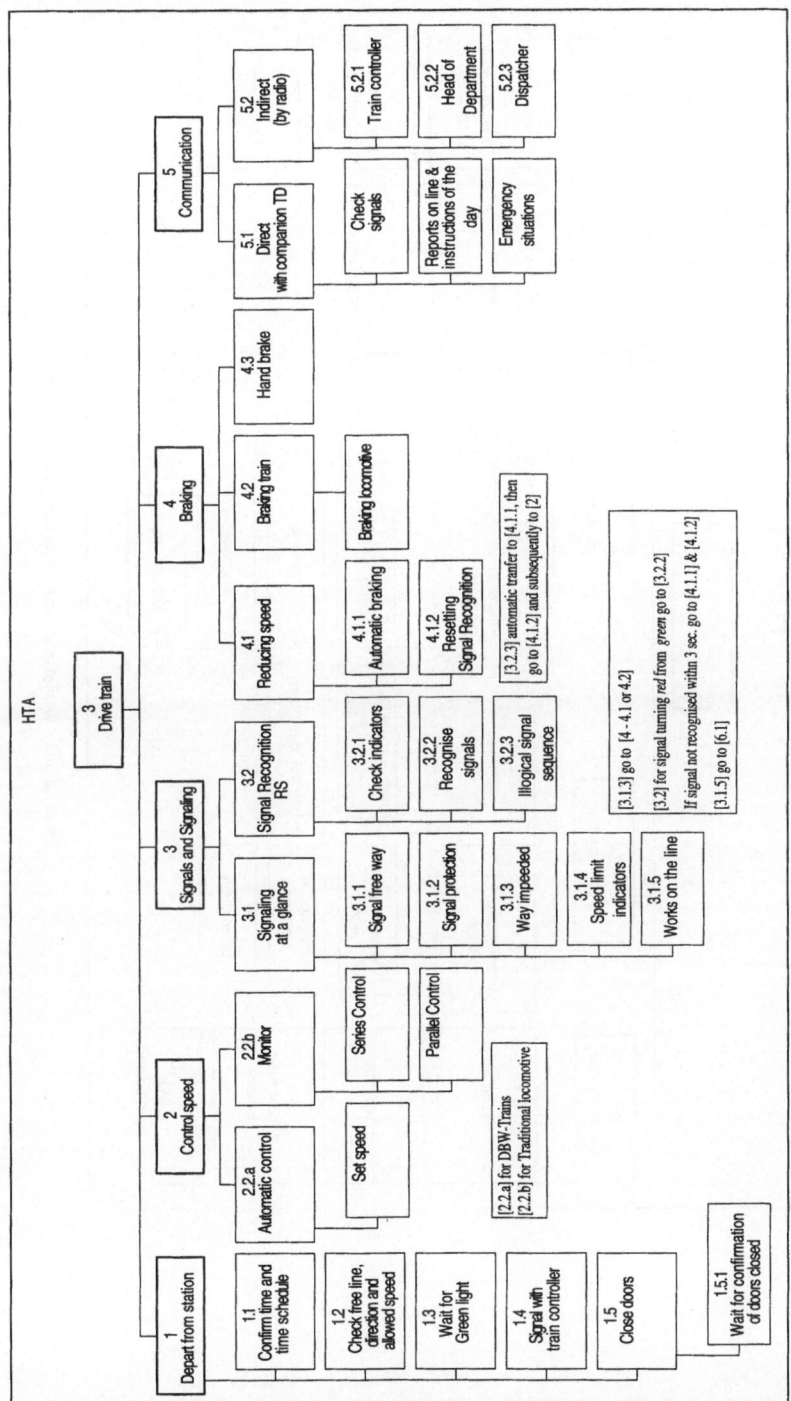

Figure 7.7 HTA – Detailed level 3. Tasks:"Drive train," subtasks 3.1–3.5.

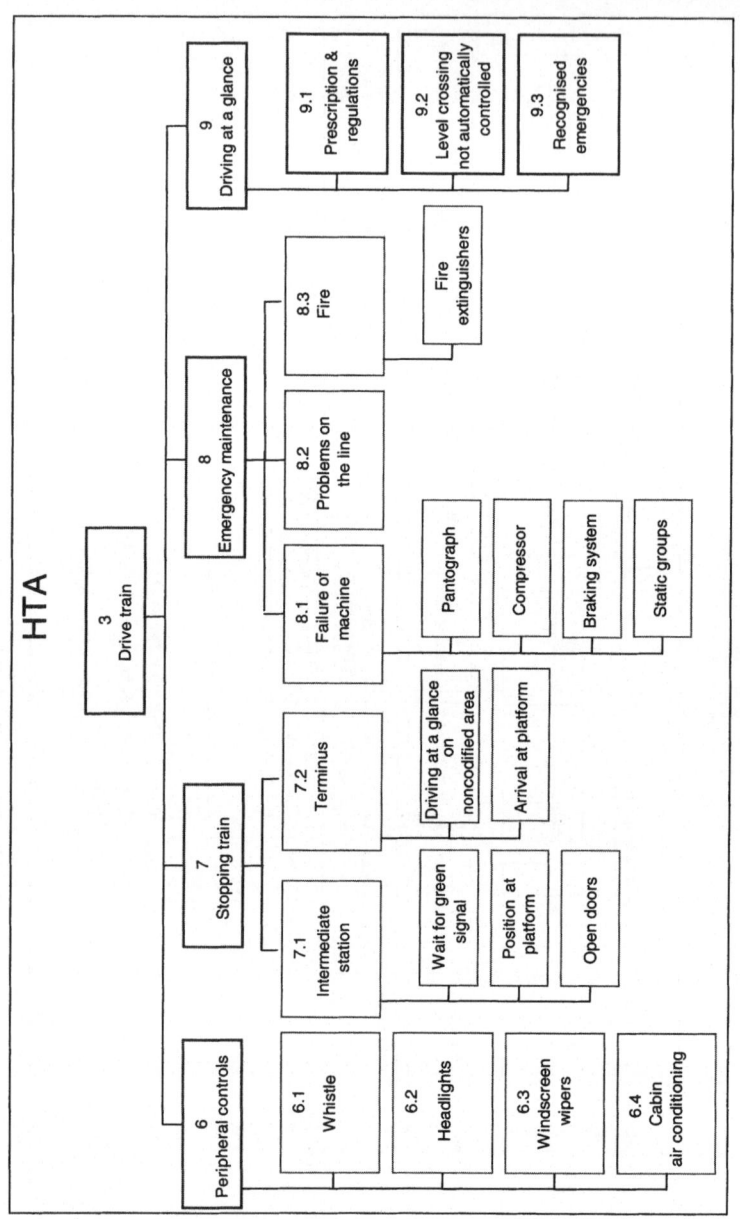

Figure 7.8 HTA – Detailed level 3. Tasks: "Drive train," subtasks 3.6–3.9.

This model, modified according to the new structure, was well accepted by all TDs of all depots where the interviews were held and served as the communication instrument between interviewers and interviewees.

7.2.2 Phase 2: Detailed Field Studies Within the ERC and Identification of Data, Causes, and Parameters

This phase of work was also structured in a number of steps. In principle, it consisted of a number of detailed field studies and data collection, and of the analysis of the collected data and information for the development of safety critical parameters, specific to the ERC organisation.

As in the previous phase, the collaboration between the HF Team and the experts of the ERC, i.e., the steering committee and the test group defined during phase 1, was of paramount importance for the successful application of the methodology.

The activity of phase 2 required the application of well-established tools and methods, such as software specifically dedicated to statistical analysis. It engaged heavily the HF Team for approximately 8 months of work before delivering the reports associated with the data analysis and the review of the applied methods and techniques.

Phase 2 consisted in four main steps. These will be discussed in detail in the following subsections:

Phase 2, Step 1. Cab rides and time line analysis.

Step 2. Detailed interviews in distributed depots.

Step 3. Definition of error analysis method.

Step 4. Analysis of data collected from questionnaires and interviews.

Cab Rides and Time Line Analysis

The cab rides were carried out immediately before and after the first set of interviews and workshop (phase 1–step 3). They helped the HF Team in refining the Task Analysis method and in preparing the study of the workload associated with the performance of the TD's task, especially with respect to available times for performing the required duties.

The theoretical instrument selected to represent graphically the time distribution of the work of TDs and of the associated workload was the so-called *Time Line Analysis* (TLA) (Kirwan and Ainsworth, 1992). TLA represented, in a simple format, the sequence of actions carried out by TDs during the performance of tasks and can be utilised to compare the workload of the TDs present in the cabin and working together.

An important aspect of the strategic work organisation of the ERC company was that two TDs are always present and collaborating on the conduction of a train. This *couple* of TDs represented a crucial team configuration in the study of safety critical factors and consequently demanded special attention. In particular, the analysis of the workload of the two collaborating TDs and their teamwork was studied in detail, including the issue of the "fixed vs. variable" couple.

The ERC utilised a "fixed couple" strategy, which foresees that a team of two train drivers always operates together, under the assumption that this is optimal for the service and safety. This strategy is the opposite to that being applied in other domains, such as aviation, where it is very rare, if not impossible, in a large company, that two pilots are assigned to the same "roster" over long periods of time (Davis and Stanton, 1988; Orasanu, 1993; Orasanu and Salas, 1993).

The time distribution of the actions and verifications carried out in three different configurations were investigated:

1. A standard working period of 9 hours, i.e., a standard shift on an advanced machine ("Drive-By-Wire–Fast Train," DBW-FT) on a track noninstrumented for high speed.
2. A period of two hours of work on an advanced machine (DBW-FT) on a track instrumented for high speed.
3. A period of 7 hours of work (traditional train).
4. A period of two hours of work on a nonautomated train with frequent stops.

For each of these periods the workload associated to each driver was estimated from (a) the analysis of the HTA (phase 1–step 4); (b) the results of the interviews with the TDs according to phase 2–step 2 that will be described in the following section; and (c) from the expert opinion elicited from managers and "line instructors" (phase 1–step 4).

The final versions of the TLAs are shown in Figure 7.9 through 7.12. In each figure, the time period of work is associated with the corresponding task, identified during the HTA carried out in phase 1 (step 4) of the study.

Detailed Interviews in Distributed Depots

This step of the study was fully dedicated to the performance of many interviews and the collection of questionnaires data.

Three main depots had been identified by the steering committee as the most representative of the organisation, and consequently extensive data collection in these depots was carried out for some months of work. In each depot, at the end of this phase of work, a workshop was organised with the TDs. This means that the workshops were held following the interviews, and the distribution, collection, and analysis of the questionnaires, developed in the previous phase of the work (phase 1–step 3).

The objective of these workshops was to further evaluate and refine the tables of Performance Influencing Factors (PIFs) already devised during the previous activity and to start discussing the issue of human error, also in connection with a preliminary evaluation of the results of the first interviews.

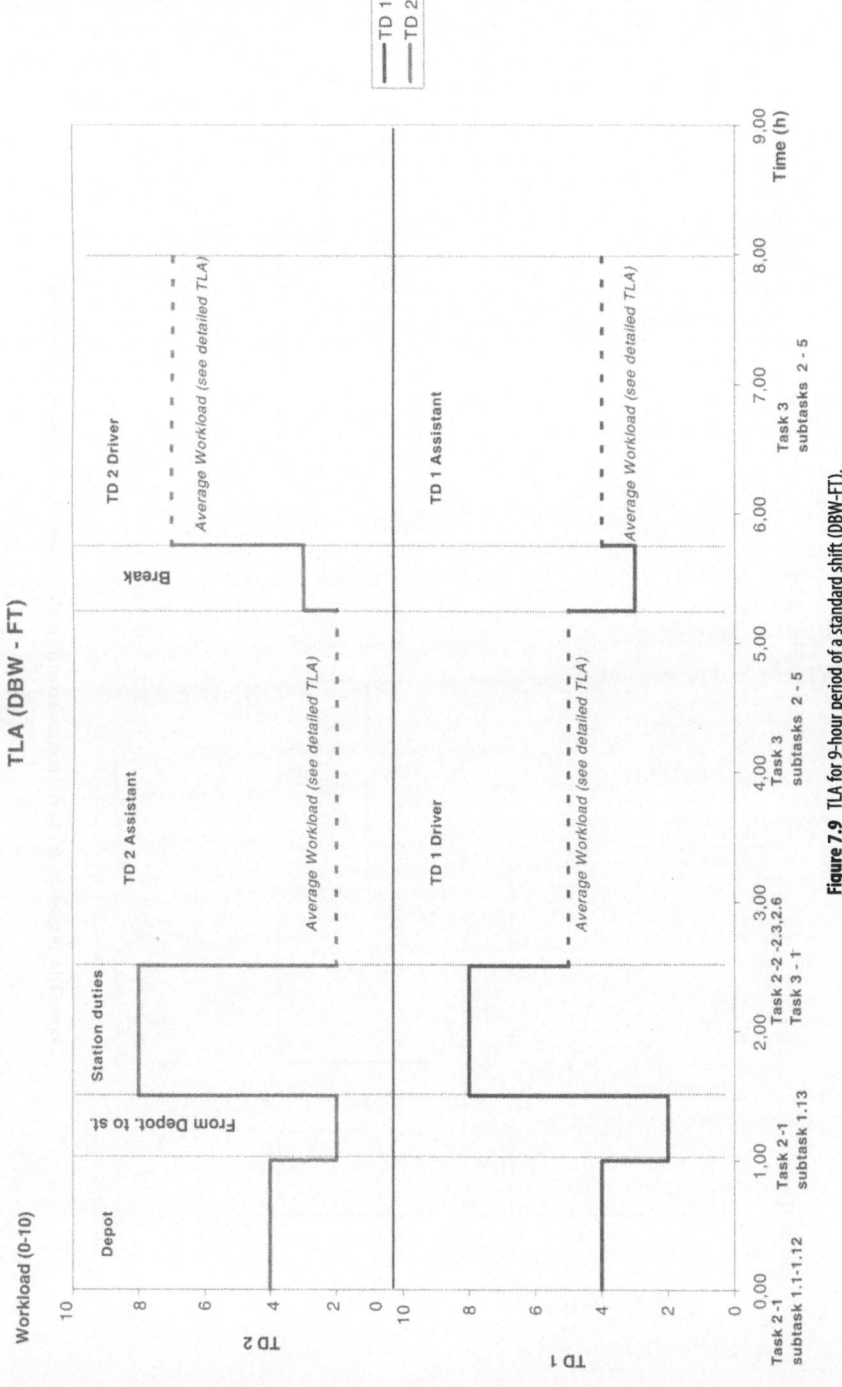

Figure 7.9 TLA for 9-hour period of a standard shift (DBW-FT).

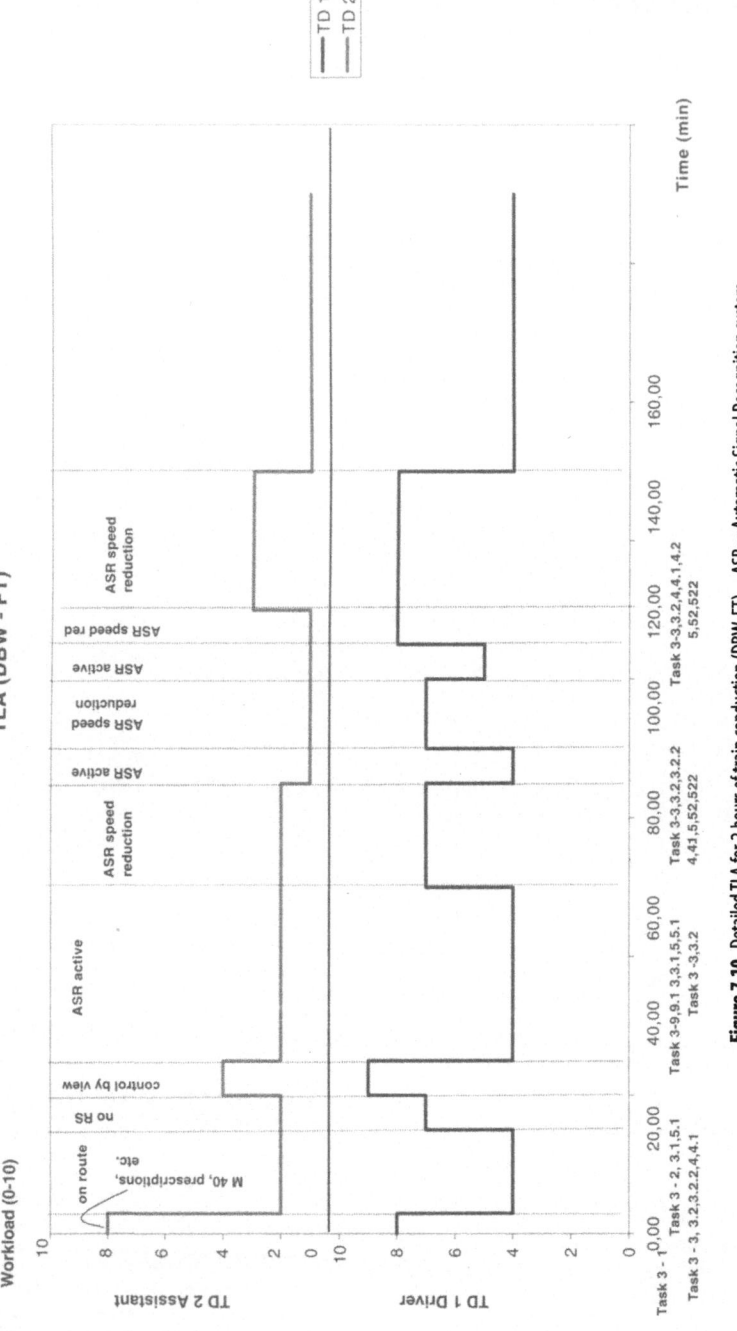

Figure 7.10 Detailed TLA for 2 hours of train conduction (DBW-FT) — ASR = Automatic Signal Recognition system.

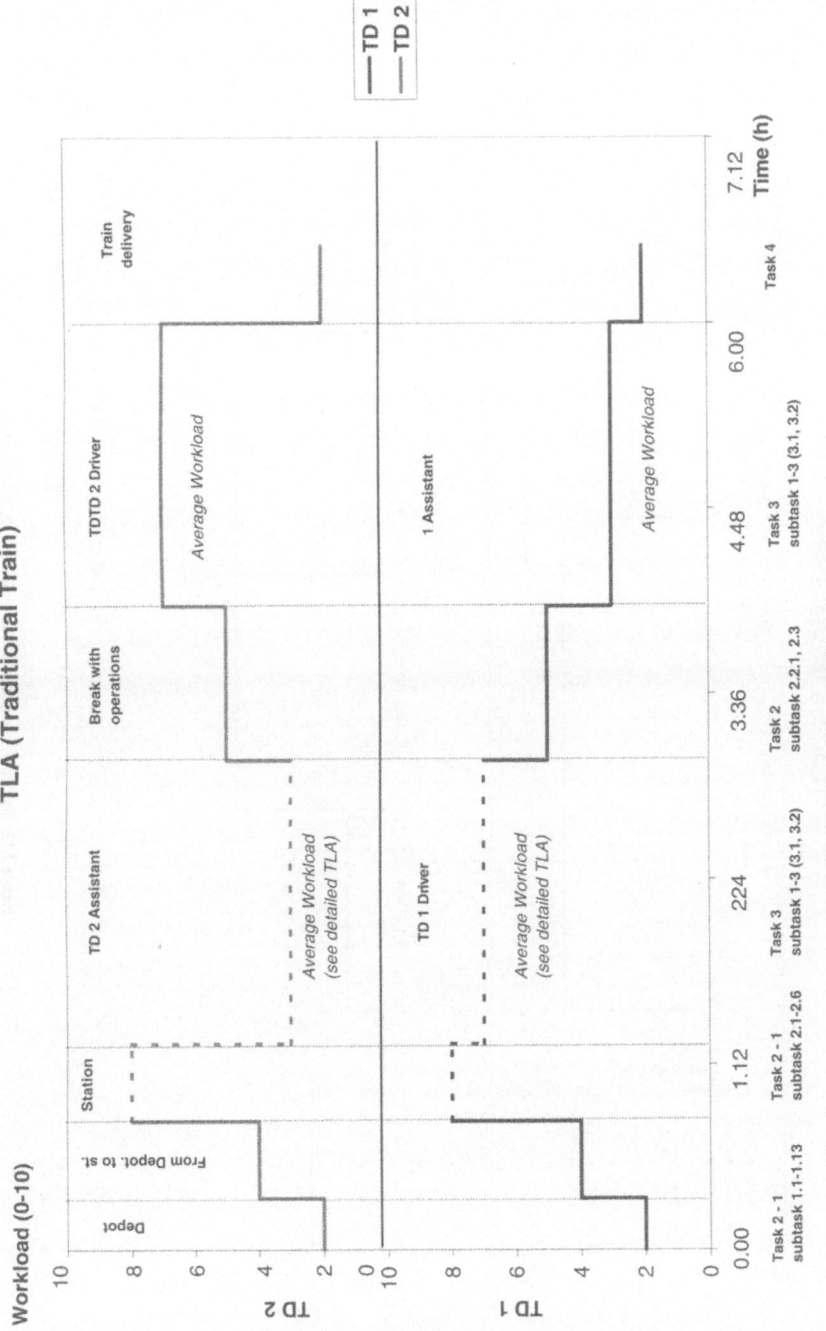

Figure 7.11 TLA for 7 hours of train conduction (traditional train).

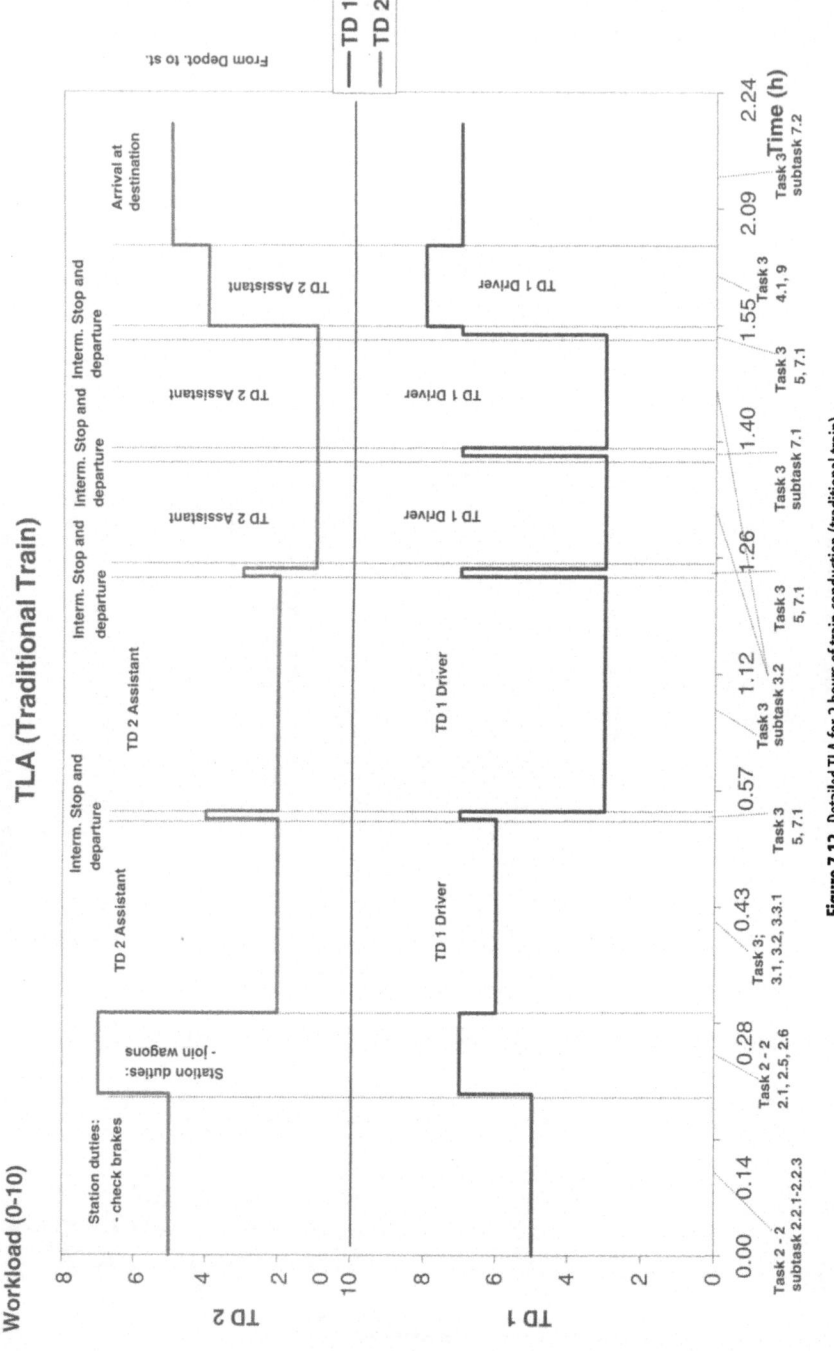

Figure 7.12 Detailed TLA for 2 hours of train conduction (traditional train).

The interviews were not structured and focused therefore on the ability of TDs to speculate on possible accident sequences and consequences of human errors, and on their likely causes. The reference model based on the SHELL paradigm (Figure 7.3) was discussed with all interviewees, covering all aspects of the TD activity, namely: the interfaces with the management and the other services of the ERC; the problems associated with the maintenance of trains and railway; the interaction with dispatchers and all other ERC personnel involved in the management of the every day activity; the ergonomics of the cabin and commands; and all other aspects of the TD activity.

More than 300 TDs were interviewed.

The questionnaire was distributed to the entire population of TDs of the three depots, i.e., approximately 2500 TDs. Altogether, 710 questionnaires were collected and the data were stored in a database.

The overall aim of the workshops and data collection was the identification of safety critical issues, i.e., the *IoS* that would finally be included in the RSA-Matrices of the Safety Audit.

Definition of Error Analysis Method

The selection of a technique for representing error generation and management during the activity of a TD has been strongly influenced by the fact that the HF Team aimed at selecting a simple theoretical configuration of human–machine interaction.

The goal was to favour the discussion with TDs for the identification of PIFs in association with possible *errors types* and *errors modes* that would be utilised in the perspective analysis to be carried out during phase 3 of application of HERMES (Figure 7.1 and Figure 7.2).

With these objectives in mind, the theoretical framework denominated THERP (Technique for Human Error Rate Prediction) (Swain and Guttmann, 1983) was selected. THERP is normally utilised in Human Reliability Assessment studies to calculate probabilities of human errors associated with the performance of tasks. However, in this case, only the basic theoretical structure and graphical representation of the method was utilised, leaving the quantification part, which is particularly critical and not-always satisfactory.

In practice, the "human error" structure of THERP ("THERP-Tree") was considered. This is based on the development of binary alternatives at each selected action point during the development of a task. Each branch generated at selected action points is then followed until either the task is successfully or unsuccessfully completed, or a recovery point is reached and the specific branch is combined with another one.

In order to assess whether this method fulfilled the objectives of the HF Team, and before applying the proposed approach for prospective analyses, a set of

THERP-Trees were developed and discussed with a number of TDs during detailed interviews.

Analysis of Data Collected from Questionnaires and Interviews

The data collected during the interviews and by means of the questionnaires that were distributed to the TDs of three different depots have contributed substantially to the development of the final list of areas of concerns with respect to safety and to the definition of the PIFs. In particular, the PSFs are the constituents of the Indicators of Safety (*IoS*) and of the *RSA-Matrices* necessary for performing reliable and coherent safety audits.

The interviews that were carried out on the basis of unstructured formats, became progressively more focused on specific aspects of TDs activities, as the HF Team became more and more experienced and familiar with certain problems of the ERC. Therefore, for this reason the last series of interviews turns out to be very different from the first ones.

Moreover, most questions and information collected during the interviews were based on "open questions" and therefore a formal analysis of the answers was more qualitative than quantitative.

Similarly, the data collected with the questionnaires were stored in a database and graphically represented on dedicated plots (Appendix 7.2). However, they were not analysed from a quantitative statistical perspective to evaluate importance, correlations, and validity of the independent variables. They were studied only from the qualitative viewpoint in conjunction with the outcome of the rest of the observations and analyses carried out by the HF Team, i.e., the interviews and direct study of the working contexts, the evaluation of past events, and the performance of task analyses.

Focusing on the content of the questionnaire and on the results of the data collected, a number of considerations could be made (Figure 7.13):

- Questions Q1–Q4 analysed the aspects of training and rules and regulations, which, in SHELL terms, represent the relationship between the TD and the "software" (L-S) (Figure 7.3). The need of more focused training and clearer regulations, particularly concerning highly automated systems, was quite clearly identified, especially amongst young TDs.
- Questions Q8–Q10 concerned the socio-technical interaction of TDs with the management and unions, and thus the L-E correlation in the SHELL format selected for the study (Figure 7.3). The results show that there is high variability within the organisation about the attitudes of the management and the unions with respect to the TDs. This showed that a problem of communication existed within the whole organisation, which needed further investigation and clarification.
- Questions Q12–Q13 concerned the practical aspects of the working contexts, including all physical environments and ergonomics conditions affecting the activity of TDs, inside as well as outside the train cabin. In SHELL terms this

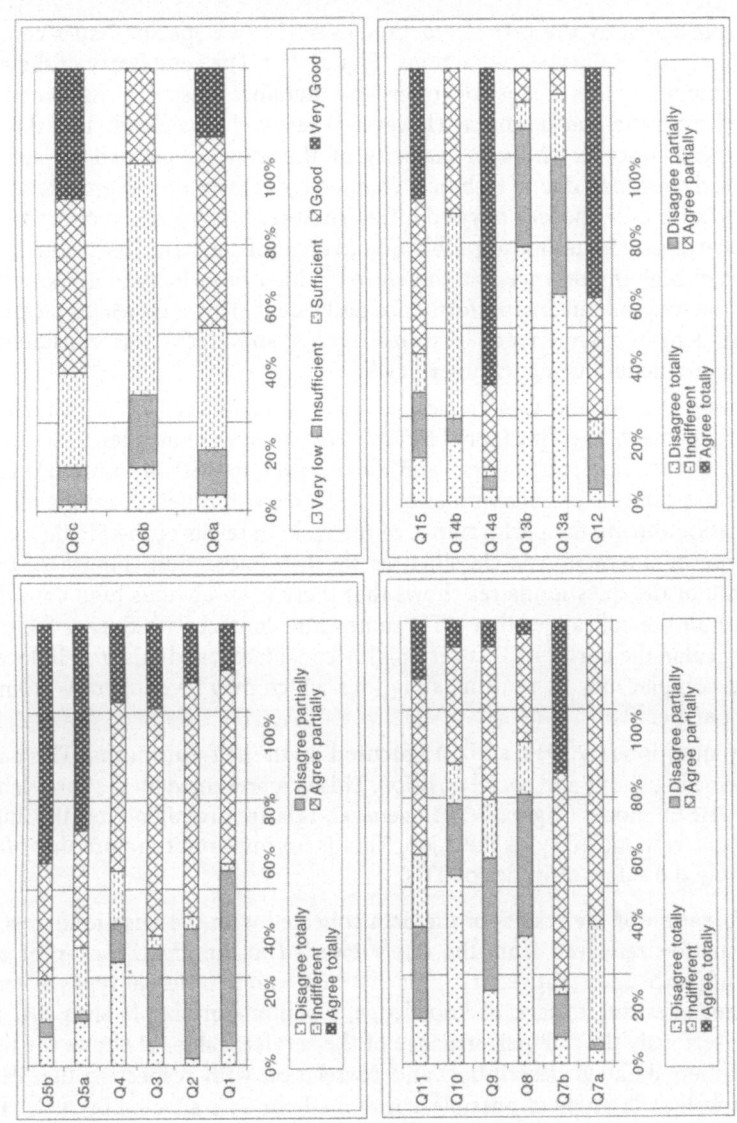

Figure 7.13 Summary result of data collected with questionnaires.

was represented by the L-H interaction, even if some specific issues could be associated with L-E type interactions (Figure 7.3). The complexity of the problems included in this area is enormous and therefore no specific input could be derived from the questionnaire. However, it resulted very clearly that this area is of great concern. The vast majority of the answers pointed towards the increased workload due to technical complexity of tasks and augmented traffic density. Moreover, the lack of good infrastructures and logistic support for TDs were also clearly identified as problems and areas of concern. This area was considered of high importance, and therefore the interviews focused in more detail on the factors influencing performance and safety critical issues. In particular, the problem of maintenance of trains, cabins, and tracks was considered in much more detail during the interviews.

- Questions Q5, Q6, and Q14 concerned the interaction of TDs with their colleagues, namely, the "partner TD," the "train crew controller," the "traffic manager," the "line instructors," and all the persons with whom a TD either cooperates directly or collaborates at a distance with different means of communication during the performance of the task. In terms of the SHELL model, this type of interaction is classified as the L-L interaction (Figure 7.3). The outcome of the questionnaires shows that there is an obvious high consideration for all the colleagues that a TD comes into contact with during the tasks. In particular, the partner TD is very highly considered, and this may induce the bias of complacency in working situations which may be dangerous from the safety perspective.

- Finally, questions Q7, Q11, and Q15 focused on the self-esteem that TDs had of their working status and social position. This is represented by the core element in the SHELL model (Figure 7.3). In general, TDs are proud and feel the importance and responsibility of their job. This is an important factors that affects positively the safety attitudes of TDs.

The combination of the results of the data collected with the questionnaires and the information retrieved from the interviews led to the identification of eight areas of influence and eight generic PIFs that may lead to different types of errors, such as mistakes due to lack of knowledge, violations or simple slips that may endanger seriously the safe performance of the service. Table 7.1 shows the list of PIFs and their detailed description and correlation with respect to the SHELL model, as well as the type of possible errors made by TDs affected by such PIFs.

The PIFs are the basic elements for performing the last phase of work for the perspective study of possible accidents induced by human errors and for the definitions of the IoS and *RSA-Matrices* for Recurrent Safety Audits.

7.2.3 Phase 3: Definition of IoS and RSA-Matrices for the Safety Audit of the ERC

This phase of work merged two essential components, namely, the information and experience gained by the HF Team during the filed studies, and the outcome of tasks, analyses, and study of reports on incidents and accidents from past operating experience.

Table 7.1 Final set of performance influencing factors (PIFs) affecting TD performance

PIF	SHELL Reference	Error types/modes
Communication within ERC	Serious problems encountered in contacting managers to discuss rules and standards (L-E)	Errors and violations of rules
Communication means	Inadequate communication technology between several actors, i.e, TDs with train crew controller, traffic manager, line instructors, etc. (L-L)	Catastrophic errors in emergency conditions and high workloads
Technological interfaces	Difficult ergonomics of signals in the cabin and inconsistent placement of signals on the railway (L-H)	Possible occurrence of signals passed at danger (SPADs)
Maintenance of trains/railway	Inadequate maintenance of trains and railway reduce reliability of system and increase difficulty of train driving (L-H, L-E)	Induced errors of traffic management and violation of procedures
Comfort of working contexts	Inadequacy of working contexts and poor logistic of rest areas increase stress and workload on long-distance shifts (L-E)	Errors due to stress
Roster and shifts planning	Too little involvement of TDs in the definition of shifts; the "fixed couple" strategy may not contribute to safety (L-E)	Errors due to stress, workload, or complacency
Rules and regulations	Too many complex rules and regulations, sometimes contrasting with each other (L-S)	Errors and violations of rules and procedures
Training methods and simulators	Existing training is inadequate to cope with advanced automation and complexity of tasks (L-S)	Errors due to lack of knowledge and training

The objective of this last phase of work was the development of the levels IoS and RSA Matrices to be applied in assessing the state of the organisation with respect to safety performances. This phase was conducted in two subsequent steps:

Phase 3, Step 1. Development of perspective THERP-Trees.

        Step 2. Generations of IoS and RSA-Matrices for Recurrent Safety Audits, and safety recommendations.

Development of Perspective THERP-Trees

The final set of PIFs (Table 7.1) were utilised for defining a number of hypothetical accidental or dangerous sequences due to human inappropriate behaviour.

The creativity of the HF Team and the opinion of ERC experts were very important for the identification of initial and boundary conditions and for the whole development of these prospective logical trees, according to the HERMES method-ology (Figure 7.1).

The TD tasks considered for the generation of the trees were the same devised during phase 1 (Step 4) (Figure 7.5–Figure 7.8), which had already been applied for the evaluation of the workload of train drivers and for the definition of the Time Line Analyses in phase 2 (step 1).

A variety of scenarios were studied, and the associated THERP-Trees were developed. For each scenario, the success or failure of the mission was evaluated in consideration of PIFs, workloads, working processes, and expected performances.

As an example, the THERP-Trees for a number of subtasks are shown in graphical and tabular form in Appendix 3. In all cases, TDs have been considered in normal mental and physical conditions at the beginning of each task. In particular, the THERP-Trees shown in Appendix 3 refer to:

- Task 2: "Prepare train." Subtasks 1.4 and 1.5: "In depot," "exterior look" and "compressors" (Figure 7.6).
  - ✓ Systemic initial and boundary conditions: One compressor only in working condition and one wheel in a bad condition (Figure 7.15, Table 7.9).
- Task 3: "Drive Train," Subtask 1.2: "Depart from station," "check free line, direction, and allowed speed" (Figure 7.7).
  - ✓ Systemic initial and boundary conditions: Signal green, and train on wrong track (Figure 7.16, Table 7.10).
  - ✓ Typical situation for an event of "Signal Passed At Danger" (SPAD).
- Task 3: "Drive train." Subtask 1.3: "Depart from station," "wait for green light" (Figure 7.7).
  - ✓ Systemic initial and boundary conditions: Signal red and train in good conditions (Figure 7.17, Table 7.11)
  - ✓ Typical situation for an event of SPAD.
- Task 3: "Drive train." Subtasks 3.1.1: "Signals and signalling," "signalling at a glance," "signal free way" (Figure 7.7).
  - ✓ Systemic initial and boundary conditions: Train in good condition, but "red light" and "signal free way" open (Figure 7.18, Table 7.12).
  - ✓ Typical situation for an event of SPAD.
- Task 3: "Drive train." Subtasks 3.1.3: "Signals and signalling," "signalling at a glance," "way impeded" (Figure 7.7).
  - ✓ Systemic initial and boundary conditions: Train in good condition; first signal "yellow" and second signal "red" (Figure 7.19, Table 7.13).
  - ✓ Typical situation for an event of SPAD.
- Task 3: "Drive train." Subtasks 3.1.4: "Signals and signalling," "signalling at a glance," "speed limit indicators" (Figure 7.7).
  - ✓ Systemic initial and boundary conditions: Train in good condition, no failures on track, but without automatic signal recognition system (Figure 7.20, Table 7.14).
  - ✓ Typical situation for an event of "spending and derailing."

- Task 3: "Drive train." Subtasks 4.1.2: "Braking," "reducing speed," "resetting automatic signal recognition" (Figure 7.7).
  - ✓ Systemic initial and boundary conditions: Train in good condition, no failures on track, and fault in automatic signal recognition system, reset necessary (Figure 7.21, Table 7.15).
- Task 3: "Drive train." Subtasks 7.1: "Stopping train," "intermediate station," case a (Figure 7.8).
  - ✓ Systemic initial and boundary conditions: Train in good condition, no failures on track (Figure 7.22, Table 7.16).
- Task 3: "Drive train." Subtasks 7.1: "Stopping train," "intermediate station," case b (Figure 7.8).
  - ✓ Systemic initial and boundary conditions: Train in good condition, red light (Figure 7.23, Table 7.17).
  - ✓ Typical situation for an event of SPAD.
- Task 3: "Drive train." Subtasks 9.2: "Driving at a glance," "level crossing not automatically controlled" (Figure 7.8).
  - ✓ Systemic initial and boundary conditions: Train in good condition, "red" signal followed by code "CD-180" which allows proceeding at a glance, but faulty Level Crossing (LC) systems and LC in open condition with no safety indication (Figure 7.24, Table 7.18).
  - ✓ Typical situation for an event of "level crossing passed at danger."

Generations of IoS and RSA-Matrices and Safety Recommendations

The final step of this study focused on the generation of a set of Indicators of Safety (*IoS*) and Recurrent Safety Audit Matrices (*RSA-Matrices*) that would allow the ERC to perform successive evaluations and audits of its own safety level. The IoS were also the guiding elements for developing safety recommendations for the railway company.

As discussed in Chapter 4, the generic format for applying HERMES in the case of a Safety Audit requires that *IoS* are identified and grouped in four "families" of indicators according to whether they can be considered *Organisational Process (OP)*, *Personal and External Factors (PEF)*, *Local Working Conditions (LWC)*, and *Defences–Barriers–Safeguards (DBA)*, (§4.4.3). In the *RSA-Matrices*, the contribution of each *IoS* is associated with its safety characteristics of *prevention*, *recovery*, or *containment* of errors and consequences.

The complete application of the approach foresees that the *RSA-Matrices* defining the current safety state of an organisation are evaluated and confronted with the corresponding *RSA-Matrices* associated with (a) the safety state of the organisation at a reference time, e.g., the initial operation for a plant; and (b) the *IoS* required by safety regulations and standards for ascertaining acceptable safety levels and therefore operability.

In the case of the study of the railway organisation ERC, only the *IoS* and associated *RSA-Matrices* relative to the current state of the organisation were performed, for a number of reasons, namely: (a) no previous evaluations of *IoS* had been performed; and (b) no comparison with the standards and regulations concerned were performed by the ERC independently. The formulation of the *RSA-Matrices* associated with the state of the organisation at the time of the study was thus the first time that the overall organisation was audited as a whole, from the human factors or Human Error and Accident Management point of view.

The complete set of *IoS* guided the generation of final safety recommendations. The *RSA-Matrices* represented the basis for future RSA. These will be briefly discussed here.

Indicators of Safety and Safety Recommendations

The Performance Influencing Factors and possible types and modes of errors identified during the previous phases of work have been considered in detail for identifying relevant *IoS*.

Table 7.2 shows the generic *IoS* that have been identified. While in Table 7.1 the PIF were correlated to types and modes of possible errors, in Table 7.2 the PIF have been associated with possible causes/effects on TD's behaviour and consequently with corresponding families of *IoS*.

In Table 7.3 the summary of the *IoS* that were utilised for performing the safety audit are reported. The safety recommendations have been developed with reference to the PIF and IoS and specific attention was dedicated to the identification of areas of concern and definition of possible interventions or improvements. The specific forms of intervention as well as their timing were entirely left to the ERC management and policy makers.

The safety recommendations started with some general considerations derived from the overall results of the investigation. In particular, the ERC was considered a company with a good overall safety record, given the amount of traffic sustained and accidents reported. Train drivers were in general proud of their role and dedicated to their job. The company was undergoing a process of renovation at organisational and technical level that aimed at improving the service for the public as well as the infrastructures and technological means (tracks, trains, communications). The burden of this process was mostly sustained by ERC personnel and management.

However, the complexity and dimension of the activity rested too heavily on personal commitment and individual intervention with little observation of the too many rules and regulations that were, in most cases, contradictory and inapplicable. Quite often, these led to the application of unwritten rules and practices that were deemed necessary for performing the tasks, but were also perceived contrary or bypassing the established safety procedures. A general revision of the policies and standards as well as the procedures applied within the company was therefore considered of primary importance in aiming at a more formal compliance with standard and emergency operating procedures (DBS_3, DBS_4).

Table 7.2 Identification of *IoS* with reference to PIF

PIF	Causes/Effects	*IoS*
Communication within ERC	• Serious problems encountered in contacting managers for discussing rules and standards • Uncertainty about future; low morale • Unions as only channel for communicating with top management level	✓ OP_1 Unwritten rules; Reporting systems ✓ PEF_1 Mental conditions ✓ OP_2 Role of unions vs. management
Communication means	• Obsolete technology for communication • Inadequate maintenance on communication means • Unclear rules for communication	✓ LWC_3 Quality of tools/actuators ✓ LWC_3 Maintenance of tools ✓ DBS_3 Policies, standards
Technological interfaces	• Poor ergonomics of interfaces of train cabins • Problems in understanding and managing automation • Inconsistency between signals on track and in cabin	✓ LWC_1 Workplace design ✓ LWC_2 Automation ✓ LWC_4 Signals: track and cabin
Maintenance of trains/railway	• Inadequate and insufficient maintenance of trains and tracks	✓ DBS_1 Safety devices ✓ PEF_2 Physical conditions, stress
Comfort of working contexts	• Obsolete technology for communication • Poor comfort of cabin, rest areas, and long-haul ERS-hotels • Lack of development plans	✓ LWC_3 Quality of tools/actuators ✓ LWC_1 Workplaces ✓ PEF_1 Mental conditions, morale
Roster and shifts planning	• Heavy and too strict TD shifts • Inadequacy of "fixed" TD couple for ensuring maximum safety	✓ PEF_2 Physical conditions ✓ LWC_5 Job planning
Rules and regulations	• Excess of rules and regulations	✓ DBS_3 Policies, standards ✓ DBS_4 Procedures
Training methods and simulators	• Inadequate training • Insufficient experience and expertise of trainers/instructors	✓ DBS_2 Training standards ✓ OP_2 Human relationships

Moreover, the development and establishment of a confidential reporting system for incidents, near misses and all other anomalies encountered while performing duties, was considered essential for improving communication with the manage-

Table 7.3 Families of *IoS* for safety assessment and audit

IoS–OP: organisational processes	IoS–PEF: personal and external factors	IoS–LWC: local working conditions	IoS–DBS: defences, barriers, safeguards
OP_1: Reporting system	PEF_1: Mental conditions	LWC_1: Workplace design; comfort	DBS_1: Safety devices
OP_2: Unions, unwritten rules, human relationships	PEF_2: Physical conditions	LWC_2: Automation LWC_3: Tools and actuators LWC_4: Signals on track and in cabin LWC_5: Job planning	DBS_2: Training DBS_3: Policies, standards DBS_4: Procedures, instructions, supervision

ment and policy makers of the company and in improving development and application of safety regulations (OP_1).

The transformation process of the company caused uncertainty about the future within the staff. Moreover, a certain number of problems at the communication level within the company aggravated the situation affecting the morale of personnel (PEF_1). In particular, the presence of "unions" as the sole communication channel was not well accepted (OP_2). The entire system for communicating with the personnel within the ERC needed revision and improvement.

Another important issue associated with the transformation of the company was the issue of job planning. In order to improve service while limiting stress and workload, the issue of the "fixed couple" train drivers was considered a very important IoS (LWC_5, PEF_2). A solution to this problem had to be studied and adequately planned, in order to compromise the requirements of modern technological means with TD's habits and expectations, in favour of safety and efficiency of performance.

From a more technical perspective, the effort of improving railway system by introducing high-speed trains and automated controls was not associated with an equivalent improvement in the quality of means for communication (LWC_3), interfaces within and outside the cabin (LWC_1, LWC_2, LWC_4). This problem was further aggravated by the fact that modern and traditional trains were very different in their layout and comfort. Therefore, the cabins of different types of trains needed revision from the ergonomic perspective. The overall signalling system needed revision in order to accommodate the coexistence of trains of different technology.

The maintenance of trains and tracks was one of the areas that needed vast amelioration and improvement. Too frequently, at least from the field observation, the existence of inadequate train or track conditions were compensated by TD actions, or even by bypassing rules and safety devices "in favour" of the service (DBS_1), causing stressful mental and physical conditions (PEF_2). This problem was considered of very high relevance and a strong recommendation was issued concerning the need to revise and renovate the whole system of maintenance.

Table 7.4 ERC RSA-matrix for the *IoS organisational processes*

IoS–OP: organisational processes	Prevention	Recovery	Containment
OP$_1$ = Reporting system • Confidential reports	φ_{rep}	φ_{rep}	–
OP$_2$ = Unions, unwritten rules, human relationships • Questionnaire/interviews	φ_{int}	φ_{int}	φ_{int}

Table 7.5 ERC RSA-matrix for the *IoS personal and external factors*

IoS–PEF: personal and external factors	Prevention	Recovery	Containment
PEF$_1$ = Mental conditions • Questionnaire/interviews	φ_{int}	φ_{int}	–
PEF$_2$ = Physical conditions • Medical checks	φ_{med}	φ_{med}	–

Finally, another very important issue, linked to the introduction of automation and modern technology, was training. Current standards and procedures for training TDs were considered insufficient and inadequate, both from the technical and human relations side (DBS$_2$, OP$_2$). The recommendation to review the entire training process was issued. In particular, the introduction of new training courses dedicated to accident and emergency management from the human factors perspective was recommended, as was done in the aviation domain with "Crew Resource Management" (CRM) courses (Chapter 6).

Recurrent Safety Audit Matrices

The *RSA-matrices* derived from the table of IoS are shown in Tables 7.4, 7.5, 7.6, and 7.7. In these matrices, each IoS is associated with three relevant elements:

● The specific safety objectives that are covered by the IoS, namely prevention, recovery, or protection from accidents/incidents.
● The definition of the approaches for evaluating the IoS. In particular, five types of assessment are identified: *Field assessment, medical checks, analysis of data, questionnaire/interviews*, and *quality and engineering checks*.
● The frequency (φ) of performance of the audit relative to each IoS.

Finally, in order to support the activity of safety engineers and experts, Table 7.8 contains for each type of assessment to be applied, i.e., medical checks, analysis of data, questionnaire/interviews, quality and engineering checks, and field assessment, the frequency of occurrence, the method utilised, the IoS to be evaluated, and the specific approach to be implemented in practice.

Table 7.6 ERC RSA-matrix for the IoS local working conditions

IoS–LWC: local working conditions	Prevention	Recovery	Containment
LWC$_1$ = Workplace design; comfort			
• Field assessment	φ_{fld}	–	–
• Questionnaire/interviews	φ_{int}	–	–
LWC$_2$ = Automation			
• Quality and engin. checks	φ_{eng}	φ_{eng}	–
LWC$_3$ = Tools and actuators			
• Quality and engin. checks	φ_{eng}	φ_{eng}	φ_{eng}
• Field assessment	φ_{fld}	φ_{fld}	φ_{fld}
LWC$_4$ = Signals: track and cabin			
• Quality and engin. checks	φ_{eng}	φ_{eng}	φ_{eng}
• Field assessment	φ_{fld}	φ_{fld}	φ_{fld}
LWC$_5$ = Job planning			
• Questionnaire/interviews	φ_{int}	φ_{int}	–

Table 7.7 ERC RSA-matrix for the IoS defences, barriers, safeguards

IoS–DBS: defences, barriers, safeguards	Prevention	Recovery	Containment
DBS$_1$ = Safety devices			
• Quality and engin. checks	φ_{eng}	φ_{eng}	φ_{eng}
DBS$_2$ = Training			
• Questionnaire/interviews	φ_{int}	φ_{int}	φ_{int}
• Field assessment	φ_{fld}	φ_{fld}	φ_{fld}
DBS$_3$ = Policies, standards			
• Quality and engin. checks	φ_{eng}	φ_{eng}	φ_{eng}
DBS$_4$ = Procedures, instructions, supervision			
• Quality and engin. checks	φ_{eng}	φ_{eng}	φ_{eng}
• Questionnaire/interviews	φ_{int}	φ_{int}	φ_{int}

In practice, Table 7.8 may be considered as a checklist to be applied by safety analysts for performing the safety audit. At the end of the safety audit process, this table can be utilised for completing the RSA-matrices and for ascertaining that all aspects and methods of the safety audit have been performed.

7.3 Summary: Application of HERMES for the Safety Audit of a Large Organisation

7.3.1 Application of HERMES

This chapter has presented a practical application of the HERMES methodology in the safety audit of a large railway company.

Table 7.8 Type of assessment and methods for the evaluation of *IoS*

Type of assessment	*IoS*	Approach

Medical Checks
✓ Method:
  Medical experience
✓ Frequency:
  $\varphi_{med} \approx 12$ months
	PEF_2 = Physical conditions	• Medical checks to monitor fitness of TDs and operation personnel

Analysis of data
✓ Method:
  Statistical analysis
  and HF experience
✓ Frequency:
  $\varphi_{rep} \approx 6$ months
	OP_1 = Reporting system	• Analysis of data from confidential and mandatory reporting system;
		• Feedback to TD and personnel

Questionnaire/interviews
✓ Method:
  Statistical analysis
  and HF experience
✓ Frequency:
  $\varphi_{int} \approx 12$ months
	DBS_2 = Training
	
	DBS_4 = Procedures, instructions, supervision
	
	OP_2 = Unions, unwritten rules, human relationships
	
	PEF_1 = Mental conditions
	
	LWC_1 = Workplace design; comfort
	LWC_5 = Job planning
		• Subjective evaluation of training
		• Subjective evaluation of procedures, etc.
		
		• Subjective evaluation:
		  a. Unions, habits;
		  b. Delay in reply from management; etc.
		• Evaluation of "mental state": motivation job satisfaction, etc.
		• Subjective evaluation of workplace design, etc.
		• Subjective evaluation of "job planning".

Quality and engineering checks
✓ Method:
  Engineering experience; standards and norms
✓ Frequency:
  $\varphi_{eng} \approx 6\text{–}12$ months
	DBS_1 = Safety devices
	
	DBS_3 = Policies, standards
	
	DBS_4 = Procedures, instructions, supervision
	LWC_2 = Automation
	
	LWC_3 = Tools and actuators
	LWC_4 = Signals along track and in cabin
		• Verify functionality, effectiveness, and maintenance
		• Verify functionality
		  a. Updates of policies
		  b. Adequate stand.; etc.
		• Verify functionality
		  c. Updates of proc.
		  d. Instr. teams; etc.
		• Verify functionality and applicability
		• Verify functionality and maintenance
		• Verify functionality and maintenance:
		  a. Delay comm.-interv.
		  b. Func. in tunnels; etc.

Table 7.8 *Continued*

Type of assessment	IoS	Approach
Field assessment common ✓ Method: Engineering and HF experience; procedures and norms ✓ Frequency: $\varphi_{fld} \approx 6$–12 months	LWC_1 = Workplace design; comfort LWC_3 = Tools and actuators LWC_4 = Signals along track and in cabin DBS_2 = Training	• Visit to cabins, hotels areas, etc. • Verify efficiency and maintenance during cab rides • Verify efficiency and maintenance during cab rides • Verify efficiency and effectiveness during training and retraining courses

The five standpoints developed in Chapter 2 to be retained in order to perform a sound HF approach have been respected:

1. The HERMES methodology was applied to the area of application "safety assessment" (standpoint 4: area of application).

2. The goals and objectives of the audit were set from the viewpoint of human error and accident management from the beginning of the study of the railway company (standpoint 1).

3. A model of reference or a frame representing human – machine interaction was selected and applied throughout the audit (standpoint 2).

4. The logical interplay between retrospective and prospective approaches was maintained, by ensuring that information and data derived from visits to depots, cabin rides, interviews, and questionnaires, and from studies of past events were taken into consideration in identifying PIFs and developing potential accident scenarios (standpoint 3).

5. The final step of the safety audit included the development of safety recommendations relative to the current sate of the company and the definition of the IoS and IoS-safety matrices typical of a safety audit. The IoS and IoS-matrices offer valuable references for performing future safety audits (standpoint 5).

Figure 7.14 summarises the whole process of application, which merged practical (field) and theoretical methods, as the HERMES methodology requires, with the aim of developing *IoS* and *RSA-matrices*.

The methodology was applied in three phases. Phase 1 of the work initially focused on the evaluation of the socio-technical context existing within the railway company. These ethnographic studies were complemented by the development of a questionnaire for the evaluation of train drivers' attitudes and the safety climate. A model representing the interactions within the company was considered.

Phase 2 focused on the collection and analysis of data and the review of reports of past events. This process led to the definition of the PIFs and the development of scenarios for potential accidents related to human factors.

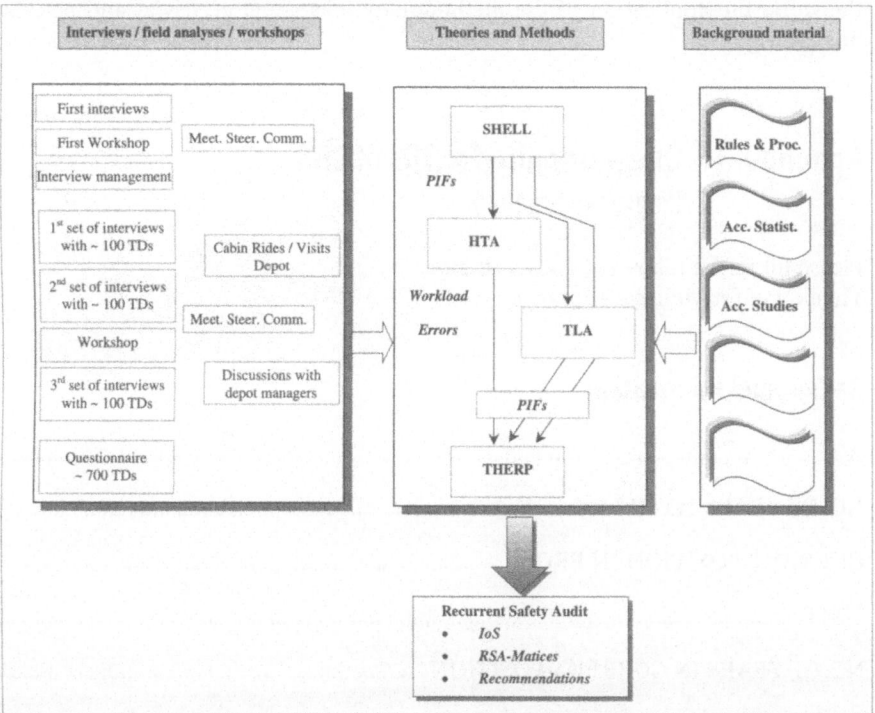

Figure 7.14 Summary of methods and techniques applied in the safety audit of a large railway company according to the HERMES methodology.

Phase 3 consisted in the actual safety audit based on the identification of IoS and the definition of IoS-matrices. Safety recommendations related to the existing safety state of the company were also developed.

7.3.2 Conclusions

The application of HERMES in this case study shows that the methodology is appropriate and can give valuable and significant results that help in improving and ascertaining safety standards.

This case study has indicated some specific formats for presenting the results and practical procedures to follow. Moreover, it has shown how the techniques can be applied and the knowledge that exists within organisations can be exploited.

At the same time, in accordance with the outcome of the other test cases discussed in previous chapters, the application of HERMES to this case study demonstrates that a substantial application of the methodology to large organisations is feasible and leads to important and useful results.

However, it requires considerable effort by human factors specialists, especially in the collection of the relevant information and data. Without this critical part of the work, the application of the methodology may generate trivial and possibly misleading results.

Appendix 1:   Questionnaire for TDs of ERC
(English translation)

Please fill in the following questionnaire.
Thank you for your collaboration.

Background Information

AGE ....

NO. OF YEARS WITH ERC ....

CURRENT POSITION IN ERC ....

UNIT ....

NO. OF YEARS IN CURRENT POSITION ....

PREVIOUS POSITIONS IN ERC:
....
....
....

POSITIONS OCCUPIED BEFORE WORKING FOR ERC:
....
....
....

TYPE OF TRAIN AND EXPERIENCE ....

TYPE OF SHIFT ....

COMMUTER ....   Y   N

Questionnaire

1. Rules and regulations are easily applied in everyday work.

Disagree totally	Disagree partially	Indifferent	Agree partially	Agree totally
☐	☐	☐	☐	☐

2. Rules and regulations deal with all real situations encountered in everyday work.

Disagree totally	Disagree partially	Indifferent	Agree partially	Agree totally
☐	☐	☐	☐	☐

3. Training and retraining processes give the best possible knowledge to the train driver.

Disagree totally	Disagree partially	Indifferent	Agree partially	Agree totally
☐	☐	☐	☐	☐

4. Certification and service on different types of trains is a valuable diversification and make everyday work more interesting.

Disagree totally	Disagree partially	Indifferent	Agree partially	Agree totally
☐	☐	☐	☐	☐

5. When I need to discuss technical problems or norms and standards the line instructor is.
   (a) The most competent person with whom to talk for support.

Disagree totally	Disagree partially	Indifferent	Agree partially	Agree totally
☐	☐	☐	☐	☐

   (b) Always available for discussion with me.

Disagree totally	Disagree partially	Indifferent	Agree partially	Agree totally
☐	☐	☐	☐	☐

6. How do you consider the professional experience of the following managers of training of ERC.

	Very bad	Insufficient	Sufficient	Good	Very Good
(a) Line instructor	☐	☐	☐	☐	☐
(b) Trainer	☐	☐	☐	☐	☐
(c) Quality manager	☐	☐	☐	☐	☐

7. The job of train driver is
   (a) Very interesting.

Disagree totally	Disagree partially	Indifferent	Agree partially	Agree totally
☐	☐	☐	☐	☐

   (b) Very satisfactory.

Disagree totally	Disagree partially	Indifferent	Agree partially	Agree totally
☐	☐	☐	☐	☐

8. ERC is a great family.

Disagree totally	Disagree partially	Indifferent	Agree partially	Agree totally
☐	☐	☐	☐	☐

9. The managers are available to listen and discuss the relevant problems and issues of train drivers.

Disagree totally	Disagree partially	Indifferent	Agree partially	Agree totally
☐	☐	☐	☐	☐

10. The unions are of great support in defending the professionalism and quality of the work of train drivers.

Disagree totally	Disagree partially	Indifferent	Agree partially	Agree totally
☐	☐	☐	☐	☐

11. Developing new competences has contributed to improving the professionalism of train drivers.

Disagree totally	Disagree partially	Indifferent	Agree partially	Agree totally
☐	☐	☐	☐	☐

12. The changes in "technical work settings," including increased traffic, number of signals, and duties, make the work of train drivers heavier.

Disagree totally	Disagree partially	Indifferent	Agree partially	Agree totally
☐	☐	☐	☐	☐

13. The logistics are adequate to support the work of train drivers.

	Disagree totally	Disagree partially	Indifferent	Agree partially	Agree totally
(a) Canteens, resting areas, hotels, etc.	☐	☐	☐	☐	☐
(b) Shifts	☐	☐	☐	☐	☐

14. The presence of a second train driver on board contributes significantly to reduce errors.

	Disagree totally	Disagree partially	Indifferent	Agree partially	Agree totally
(a) Usual buddy	☐	☐	☐	☐	☐
(b) Available TD	☐	☐	☐	☐	☐

15. The train driver is a professional figure in modern society.

Disagree totally	Disagree partially	Indifferent	Agree partially	Agree totally
☐	☐	☐	☐	☐

Appendix 2: Graphs of Results from Questionnaires

The figures that follow contain the results of the questionnaires collected in the three depots of the ERC where they were distributed.

The questionnaire contained 15 questions (see Appendix 1), some of which were split in two or three parts, and covered the general issues of the reference model

SHELL selected for describing the tasks and relations of TDs with their working context. In total, 2500 questionnaires were distributed and 710 were returned and stored in the database. The data analysis focused on a qualitative study rather that the quantitative assessment of the statistical correlations and validity of the variables. The graphical results for each question are reported here.

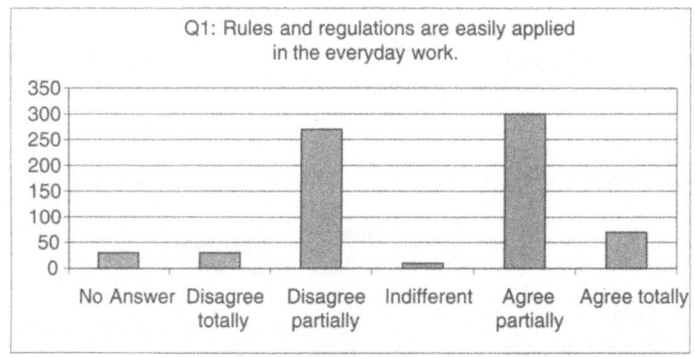

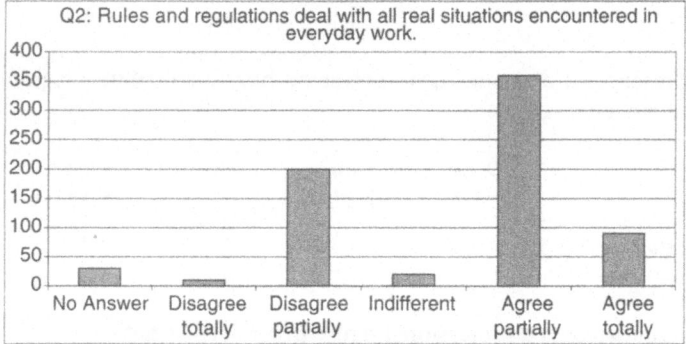

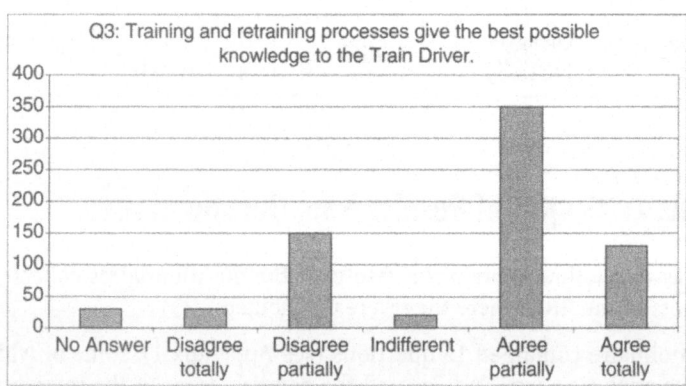

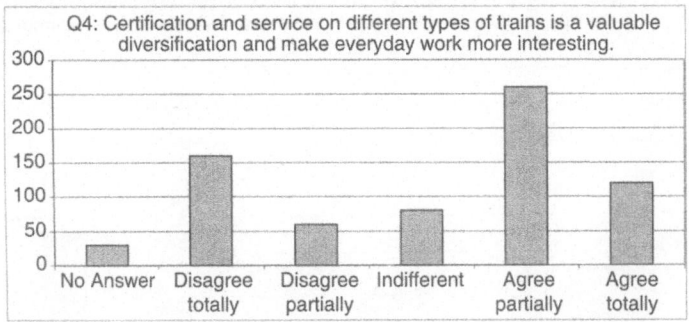

Q4: Certification and service on different types of trains is a valuable diversification and make everyday work more interesting.

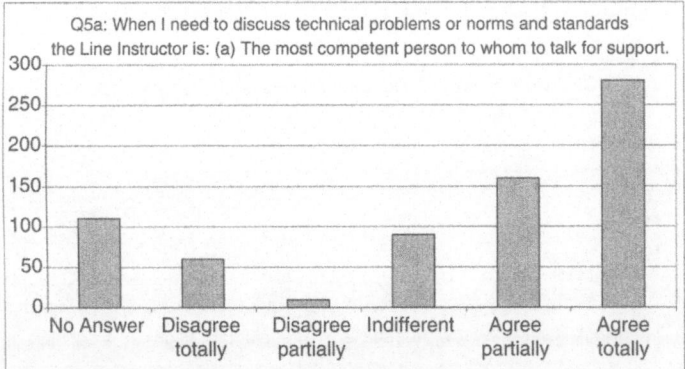

Q5a: When I need to discuss technical problems or norms and standards the Line Instructor is: (a) The most competent person to whom to talk for support.

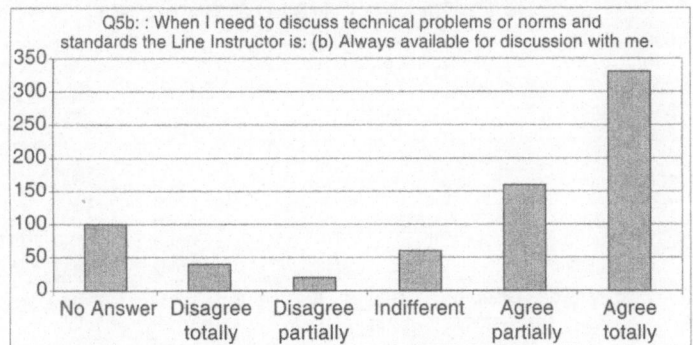

Q5b: : When I need to discuss technical problems or norms and standards the Line Instructor is: (b) Always available for discussion with me.

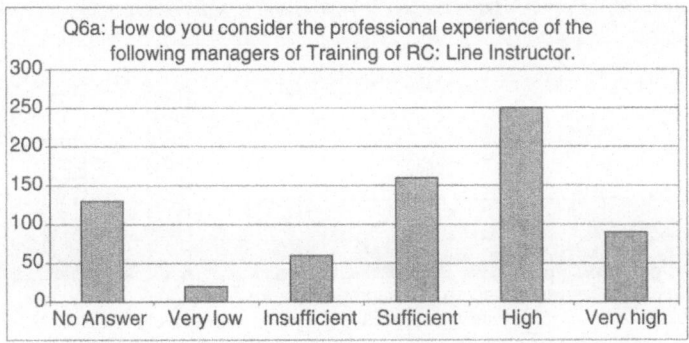

Q6a: How do you consider the professional experience of the following managers of Training of RC: Line Instructor.

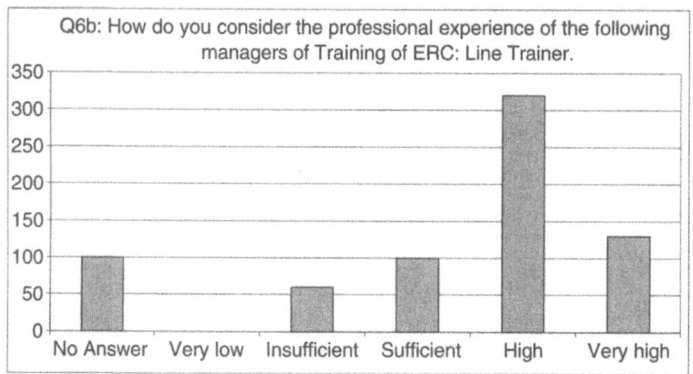

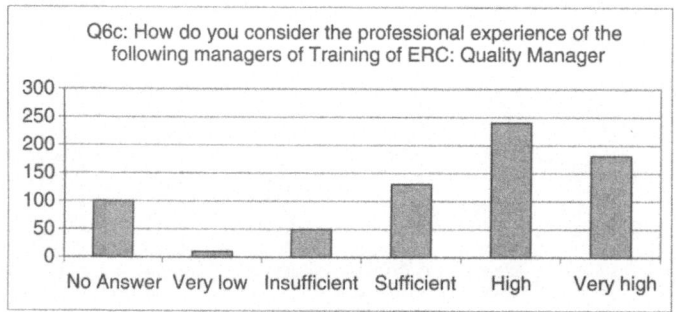

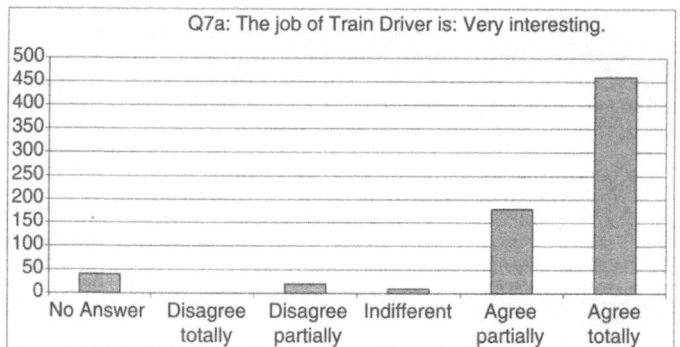

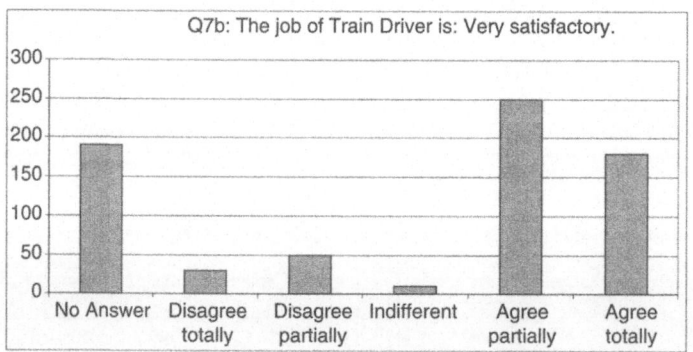

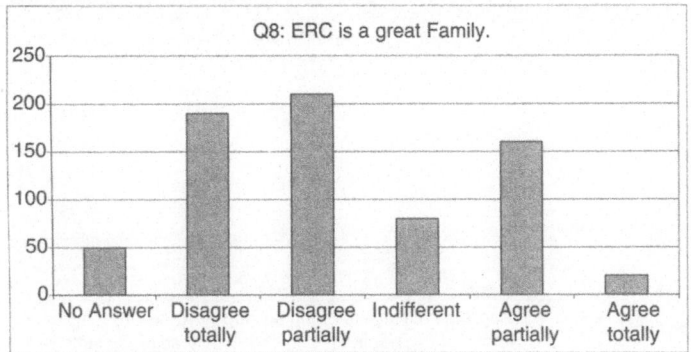

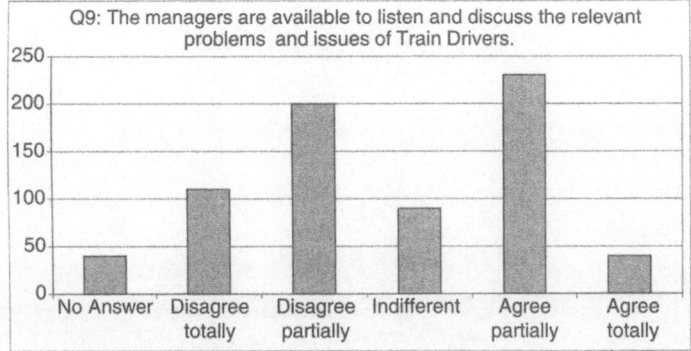

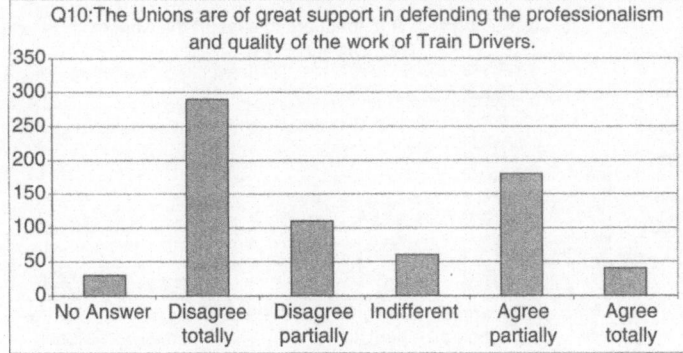

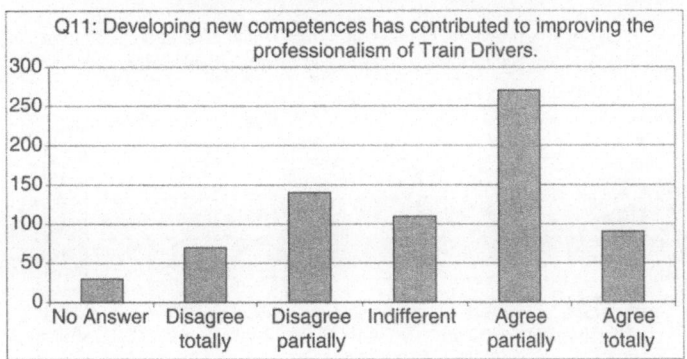

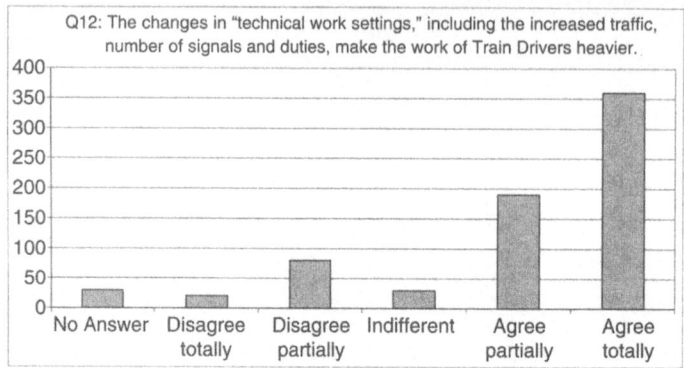

Q12: The changes in "technical work settings," including the increased traffic, number of signals and duties, make the work of Train Drivers heavier.

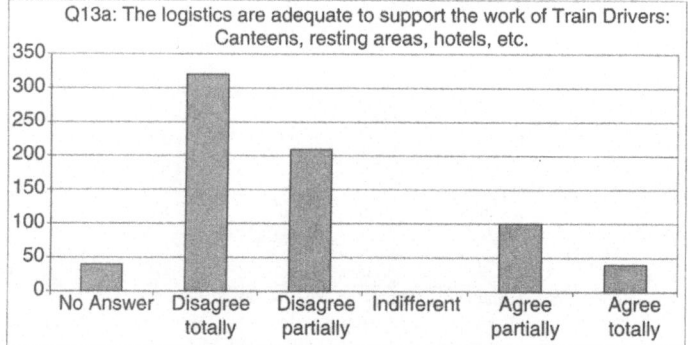

Q13a: The logistics are adequate to support the work of Train Drivers: Canteens, resting areas, hotels, etc.

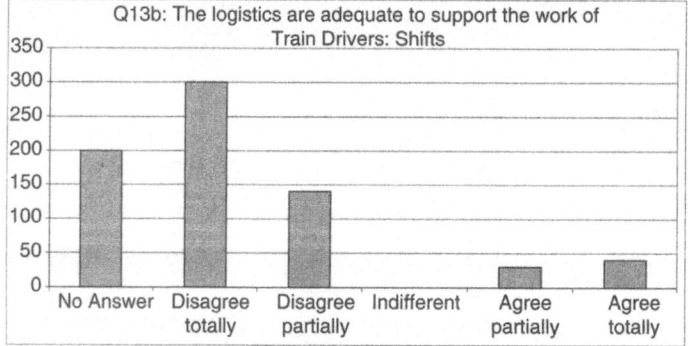

Q13b: The logistics are adequate to support the work of Train Drivers: Shifts

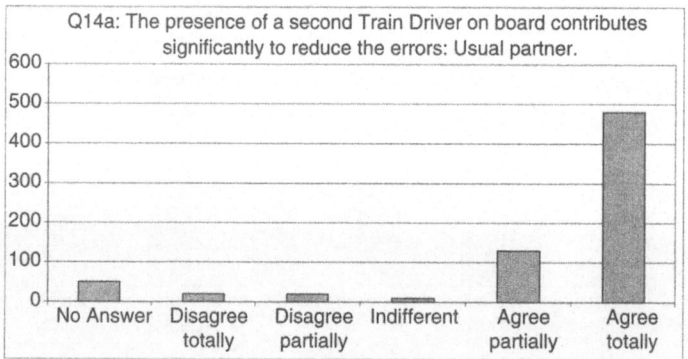

Q14a: The presence of a second Train Driver on board contributes significantly to reduce the errors: Usual partner.

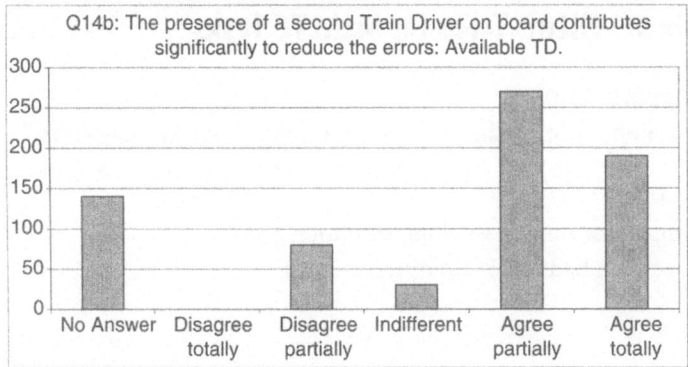

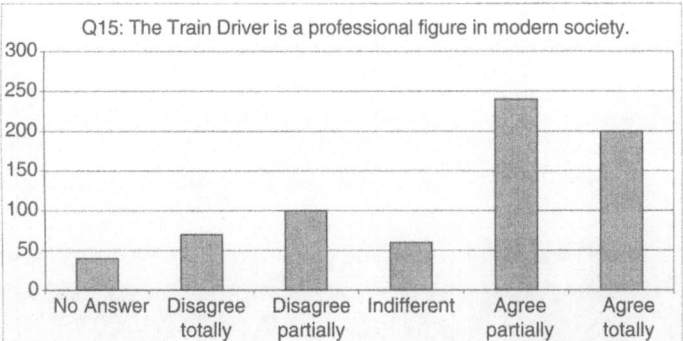

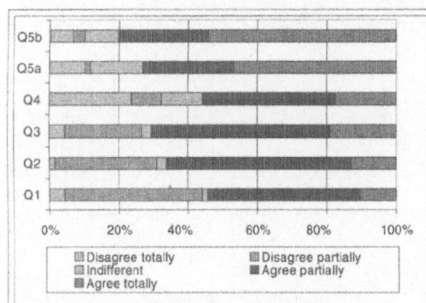

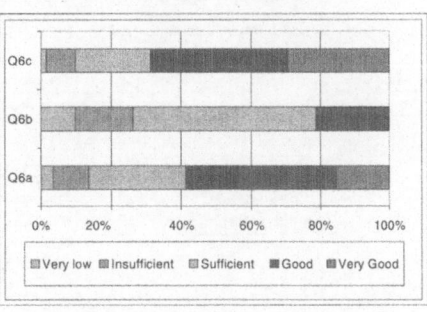

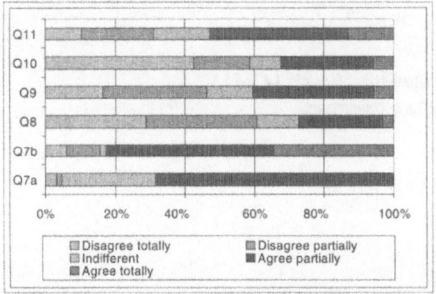

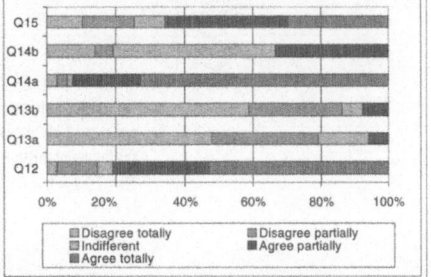

Appendix 3: THERP-Trees for Main TD Tasks

Task 2: "Prepare train"
Subtasks 1.4 and 1.5: "In depot," "exterior look," and "compressors."

Initial Conditions:

- One compressor only in working conditions; and
- One wheel in a bad "oval" conditions.

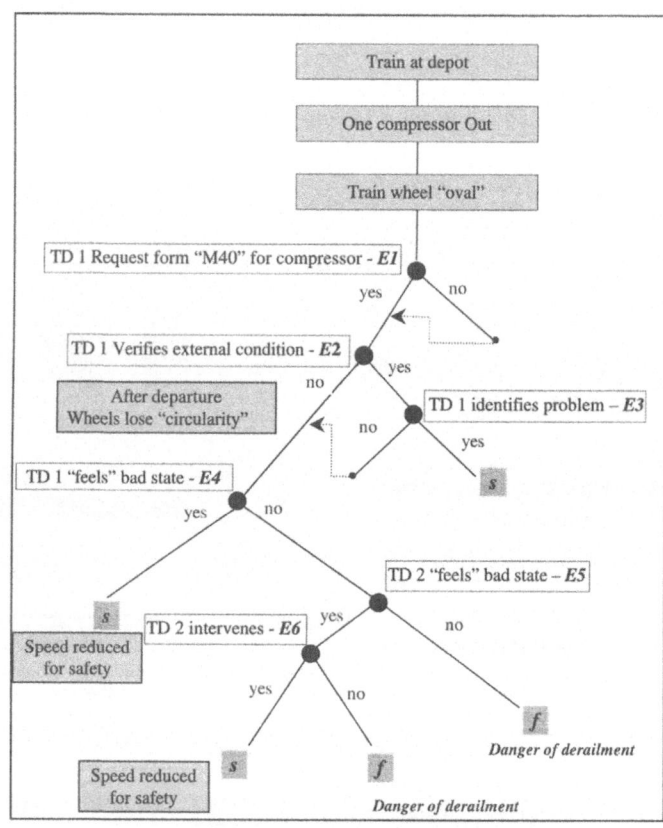

Figure 7.15 THERP-tree for task 2: "Prepare train." Subtasks 1.4 and 1.5:
"In depot," "exterior look," and "compressors."
s = success of mission
f = failure of mission
TD 1, TD 2 = train driver 1, train driver 2.

Table 7.9 Error modes, PIFs, and SHELL reference for Task 2: "Prepare train." Subtasks 1.4 and 1.5: "In depot," "exterior look," and "compressors."

Error mode	PIF – error causes	SHELL reference
E1: TD does not request form "M40" prior to departure	• Inappropriate knowledge of rules	• L-S: rules
E2: TD does not verify properly external train condition	• Lack of attention, commitment • Distraction • Time pressure • Dependency from previous event	• L-E: relationship with v • L-H: stress • L-E: relationship ERC • L-H: habit
E3: TD does not identify problem with wheel state, in depot	• Lack of attention, commitment • Time pressure • Lack of experience	• L-E: relationship with ERC • L-E: relationship ERC • L-S: training
E4: TD 1 does not identify problem with wheel state, during conduction (Task 3)	• Lack of attention, commitment • Time pressure • Lack of experience	• L-E: relationship with ERC • L-E: relationship ERC • L-S: training
E5: TD 2 does not "feel" problem with wheel	• Stress • Complacency with companion	• L-E: shifts • L-L: relationship TD – partner TD
E6: TD 2 does not intervene	• Complacency with companion • Misinterpretation	• L-L: relationship TD – partner TD • L-H: misinterpretation

Task 3: "Drive train"
Subtask 1.2: "Depart from station," "check free line, direction, and allowed speed."

Initial Conditions:
- Train on wrong track.
- Signal "green."

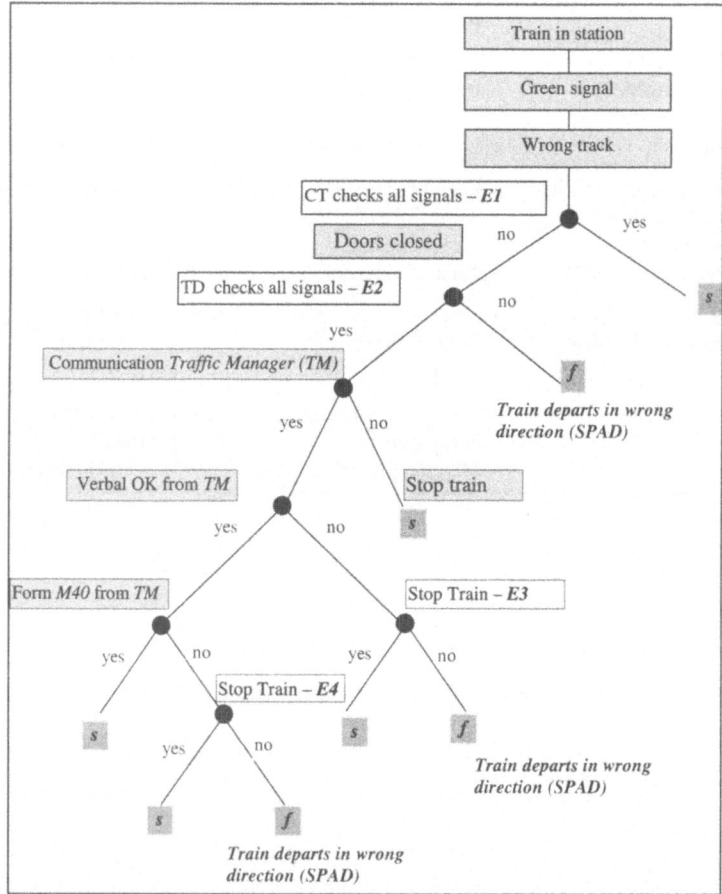

Figure 7.16 THERP-tree for Task 3: "Drive train." Subtasks 1.2: "Depart from station," "check free line, direction, and allowed speed."
s = success of mission
f = failure of mission
M40 = written form for change of normal procedure
TD = train driver 1
TM = traffic manager
SPAD = signal passed at danger
CT = crew train controller

Table 7.10 Error modes, PIFs, and SHELL reference for Task 3:"Drive train." Subtasks 1.2:"Depart from station,""check free line, direction, and allowed speed"

Error mode	PIF – error causes	SHELL reference
E1: CT does not check all signals (including line)	• Lack of attention • Does not know need to do so • Signal not well visible	• L: personality • L-S: lack of training, knowledge • L-H: environment, maintenance line
E2: TD does not check all signals (including line)	• Distraction • Tired and affected by "get-home, itis" • Signal not well visible	• L: attitudes • L-H: workload, heavy shifts • L-H: environment, maintenance line
E3: TD 1 Does not stop train (only verbal contact, but no verbal OK from *TM*)	• Ignore rules • Time pressure • Complacency • Machismo • Invulnerability	• L-S: lack of training • L-E: pressure of ERC • L-L: trust on *TM* • L: personality, dangerous attitudes
E4: TD 1 Does not stop train (in absence of written form, with only verbal OK from *TM*)	• Ignore rules • Time pressure • Complacency	• L-S: lack of training • L-E: pressure of ERC • L-L: trust on *TM*

Task 3: "Drive train"
Subtask 1.3: "Depart from station," "wait for green light."

Initial Conditions:

- No specific failures or malfunctions.
- Signal "red."

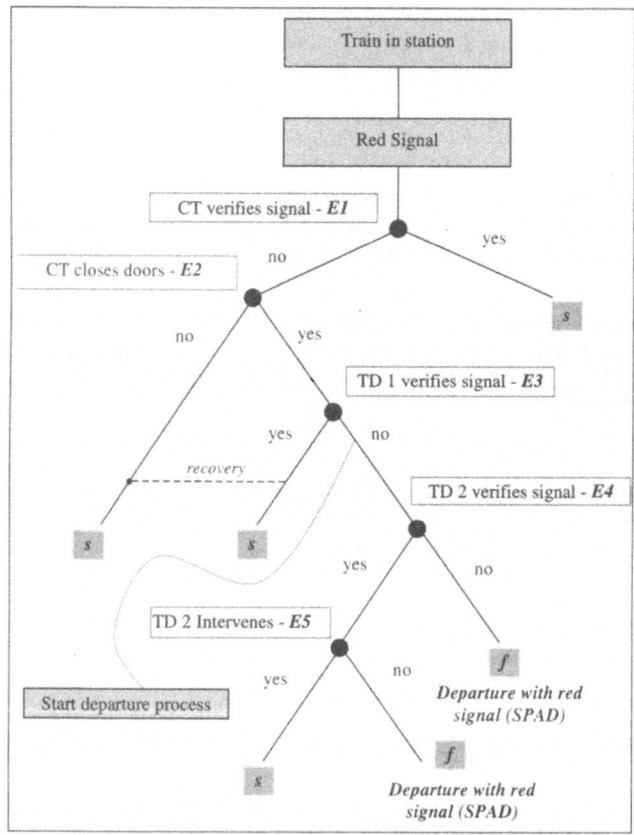

Figure 7.17 THERP-tree for Task 3: "Drive train." Subtasks 1.3: "Depart from station," "wait for green light."
s = success of mission
f = failure of mission
TD 1, TD 2 = train driver 1, train driver 2
CT = crew train controller
SPAD = signal passed at danger

Table 7.11 Error modes, PIFs, and SHELL reference for Task 3: "Drive train." Subtask 1.3: "Depart from station," "wait for green light"

Error mode	PIF – error causes	SHELL reference
E1: CT does not verify signal	• Distraction • Lack of concentration, boredom, • Job dissatisfaction	• L: personality • L-E: relationship with ERC • L-E: relationship with ERC
E2: CT closes doors	• Habit • Lack of responsibility	• L: attitudes • L-E: relationship with ERC
E3: TD 1 does not verify signal	• Habit • Lack of responsibility • Tired	• L: attitudes • L-E: relationship with ERC • L-H: heavy shifts
E4: TD 2 does not verify signal	• Lack of responsibility • Complacency with companion	• L-E: relationship ERC • L-L: relationship with TD, partner TD
E5: TD 2 does not intervene	• Stress • Complacency with companion	• L-E: shifts • L-L: relationships, TD–TD
E6: TD 2 does not intervene	• Complacency with companion • Misinterpretation of rules	• L-L: relationship TD with partner TD • L-S: rules

Task 3: "Drive train"
Subtasks 3.1.1: "Signals and signalling," "signalling at a glance," "signal free way."

Initial Conditions:

- No specific failures or malfunctions.
- Signal "red" but "signal free way" open.

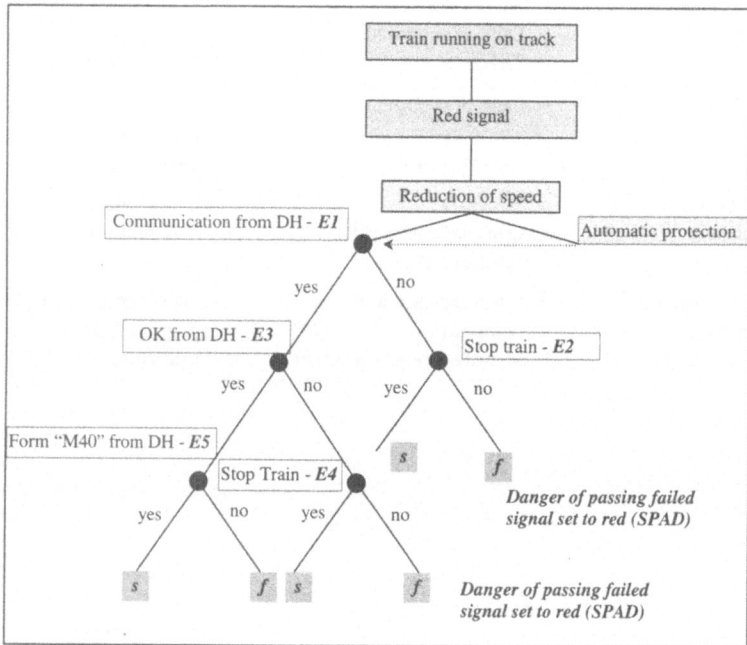

Figure 7.18 THERP-tree for Task 3: "Drive train." Subtasks 3.1.1: "Signals and signalling," "signalling at a glance," "signal free way."
s = success of mission
f = failure of mission
M40 = written form for change of normal procedure
DH = department head
SPAD = signal passed at danger

Table 7.12 Error modes, PIFs, and SHELL reference for Task 3: "Drive train." Subtasks 3.1.1: "Signals and signalling," "signalling at a glance," "signal free way"

Error mode	PIF – error causes	SHELL reference
E1: TD does not communicate with DH	• Lack of trust or confidence in DH* • Job dissatisfaction • Does not know/ recognise need to do so • System rarely works	• L-E: relationship with ERC • L-E: relationship with ERC • L-S: rules • L-H: maintenance
E2: TD does not stop train	• Lack of knowledge of • Macho + lack of trust in DH*	• L-S: rules procedures • L-L-E: attitude, relationship with ERC
E3: TD does not wait for OK from DH	• Lack of trust or confidence in DH* • Time pressure • System rarely works	• L-E: relationship with ERC • L-E: relationship with ERC • L-H: maintenance
E4: TD does not stop train	• Ignores rules • Machismo • Invulnerability • Time pressure	• L-S: rules • L: personality attitudes • L-E: relationship with ERC
E5: Does not wait for form "M40" from DH	• Annoyed by too many formalities • Time pressure	• L-S: number and congruence of rules • L-E: relationship with ERC

* DH = department head.

Task 3: "Drive train"
Subtasks 3.1.3: "Signals and signalling," "signalling at a glance," "way impeded."

Initial Conditions:

- No specific failures or malfunctions.
- First signal "yellow" and second signal "red."

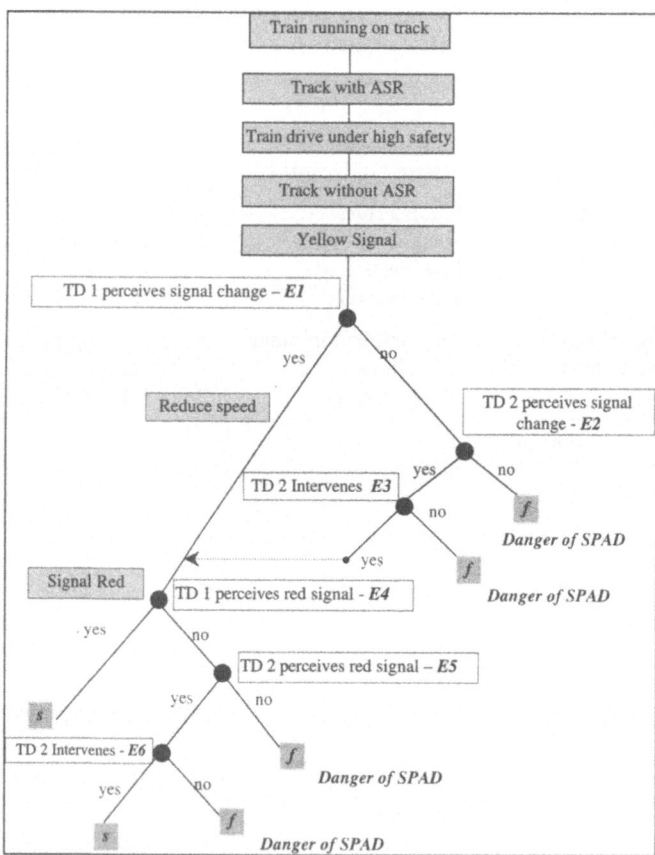

Figure 7.19 THERP-tree for Task 3: "Drive train." Subtasks 3.1.3: "Signals and signalling," "signalling at a glance," "way impeded."
s = success of mission
f = failure of mission
ASR = automatic signal recognition (on board automatic safety train stop)
SPAD = signal passed at danger
TD 1, TD 2 = train driver 1, train driver 2

Table 7.13 Error modes, PIFs, and SHELL reference for Task 3: "Drive train." Subtasks 3.1.3: "Signals and signalling," "signalling at a glance," "way impeded"

Error mode	PIF – error causes	SHELL reference
E1: TD 1 does not perceive change in track automation	• Distraction • Lack of attention • Bad visibility	• L: personality • L: habit to SAC* • L-H: maintenance
E2: TD 2 does not perceive change in track automation	• Lack of attention • Bad visibility • Complacency • Lack of responsibility	• L: habit to SAC* • L-H: maintenance • L-L: trust in partner • L-E: relationship with ERC
E3: TD 2 does not intervene	• Lack of attention as second TD • Complacency • Ignore rules • Misinterpretation	• L: attitudes • L-L: trust in partner • L-S: rules • L-H: cabin ergonomics
E4: TD 1 does not perceive red signal	• Lack of attention • Forgets track low automation • Bad visibility	• L: personality • L-H: complacency with SAC* • L-H: maintenance
E5: TD 2 does not perceive red signal	• Lack of attention as second TD • Forgets track low automation • Misinterpretation • Bad visibility	• L: attitudes • L-H: complacency with SAC* • L-H: cabin ergonomics • L-H: maintenance
E6: TD 2 does not intervene	• Complacency with companion • Misinterpretation of rules • Misinterpretation	• L-L: relationship with TD, partner TD • L-S: rules • L-H: cabin ergonomics

* SAC = signal automatic control.

Task 3: "Drive train"
Subtasks 3.1.4: "Signals and signalling," "signalling at a glance," "speed limit indicators."

Initial Conditions:

- No specific failures or malfunctions.
- Track without automatic signal recognition system.

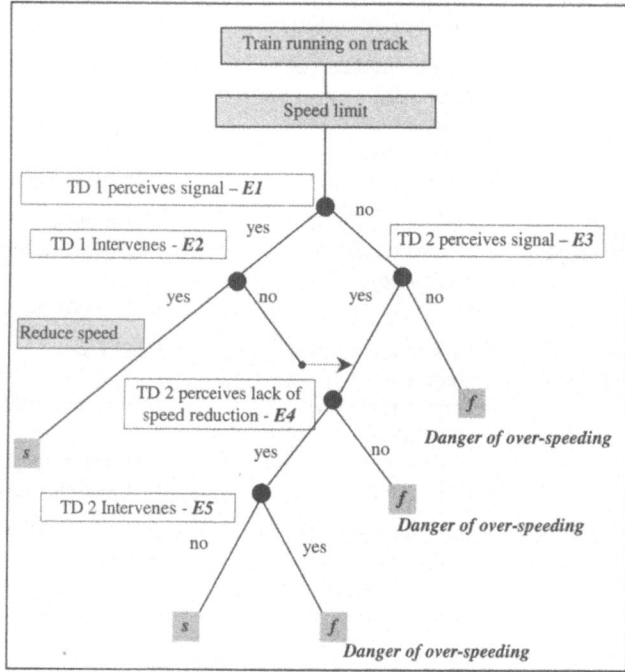

Figure 7.20 THERP-tree for Task 3: "Drive train." Subtasks 3.1.4: "Signals and signalling," "signalling at a glance," "speed limit indicators."
s = success of mission
f = failure of mission
TD 1, TD 2 = train driver 1, train driver 2

Table 7.14 Error modes, PIFs, and SHELL reference for Task 3: "Drive train." Subtasks 3.1.4: "Signals and signalling," "signalling at a glance," "speed limit indicators"

Error mode	PIF – error causes	SHELL reference
E1: TD 1 does not perceive speed limit	• Lack attention • Stressed • Time pressure • Bad visibility • Bad position of signal	• L: personality • L-H: heavy shifts • L-E: relationship with ERC • L-H: weather • L-H: interfaces
E2: TD 1 does not intervene	• Lack attention • Stressed • Bad visibility • Misinterpretation	• L: personality • L-H: heavy shifts • L-H: weather • L-H: cabin ergonomics
E3: TD 2 does not perceive speed limit	• Lack of attention as second TD • Complacency with companion • Bad position of signal • Bad visibility	• L: attitude • L-L: relationship with TD, partner TD • L-H: Interfaces • L-H: Weather
E4: TD 2 does not perceive lack of speed reduction	• Lack of attention as second TD • Complacency with companion • Stressed	• L: attitudes • L-L: relationship with TD, partner TD • L-H: heavy shifts
E5: TD 2 does not intervene	• Complacency with companion • Lack of communication • Misinterpretation	• L-L: relationship with TD, partner TD • L-E: noise, cabin ergonomics • L-E: cabin ergonomics

Task 3: "Drive train"
Subtasks 4.1.2: "Braking," "reducing speed," "resetting automatic signal recognition."

Initial Conditions:

- No specific failures or malfunctions.
- Fault in automatic signal recognition system; reset necessary.

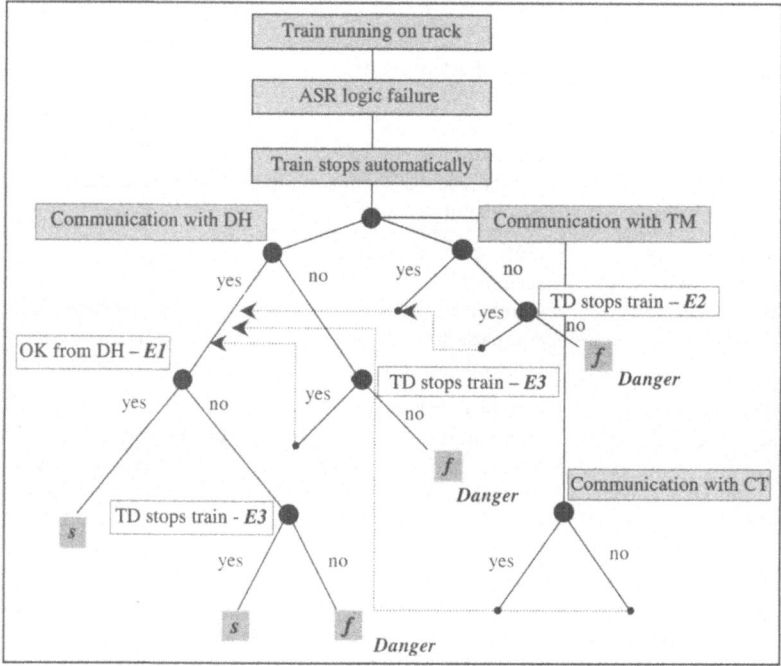

Figure 7.21 THERP-tree for Task 3:"Drive train." Subtasks 4.1.2:"Braking," "reducing speed," "resetting automatic signal recognition."
s = success of mission
f = failure of mission
DH = department head
TM = traffic manager
CT = crew train – controller
TD = train driver
ASR = automatic signal recognition (on board automatic safety train stop)

Table 7.15 Error modes, PIFs, and SHELL reference for Task 3: "Drive train." Subtasks 4.1.2: "Braking," "reducing speed," "resetting automatic signal recognition"

Error mode	PIF – error causes	SHELL reference
E1: DH does not give OK	• Lack of trust in TD • Stress • System rarely works	• L-E: relationship with ERC • L-H: workload • L-H: poor interface
E2: TD does not stop train in absence of communication with TM	• Lack of trust in TM • System rarely works • Poor rule knowledge • Time pressure • Machismo	• L-E: relationship with ERC • L-H: poor interface • L-S: training • L-E: work demands • L: personality
E3: TD does not wait for communication with DH	• Lack of trust or confidence in DH • Time pressure • System rarely works • Poor rule knowledge • Machismo	• L-E: relationship with ERC • L-E: relationship with ERC • L-H: poor interface • L-S: training • L: personality
E4: TD starts train without OK from DC	• System rarely works • Ignores rules • Machismo • Invulnerability • Time pressure	• L-H: poor interface • L-S: rules • L: personality attitudes • L-E: relationship with ERC

Task 3: "Drive train"
Subtasks 7.1: "Stopping train," "intermediate station," case a.

Initial Conditions:
- No specific failures or malfunctions.

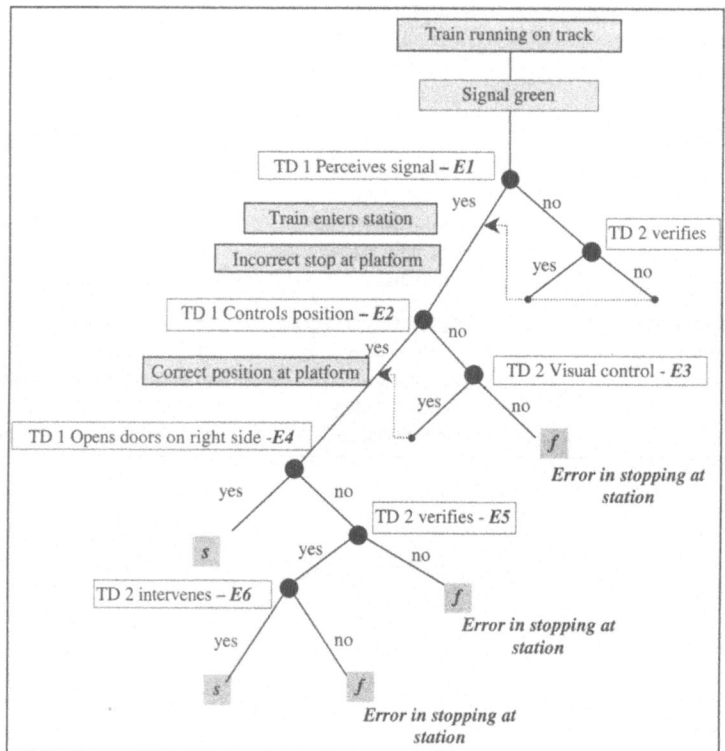

Figure 7.22 THERP-tree for Task 3:"Drive train." Subtasks 7.1:"Stopping train,""intermediate station."
s = success of mission
f = failure of mission
TD 1, TD 2 = train driver 1, train driver 2

Table 7.16 Error modes, PIFs, and SHELL reference for Task 3: "Drive train." Subtasks 7.1: "Stopping train," "intermediate station," case a

Error mode	PIF – error causes	SHELL reference
E1: TD 1 does not perceive green signal	• Bad visibility • Stress	• L-E: weather, maintenance • L-H: shifts
E2: TD 1 does not stop at right place on the platform	• Distraction	• L: attitude • L-H: shift workload
E3: TD 2 does not pay attention to train position	• Attention failure • Complacency with companion	• L-E: relationship with ERC • L-L: relationship with TD, partner TD
E4: TD 1 opens door on wrong side	• Habit • Lack of attention • Misinterpretation	• L: absent minded • L-E: shifts workload • L-E: cabin ergonomics
E5: TD 2 does not pay attention to door opening	• Attention failure • Complacency with companion	• L-E: relationship with ERC • L-L: relationship with TD, partner TD
E6: TD 2 does not intervene	• Complacency with companion • Lack of competence	• L-L: relationship with TD, partner TD • L-S: rules, training

Task 3: "Drive train"
Subtasks 7.1: "Stopping train," "intermediate station," case b.

Initial Conditions:

- No specific failures or malfunctions.
- "Red" signal on track.

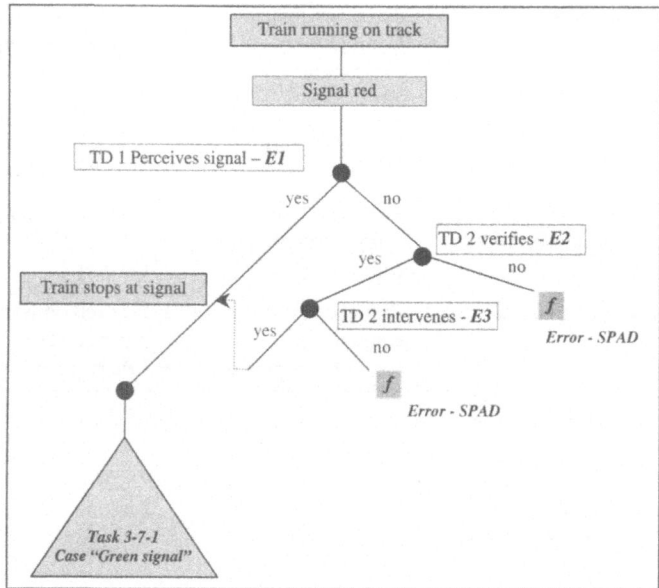

Figure 7.23 THERP-tree for Task 3: "Drive train." Subtasks 7.1: "Stopping train," "intermediate station."
s = success of mission
f = failure of mission
TD 1, TD 2 = train driver 1, train driver 2
SPAD = signal passed at danger

Table 7.17 Error modes, PIFs, and SHELL reference for Task 3: "Drive train." Subtasks 7.1: "Stopping train," "intermediate station," case b

Error mode	PIF – error causes	SHELL reference
E1: TD 1 does not perceive red signal	• Bad visibility • Stress	• L-E: weather, maintenance • L-H: shifts
E2: TD 2 does not verify signal	• Attention failure • Complacency with companion • Habit	• L-E: relationship with ERC • L-L: relationship with TD, partner TD • L: attitude
E3: TD 2 does not intervene	• Complacency with companion	• L-L: relationship with TD, partner TD

Task 3: "Drive train"
Subtasks 9.2: "Driving at a glance," "level crossing not automatically controlled."

Initial Conditions:
- No specific failures or malfunctions.
- "Red" signal followed by code "CD-180" = allowed to proceed at glance.
- Faulty level crossing (LC) systems. LC is open with no safety indication.

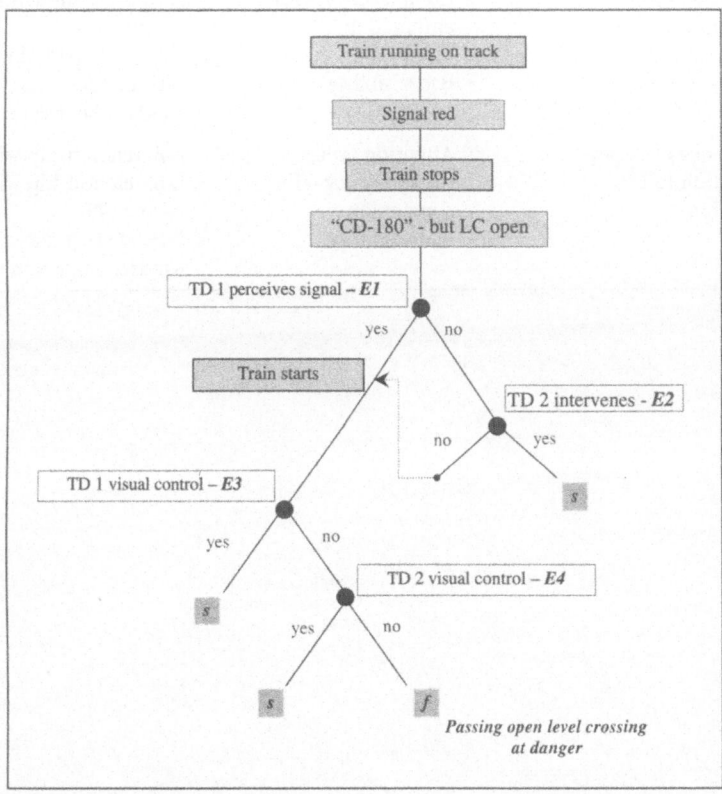

Figure 7.24 THERP-tree for Task 3: "Drive train." Subtasks 9.2: "Driving at a glance," "level crossing not automatically controlled."
s = success of mission
f = failure of mission
TD 1, TD 2 = train driver 1, train driver 2
CD-180 = Code 180 – "proceed at glance"
LC = level crossing

Table 7.18 Error modes, PIFs, and SHELL reference for Task 3:"Drive train." Subtasks 9.2:"Driving at a glance,""level crossing not automatically controlled"

Error mode	PIF – error causes	SHELL reference
E1: TD 1 does not perceive signal code 180	• Attention failure	• L: attitude • L-H: workload, shifts
E2: TD 2 intervenes and starts train	• Distraction • Complacency with companion	• L: attitude • L-H: shift workload • L-L: relationship with TD, partner TD
E3: TD 1 does not pay attention to LC position	• Time pressure • Complacency with automation • Tired • Bad visibility	• L-E: relationship with ERC • L-H: overconfidence in automation • L-H: workload, shifts • L-H: maintenance windscreen, weather
E4: TD 2 does not pay attention to LC position	• Attention failure • Complacency with companion • Bad visibility	• L-E: relationship with ERC • L-L: relationship with TD, partner TD • L-H: maintenance windscreen, weather

8

Application of HERMES for Accident Investigation: The Case of a Thermoelectric Power Plant

This chapter focuses on the application of the HERMES methodology in the performance of the investigation that followed an accident in an energy production plant.

Accident investigation in modern industrial environments is a very well established practice, especially when the impact of serious events on the environment and society is relevant. In these cases, safety authorities require that an accurate estimation be performed on the root causes, followed by recommendations and implementation of appropriate measures to improve safety.

In domains such as aviation and, in general, in all transport systems, with the exception of automotive road transport, in energy production and process industry, any major accident is followed by an internal investigation within the company and an external one performed by the safety authorities. In more serous cases, a specific "commission of enquiry" is set up of independent experts in the domain, called in to perform the investigation.

National and international organisations have developed guidelines for carrying out accident investigations (ICAO, 1986, 1988), which tend to focus primarily on the identification of root causes and on the way to classify events and reporting occurrences (ICAO, 1987, 1997). Rarely, the accurate analysis of the socio-technical context and familiarisation with the habits and attitudes of the organisation in which the accident has developed are explicitly identified as necessary preliminary steps of the investigation.

The HERMES methodology formalises the whole process of accident investigation and distinguishes clearly between two phases:

1. A preliminary phase that requires the definition of an adequate theoretical framework supporting the experts for the whole investigation, and the performance of ethnographic analyses on the working environments involved in the accident.
2. A second phase dedicated to the analysis of the actual accident, the recognition of root causes, and development of recommendations.

In the following sections, the problem statement for the specific accident investigation will be presented. Then, the way in which the HERMES methodology was applied for generating a consistent set of data, and for identifying a reference method to guide the root cause analysis and classification of events, will be described.

The selected method and data collected will be applied for the identification of the root causes of the specific accident and for the development of recommendations aimed at preventing the occurrence of similar accidents, as well offering means for recovery actions and mitigating the consequences in the case of similar occurrences.

Finally, the effectiveness of the recommendations will be evaluated in terms of HEAM, according to the requirements of the DSME (Define, Select and Implement, Monitor, and Evaluate) process discussed in Chapter 4 (§4.5) of this book.

As in the case of the previous chapters, in order to preserve confidentiality with respect to the organisation involved in the accident, some of the data and references have been slightly modified or omitted. This has not altered the overall process of application of the methodology nor has it reduced the practical implications involved in the performance of a real accident investigation.

8.1 Problem Statement

A serious accident occurred in a thermoelectric cogeneration plant, which provides electricity and hot pressurised water for heating purposes to a population of several hundred thousand people concentrated in a vast urban area.

The accident did not cause loss of lives or serious environmental damage, mainly because the release of energy and pollution was generally contained within the structures of the plant. However, the dimension of the events could easily have generated a much more severe impact on the population and environment surrounding the plant.

The investigation that immediately followed the accident was limited to the internal assessment of technical causes, and many human factors implications were only generically identified, with no deeper search for their root causes. This led to serious shortcomings in the recommendations concerning human factors aspects and the organisation owning the plant requested a further phase of investigation with the following goal:

- *Ascertaining detail HF aspects and defining possible improvements and ameliorations in the control process to enhance safety.*

In general, the performance of an accurate accident investigation is quite complex and requires a variety of competencies, especially in the technological domain of the accident, as well as in accident investigation, data collection, statistical analysis, field assessment, and human factors.

The need to include HF within the generic competencies of accident investigation is due to the fact that human factors are always present in any major accident. Moreover, in the case of a specific analysis focused on HF and their root causes, a particular competence and experience is required in this domain. The HERMES methodology is principally suited for guiding this type of investigation.

In the case of the accident discussed in this chapter, the active collaboration of the organisation managing the plant was of paramount importance to the success of the entire investigation, especially for the preliminary phase of work. In particular, top managers, operators, and primary actors involved in the accident were very cooperative.

This allowed the set-up of a limited team of human factors investigators (HF Team) made up of a few HF specialists with engineering know-how of the specific technological domain (energy production and thermohydraulics). In particular, the HF Team easily developed further experience of the plant performances and operations, and, in a relatively short time gained acquaintance with the habits and everyday practices, and knowledge about past events, relative to the specific plant. Moreover, the HF Team could access a variety of information and data concerning the accident under investigation in a concentrated and specific format for assessment.

The development of the accident analysis and the findings of the HF Team will be discussed in the following sections.

8.2 The Investigation of a Thermoelectric Power Plant Accident

The procedure discussed in Chapter 4 that describes the specific instantiation of HERMES for accident investigation was followed in detail.

The goals of the activity were clearly defined from the start of the work as discussed in the previous section. Then, a method judged most suited for the specific accident investigation was selected and applied. During the analysis of causes associated with the accident, an adequate monitoring of the attainment of the goals was performed. Finally, once the recommendations were developed and submitted to the management of the power plant, a plan for implementation and cost/benefit analysis was also provided to evaluate the effectiveness of the proposed safety measures.

The existence of a previous investigation report and the openness and willingness of the management and operators of the plant in supporting the specific HF investigation were extremely helpful. Therefore, even though the system was quite complex and the accident involved several operators, the Human Factors Team could be limited to the following members:

- A group of operators and shift supervisors, with considerable experience and expertise in the management of the plant.

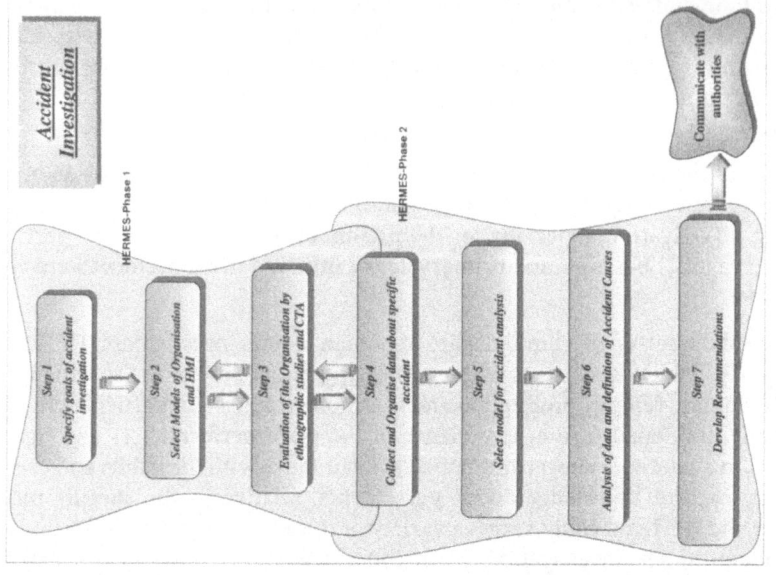

Figure 8.2 Stepwise procedure for accident investigation and HERMES phases.

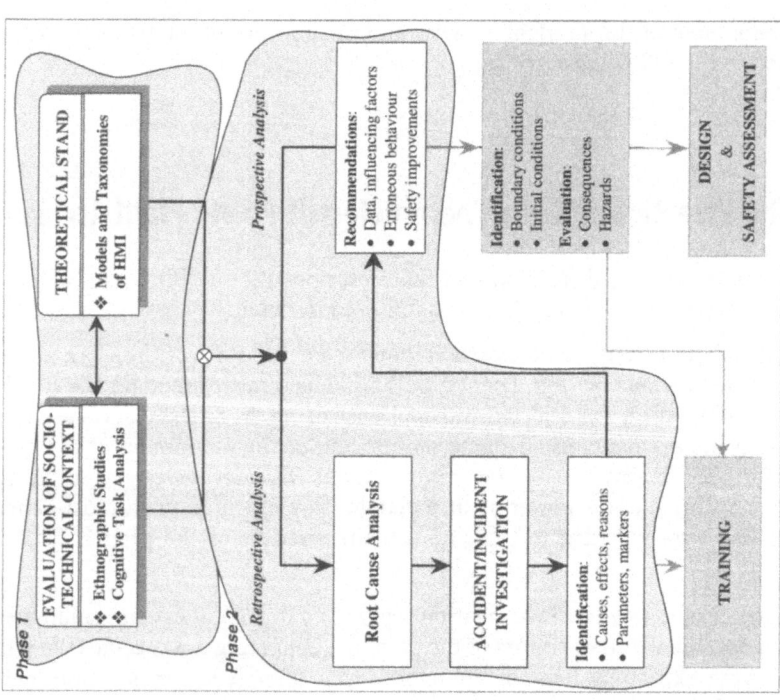

Figure 8.1 Application of HERMES for the accident investigation of a power plant.

- Two electrical and energy production engineers, with experience in human factors and knowledge in psychology and field studies.
- A human factors specialist, who supported the activity mainly from the methodological viewpoint.

The application of HERMES for defining causes and specific HF-related issues and for developing recommendations covered several months of work by the HF Team.

As discussed above, the accident investigation based on HERMES pivots on the actual occurrence and on the assessment of the socio-technical context in which the accident developed. Therefore, the application of the HERMES methodology was performed in two formal and time distributed phases (Figure 8.1). This process allows the integration of the HERMES methodology with the stepwise process for accident investigation discussed in Chapter 4 (Figure 8.2).

The two phases of application of the HERMES methodology were:

Phase 1. The evaluation of the socio-technical context of the work, and familiarisation with the practices, and habits existing within the power plant. This process was coupled with a more theoretical analysis of philosophies, policies, and procedures promoted by the management along with a definition of methods for root-cause analysis.

Phase 2. The detailed evaluation of data collected during *phase 1* and of information about the accident, and the performance of the actual root cause analysis and accident investigation with the identification of the causes, effects, and reasons for the accident and relative recommendations for safety improvement.

As all accident investigations always pivot around the accident itself, the description of the phases of application of HERMES will follow the detailed description of the plant and accident sequence, as reported by the formal investigation, which immediately followed the event. This preliminary step has been clearly identified as necessary for a practical approach to accident investigation in Chapter 4 (section 4.5.1) and implies the collection of data about the accident and the structuring of such data in a temporal sequence (Event Time Line, ETL) that gives an overall sequential picture of the accident and its dynamic development.

In this preliminary phase, it is very important that the team of investigators familiarises themselves with the plant and working situations, even before beginning the ethnographic analysis and field assessment. In this case, the engineering knowledge and experience in human factors of the HF Team is very important in generating an initial expectation of a possible modelling structure and of a specific method to apply in the accident investigation.

8.2.1 Plant Characteristics and Accident Sequence

Plant Characteristics

Plant Description

The plant under consideration combines electrical and thermal energy production, providing electricity and hot pressurised water for heating purposes for a population of several hundred thousand concentrated in an urban area.

The system comprises the necessary apparatuses for heat generation and distribution. Transport to subsystems for the heat exchange and distribution to the single users is also included. The major technical characteristics of the system existing at the time of accident were:

1. A *power generation group,* named "*GT-2,*" generating 136 MW of electrical power, fuelled by either methane gas or oil, giving a combined thermal power of 162 MW.
2. A *turbo-gas group,* fuelled by methane gas or gasoline, producing a nominal electrical power of 35 MW and a thermal power of 63 MW, obtained from the discharge gas.
3. A *thermal power station* for integration and reserve, able to provide a thermal power of 141 MW, produced by three steam generators of 47 MW each, fuelled by either methane gas or oil.
4. A *hydroelectric generating station* of 4.5 MW.

In Figure 8.3, a part of the control system of the turbine/alternator complex is shown (Piccini, 1998).

The distribution network consists of a dual piping system first connecting the power generation stations with three main-pumping systems, and then supplying the different main-users distributed along the whole network. These main-users consist of heat exchangers, which essentially provide hot water to each single user, i.e., private houses, shops, offices, etc.

The fluid utilised for thermal distribution is pressurised water, operating at a maximum temperature of 120°C in the feed-line and returning at 60–70°C in the return-line. In some special cases the water temperature could reach 100°C. The system operates at high pressure (approximately 0.95 MPa).

Altogether, the plant provides heat and electricity for an area that covers approximately 12 km^2, and the volume of buildings heated is approximately 22.5 10^6 m^3.

Control System

The power station could be configured in four "working modes": not available; stand-by; ready to operate; and in operation.

The power station and the three main-pumping systems could run in three different control modes:

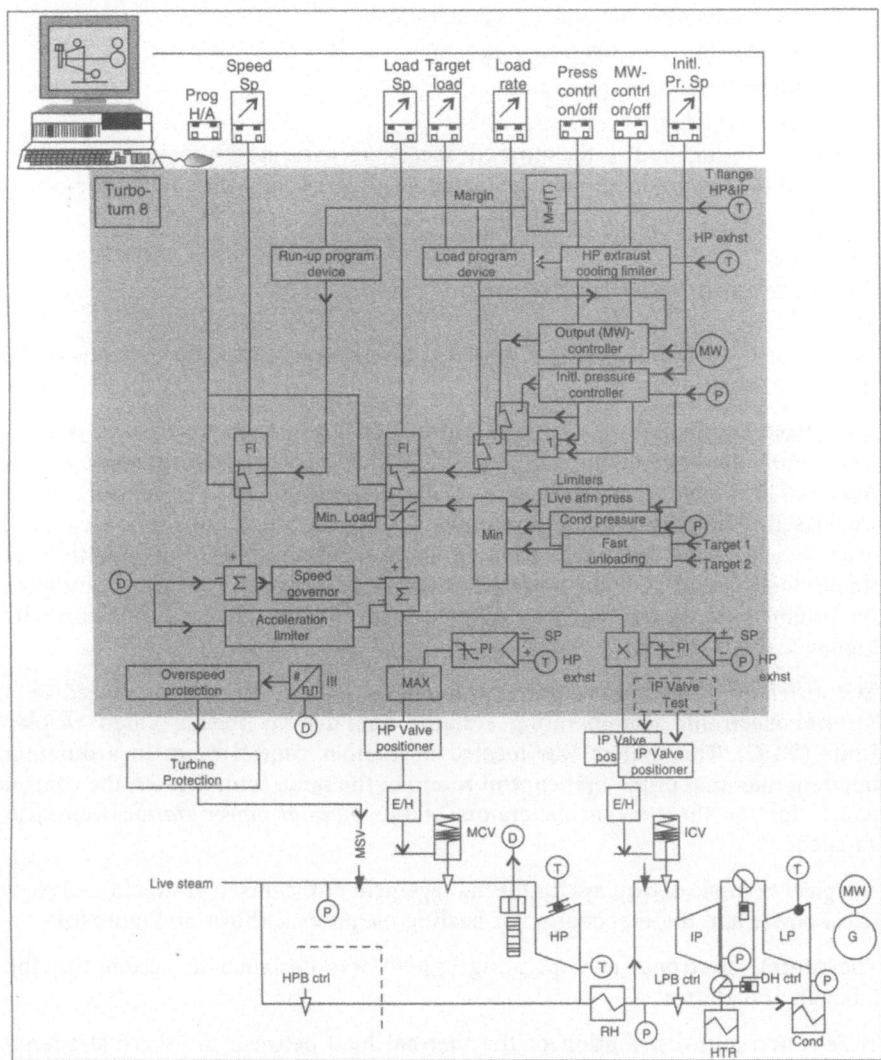

Figure 8.3 Portion of the alternator/turbine control system.

1. *Manual operations (M).*
   The manual operations are carried out locally by operators on all motor-operated valves, regulators, and all pumps. This control mode is considered as an emergency and backup system to be rarely utilised for obvious reasons of complexity and difficulty of performance.

2. *Automatic-local operations (AL).*
   This form of control implies that operators define and input all set points locally before starting automatic system operation. In particular, the following important set points have to be assigned:

- Thermal energy for each operating boiler.
- By-pass flow for each heat generator.
- Exit water temperature of boilers.

3. *Automatic-distant operations* (*AD*).

This operating mode is the most advanced one, in terms of supervisory control, and allows operators to manage the whole process from the control room.

Working Environment and Tele-operation System

The normal operations of the power station were performed in physically separated areas.

The mixed production of electrical power and thermal energy (*power generation group, thermal power station*, and *hydroelectric generating station*) was managed in a control room located in the same building. The control system was based mainly on synoptic instruments, and only some functions were governed by electronic instrumentation. In the same room, some of the functions of the *turbo-gas group* could be managed. However, the majority of the operations of this group were carried out in a different control room that was not normally manned.

The *distribution* of thermal power throughout the network was managed by a "central electronic tele-operating system," operated by means Video Display Units (VDU). This system was located in another control room in a different building than that of the first control room. In the same control room the control panels for the three steam generators of the *thermal power station* were also situated.

A typical example of displays for the management of the mixed production of electrical power and thermal energy for heating purposes is shown in Figure 8.4.

The central electronic tele-operating system was designed to accomplish the following functions:

1. Selection and distribution of the thermal load between the thermal energy generation groups.
2. Tele-control of the various supporting systems: pumping stations, back-up and pressurisation systems, regulation and throttling valves, local heat-exchange stations, and pressure losses.

Other instrumentation and management systems were present in the same control room of the central electronic tele-operating system. These were namely:

- Two VDUs for the management and control of the entire network.
- A personal-computer-based position for configuring the process control.
- Two redundant systems to the central electronic tele-operating system, for the acquisition of data and for the management of several control subsystems, called *Series-9000*.

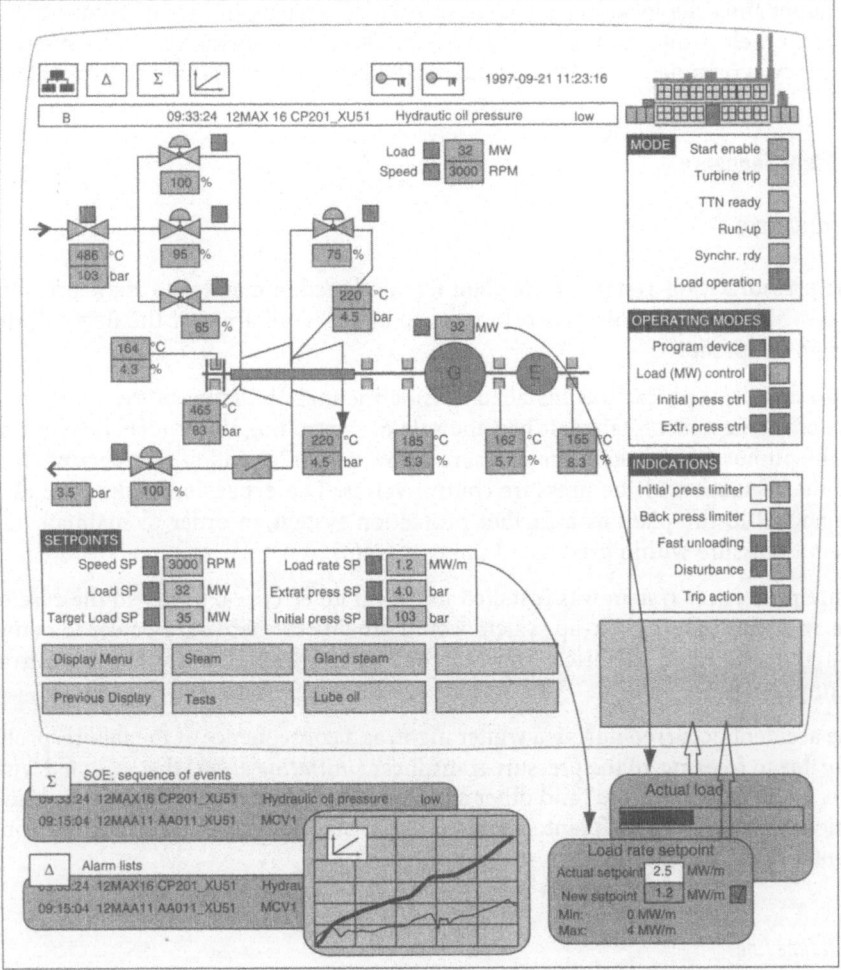

Figure 8.4 Typical human–machine interface for plant management by the central electronic tele-operating system.

- Two printers for presentation of data and alarms.
- Optical data system for the conversion of signals, through optical fibres, between the central electronic tele-operating system and the *Series-9000* control systems.

These management systems acted as primary nodes in the control process. They had two functions:

1. Connecting the different subsystems with the central electronic tele-operating system.

2. Supporting the local management of each subsystem, in case of failure of the central electronic tele-operating system. The actual operations on each emergency system had to be carried out locally, by means of specific control panels.

Accident Sequence

The Occurrence

The pressurisation system of the plant was designed to maintain a static pressure of 0.95 MPa in the whole network, so as to attain in all areas of the network the design temperature.

A pressure transducer was installed outside the main building of the plant, near one of the expansion tanks. It had the role of controlling the regular revolutions of the pumps (to maintain the pressure below 0.95 MPa) and of transferring data for the regulation of the pressure control valves. Two expansion tanks were also connected to the plant as a further protection system, in order to maintain the overall pressure within predefined safety margins.

A fire protection system was installed to cut off all electrical power in the case of fire. In such a case, a back-up system would provide the necessary power to maintain the minimum electrical power required to retain the automatic control settings.

The accident occurred during a winter night, as a consequence of the failure, probably due to freezing, of the pressure transducer (*initiating event*) that, coupled with a series of "human errors" and other system faults, led to a water hammer causing serious damages to the plant. There were no casualties and irrelevant environmental consequences.

Event Time Line

The evolution of the accident has been structured on a time sequence of events, called the Event Time Line (ETL) (Piccini, 1998; Carpignano et al., 1998, 1999).

Each event contains data and information for further analysis and represents the input to the methods and approaches selected for classifying the accident and identifying its root causes.

Figure 8.5 shows the ETL, where each event is shortly described and associated with the time of occurrence. The short description of each event is as follows:

Event 1. The pressure transducer failed due to a very low temperature (however, not exceptional) reached during a winter night, which also caused the freezing of the signal transmission piping. As a consequence of this failure, an erroneous pressure indication of 1.76 MPa, outside the design values, was reached, while, in reality, the pressure in the system was normal. The entire automatic control system started reacting to the (erroneous) signal, attempting to reduce the pressure to the design value.

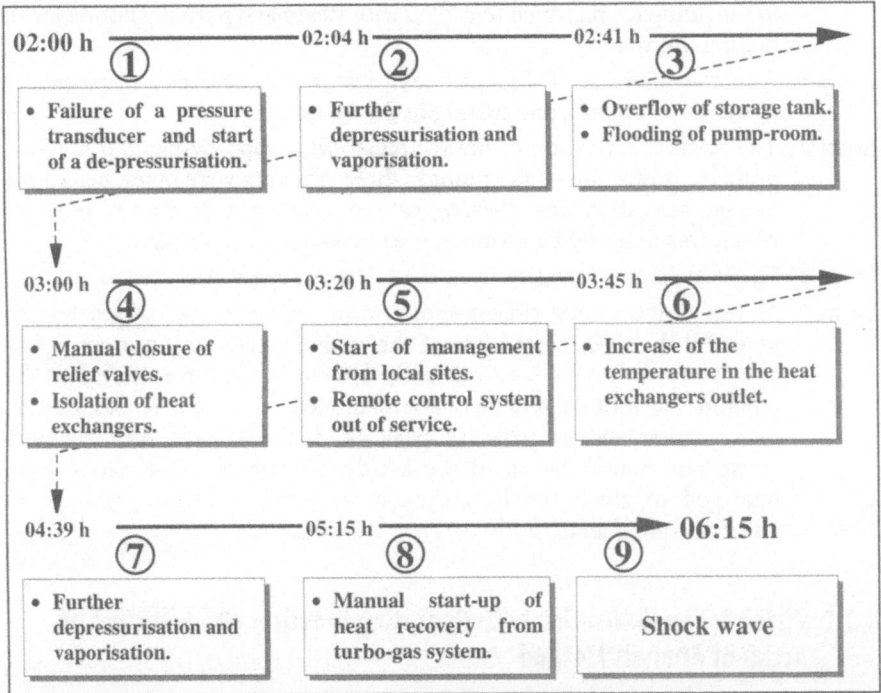

Figure 8.5 Event time line of the accident in the power plant.

The pressurising pump revolutions were reduced, the relief valves of the pressurising system and the valves of the expansion tanks were opened. The main feed pumps were also stopped. A rapid transient de-pressurisation started.

Event 2. Even though the pressure in the system was low, the main recirculation pumps were manually started from the control room, but at minimum revolution. This resulted in a further trend of water vaporisation.

Event 3. The attainment of the maximum level in the storage tank led to water discharge through the relief valves. Unfortunately, the emergency system was not of a sufficient dimension to fully discharge the water flow and some hot boiling water started to flood the pump room.

Event 4. The relief valves were manually closed and the heat exchangers were bypassed. These essential protective actions were performed, however, with a very relevant delay.

Event 5. The flooding of the pump room with boiling hot water caused the saturation of the atmosphere with vapour. This triggered the intervention of the fire protection system, which was unable to distinguish between vapour and smoke. The electrical power was completely cut off and, consequently, all pumps stopped working. The remote control system also went out of service and, from this moment forward, the most relevant control operations had to be performed locally, including the operations in the pump room, which was filled with steam and partially flooded with boiling hot water.

Event 6. A further increase of the water temperature occurred as a consequence of the water spilling and mixing in the system.

Event 7. The manual activation of the recirculation pumps, by-passing the low-pressure protection of the pumps (three attempts were made before the manual activation was effective) caused a further reduction of pressure which was followed by a violent pressurisation of the system.

Event 8. The heat recovery system from the *turbo-gas group* was started.

Event 9. At this point a very violent shock wave was generated which was so powerful that it displaced one of the anchor points of the main piping line of 90 mm, leaving it permanently deformed. The force exerted on the pipeline was estimated to be of approximately 35–45 tons corresponding to an instantaneous pressure peak of 3–4 MPa. This was the only immediate manifestation of the accident, however – the shock wave managed to reach the heat exchangers, causing serious damage to internal components.

8.2.2 Phase 1: Familiarisation with Operating Practices and Selection of Accident Analysis Method

This first phase of work is essential in the application of HERMES, as it leads to the selection of the basic instruments for performing the human factors analysis based on the ethnographic study of the organisation for assessing procedures and practices during normal and emergency conditions.

The work was highly simplified by the availability of operators to discuss problems and practical solutions during the management of the plant.

The combination of this unbiased acquaintance with real working context and the knowledge of the history of the accident strongly affected the identification of the most suitable approach for carrying out the root cause analysis and accident investigation process (Cacciabue et al., 1998).

In practice, phase 1 of work was structured into two correlated steps performed in sequence by the HF Team:

Phase 1, Step 1. Familiarisation with the working context and practices of plant control and emergency management.

Step 2. Selection of the most suitable method for root cause analysis and accident investigation.

Familiarisation with Working Context and Practices

A period of familiarization with the plant was performed, based on "walk-through" and "talk-through" activities. These required several weeks of visits and observation of the working contexts, as well as interviews with operators, based mostly on open and unstructured formats.

These field investigations were coupled with the study of the plant characteristics. These focused on the design of the control and emergency systems, on the structural and geometrical layout of the plant and its control rooms, on the procedures for normal and emergency operations, and, in general, on the overall plant performance.

The objective of these combined activities was to become as familiar as possible with the actual working contexts, and to understand the amount of empirical actions that operators had to put into practice in order to comply with the required plant performances and efficiency of production.

From reading the accident report and from the first visits to the plant, it became immediately clear that the system was not sufficiently regulated by formal procedures, given the variety and complexity of phenomena involved, the dimension of the plant, and the possible hazards involved. Too many control activities were assigned or left to the practical know-how and experience of operators. Two aspects were particularly critical in the management of processes:

1. The lack of written emergency operating procedures.
2. The complexity of the central electronic tele-operating system.

Emergency Operating Procedures

The Emergency Operating Procedures (EOP), as well as the Standard Operating Procedures (SOP), were developed with reference to the preliminary design of the plant, but had never been revised or adapted to the control systems actually implemented.

A number of drawbacks were identified, such as: only few situations were contemplated; no account was given for a number of subsystems actually operating in the plant; no reference was made to the HMI in the control rooms; no description was given of the logistic distribution of control means; no account was made of the empirical actions required in order to cope with production requirements.

These problems led operators to consider the written procedures with extreme parsimony and prudence, as they were presumed not to correspond to real system configurations and conditions. Written procedures were actually thought as being more confusing than helpful in the management process.

Another element of risk associated with plant management was identified in the lack of training of the novices or the retraining of experienced operators. Normally, the novices received only "on-the-job" training, performed by expert operators. Therefore, very little grounding was given on emergency management.

Emergencies were to be dealt with by most experienced operators, with minimum guidance from existing procedures.

Consequently, a simple and straightforward recommendation could be developed from this first analysis:

- *To develop written procedures for normal and emergency conditions (SOP and EOP) to be applied by operators in a wide variety of circumstances and dynamic situations.*

Central Electronic Tele-operating System

The central electronic tele-operating system was designed to exploit the supervisory control power of modern electronic systems. The extensive use of VDUs helps in reducing the amount of space and distribution of controls and offers the possibility to concentrate the information on the whole system performance in a few screens and pages of computer presentation.

However, several problems exist and may be aggravated if all aspects of supervisory control, and especially the drawbacks, are not dealt with appropriately. Firstly, the management of the alarms was carried out in a dedicated "page" of a VDU, and overlapped with the normal indicators of crucial system variables. This required continuous shifts from one page to the other during an emergency situation in order to develop a clear awareness of the plant conditions. Moreover, many indicators were unable to manage the extreme values recorded by sensors during an emergency and they simply showed the indication "out-of-range," which was of little help.

Moreover, no hierarchy was planned in the presentation of the alarms and also the written information reported on the printers associated with trend and dynamic recording of data about system performance were cumulative. The operators had to differentiate between various indications and values.

Other difficulties with the management system both at the centralised and local levels were identified and examined, such as:

- issues of syntactic and semantic nature in the displays;
- inappropriate grouping in the representation of important parameters and process variables;
- incoherencies of graphical and qualitative data presentation, as well as of representation of devices;
- erroneous grouping in one single interface of controls pertaining to different systems.

Consequently, from the evaluation of the central electronic tele-operating system a second major recommendation could be developed:

- *To redesign the computerised control unit and associated interface in consideration of the fundamental principles of supervisory control and human-centred design.*

Finally, the evaluation of the EOP and central electronic tele-operating system clearly indicated the existence of several problems at the organisational level. These represented contributing factors and latent conditions favouring inadequate behaviour of operators during the management of an accident.

Selection of the Most Suitable Method for Root Cause Analysis

The step of the selection of the most suitable method for performing the accident investigation and discovering root causes was very strongly influenced by the experience and findings of the walk-through and talk-through activities and by the analysis of the accident report.

From the early stages of the investigation, it became clear that the plant, although managed very successfully and efficiently, was affected by a variety of organisational issues and latent failures that had a paramount effect in the development of the events and were fundamental in the performance an accurate etiology of the accident.

Therefore the primary goal of the investigation became the identification of these organisational factors and latent failures at systemic and human level, while the definition of human errors and inadequate functions of cognition of front line operators or specific deficiencies of components affecting local events were of lesser importance. Consequently, the HF Team focused towards the selection of a "macroscopic" type RCA approach (section 3.6).

The Integrated Systemic Approach for Accident Causation (ISAAC) (§3.6.1) seemed the most suitable method to apply in this case. The ISAAC method has been described in detailed in Chapter 3. However, it will be briefly recalled here for completeness (Figure 8.6).

The objective to concentrate on organisational and latent factors favoured the selection of a fairly simplified and generic taxonomy of human errors and system failures.

For human errors, the taxonomy based on Reason's model (1990, 1997), considers the following items:

- *slips*, *lapses*, *mistakes*, and *violations* for erroneous behaviours;
- *active* and *latent* errors for defining inappropriate actions performed either during the actual development of the accident or at an earlier time; and
- *management* or *organisational* errors for distinguishing between errors at the organisational level from *active* errors at the *individual* level.

System failures entail any malfunction or inappropriate performance of physical components. However, as the goal of the investigation was to identify human-related root causes, no specific taxonomy of system failures was considered.

The ISAAC method offers the possibility of associating with a system failure either a "latent or distant human" root cause, such as, for example, a "design error" or "maintenance error," or a contextual condition, or even a random cause. Any occurrence, i.e., a system failure, an error, or a negative or inadequate working context or condition, is connected to other occurrences by a set of logical links.

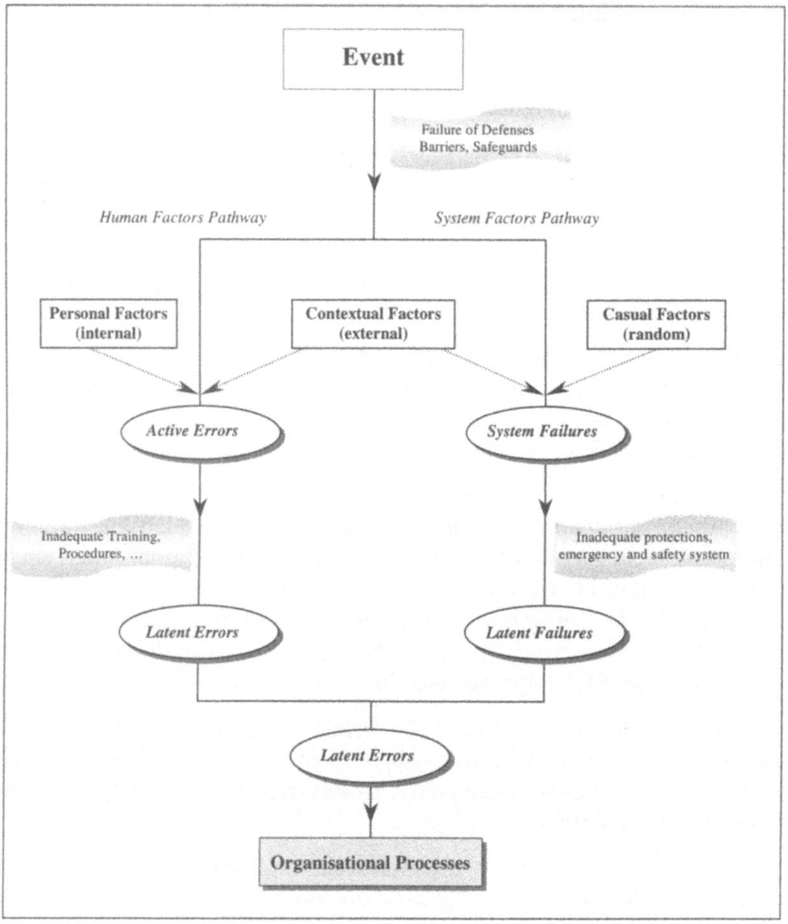

Event

Failure of Defenses
Barriers, Safeguards

Human Factors Pathway   *System Factors Pathway*

Personal Factors
(internal)

Contextual Factors
(external)

Casual Factors
(random)

Active Errors   *System Failures*

Inadequate Training,
Procedures, ...

Inadequate protections,
emergency and safety system

Latent Errors   *Latent Failures*

Latent Errors

Organisational Processes

Figure 8.6 Integrated systemic approach for accident causation (ISAAC) method (section3.6.1).

In order to enable the graphical representation of dependencies between occurrences, a certain formalism of connection is adopted:

- The *arrowed full line* indicates "backwards logical links" and implies that an occurrence "*A*" was generated by or favoured by occurrence "*B*";

The dependence of active errors and system failures on latent failures and of these on organisational processes can be depicted by "backwards connections," to show the complete chain of logical connections (Figure 8.6).

- The *arrowed dotted line* indicates "forwards logical link" and implies that an occurrence "*C*" engendered or favoured occurrence "*D*";

The identification of contextual, personal, and random factors that may affect specific failures or errors are represented by "forward connections" (Figure 8.6).

Moreover, in a specific root cause process, it is possible that an active failure may engender the failure of certain components (or vice-versa). These are typical links between occurrences that can also be identified by "forward connections."

In this way the complete set of connections can be developed in a single ISAAC tree. The method can be applied in retrospective or prospective mode, according to whether it is utilised to perform an accident investigation or to develop a safety assessment or to design a safety system. For the case under development, the method was applied in a retrospective mode to support the accident investigation.

The application starts by considering the sequence of *events* distributed in time, identified by a specific Event Time Line (ETL) analysis.

Events are usually associated with the failure of a protection or safety system, i.e., a defence, barrier, or safeguard, and can be the result of a combined effect of systemic and/or human-related factors.

For each *event*, the (RCA) is then developed according to the two following pathways:

1. The *human factors pathway*, which is applied to assess the causes associated to human nature, e.g., an operator's inappropriate action or inadequate of behaviour. The process of RCA in a *human factors pathway* develops as follows:

   • The contextual and personal factors that may have affected the specific behaviour, or inappropriate action, are identified. The forward link (*arrowed dotted line*) in Figure 8.6 indicates that certain *personal* or *contextual factors* may have affected, favoured, or engendered the specific inadequate behaviour.

   • In the case that the action is actually considered erroneous, then an *active error* is defined.

   • In certain cases an erroneous action may be considered not only dependent on personal or contextual factors, but may be the result of certain circumstances or be caused by reasons at the organisational level. Therefore, the analysis is expanded to look for the *latent errors* and/or *organisational processes* that engendered the inappropriate behaviour.

2. The *system factors pathway*, which is applied for evaluating the technical aspects of the *event*. An equivalent analysis to the one above is performed in order to discover whether a system failure is due to local random and/or contextual factors or resides in *latent failures* or *organisational processes*.

Latent errors and failures can be specific to the human factors pathway or to the system pathway, or can be common to both, as they are directly related to higher organisational processes that may have affected both types of latent conditions.

The outcome of the ISAAC analysis shows the combination of human-related and system-dependent factors that led to the various *events* and eventually to the accident as a whole.

The application of this procedure has been applied to the case of the power plant and will now be discussed in detail.

8.2.3 Phase 2: Performance of Accident Investigation and Development of Recommendations

Phase 2 essentially consisted in the detailed analysis of the nine events that had been identified in the ETL analysis using the ISAAC method, and the development of the recommendations for the company owning the plant, aimed at improving the safety level of the whole plant.

Phase 2 was therefore clearly divided into two correlated steps, namely:

Phase 2, Step 1. Performance of the RCA and accident investigation using the ISAAC method.

Step 2. Development of recommendations to improve safety.

Root Cause Analysis and Accident Investigation

Each event of the ETL (Figure 8.5) was classified and analysed according to the ISAAC method.

Event 1: Initiating Event

The initiating event of the accident consisted of the failure of the pressure transducer and the consequent depressurisation, which was automatically initiated by the protection system.

The detailed analysis of *event 1* and the single contributors to this event have been classified as follows (Figure 8.7):

1. *Contextual Factor:*
   - The external temperature was −5°C, well below freezing, but not unusual for the time of the year in that particular region.

2. *Local Working Conditions:*
   - The automatic control system did not have an alarm system for "over-range" of important physical variables, such as pressure or temperature.

3. *Contextual and Personal Factors:*
   - The event occurred at night, in conditions of low thermal power production and relatively small water flow. These factors caused a low level of attention by the operators in the first phase of the transient depressurisation.

4. *Active Error:*
   - During the first phase of the transient depressurisation, although correct indications were available on the ongoing processes, the operators did not understand the real conditions of the plant and did not undertake the adequate actions that would have avoided the rapid depressurisation. (*Mistake*)

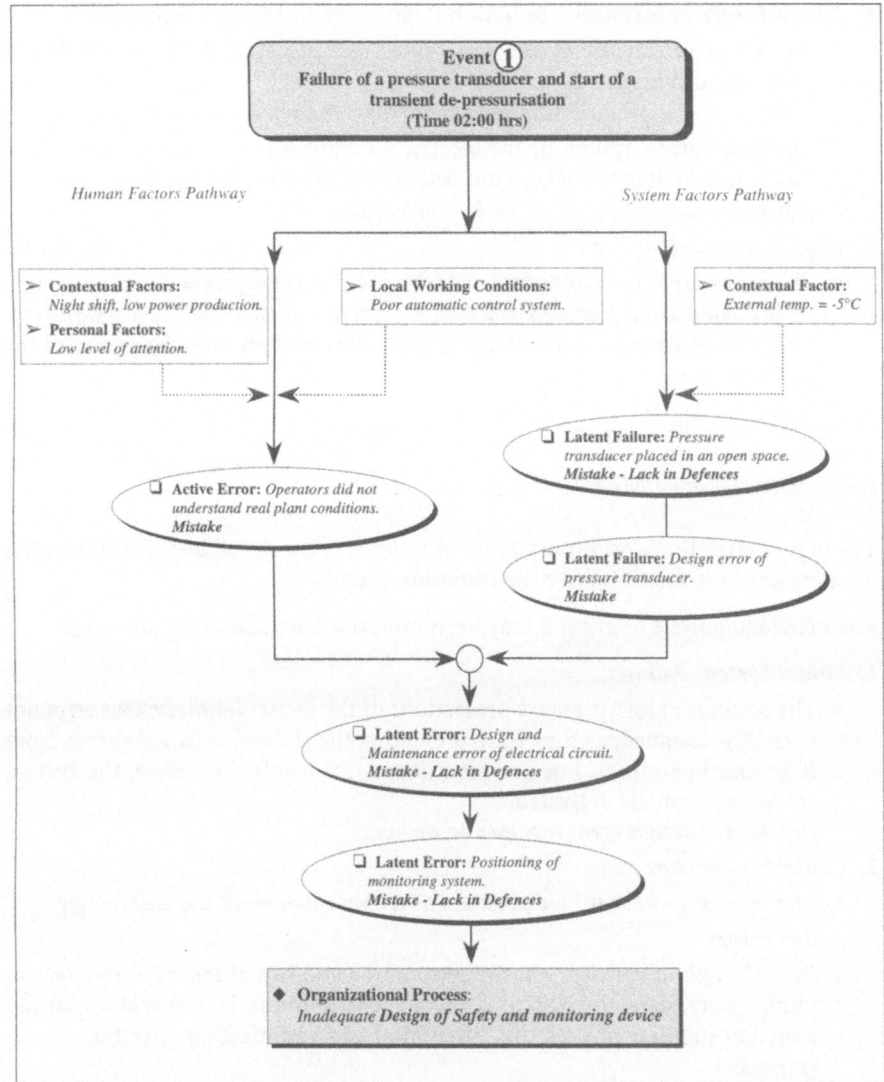

Figure 8.7 Analysis of *event 1*: Initiating event: failure of the pressure transducer and start of depressurisation.

5. *Latent System Failures and Lack in Defences:*
   - The pressure transducer was placed in an open space.
   (*Mistake – design error = > lack of defences*)
   - This transducer performed too many functions at the same time, namely: (a) control of the pressure level of the pumps; (b) activation of the safety and release valves; (c) presentation of indications and alarms to operators. Due to this "common cause failure," all these functions were lost.
   (*Mistake – design and management error*)

6. *Latent Errors*, contributors to both human error and system failure:
   - The electrical circuit of the transducer heating system was not properly designed and installed.
     (Mistake – Design and Maintenance error = > Lack in Defences)
   - The monitoring system of the electrical circuit was wrongly positioned, so that it was unable to register any anomaly and warn the operators.
     (Mistake – design error = > lack of defences)
7. *Organisational Process:*
   - The main organisation process affecting the initiating event, independently of the contextual and environmental conditions that favoured the event, was the inadequacy in the design of the safety and monitoring device of the pressure transducer.

Event 2: Further Depressurisation and Vaporisation

The depressurisation and vaporisation of water was enhanced by the intervention of operators four minutes after the initiating event.

The detailed analysis of *event 2* may be summarised as follows (Figure 8.8):

1. *Latent System Failure:*
   - The settings of low-pressure protections of the recirculation system were not correctly designed, as they were protecting the individual components from improper operations, but not the system as a whole. Therefore, the system could be manually activated.
     (Mistake – design error = > lack in defences)
2. *Contextual Factor:*
   - The operators were under pressure as a consequence of the emergency.
3. *Active Error:*
   - Even though the pressure in the system was low, two of the main circulation pumps were manually started from the control room. This operation caused a further increase of pressure differential and vaporisation of water.
     (Mistake)
4. *Latent Management Errors:*
   - No training programme was provided for the operators, who learnt directly by experience. They were unprepared to face an emergency situation. In particular, they were not sufficiently informed about causes and consequences of shock waves.
     (Mistake – management error = > lack in defences)
   - No written emergency procedures were available.
     (Mistake – management error = > lack in defences)
5. *Organisational Process:*
   - The organisation as a whole could be considered inadequate from the safety perspective, especially concerning training, design of safety devices, and emergency procedures.

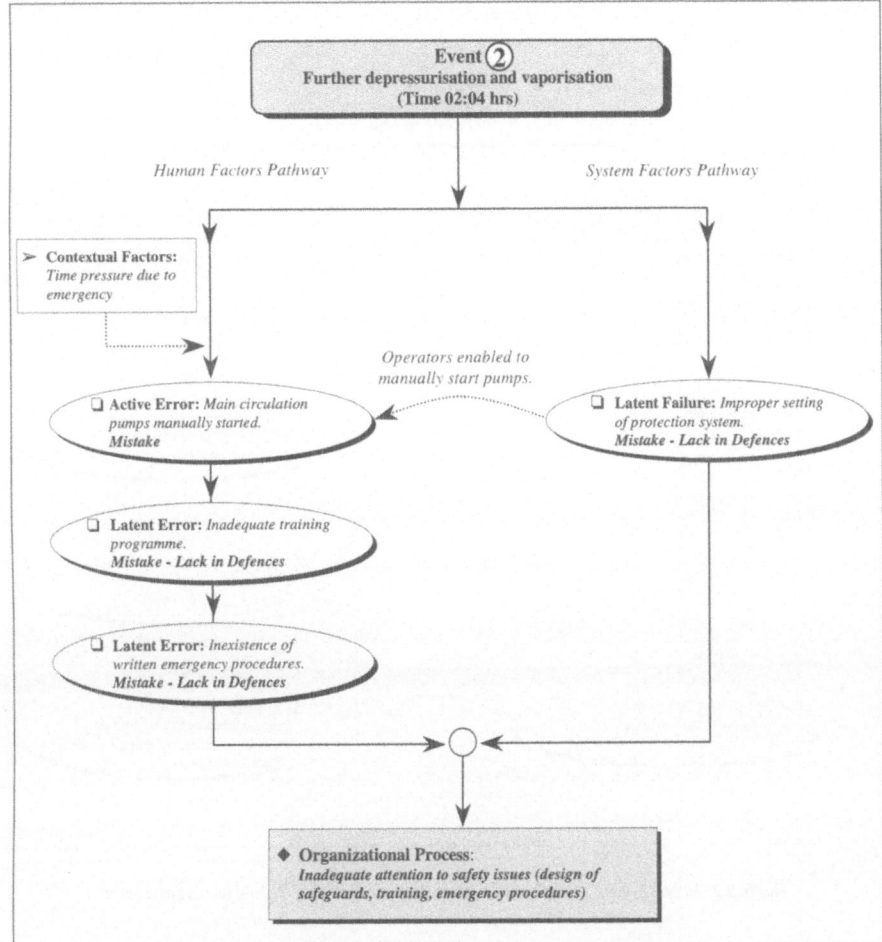

Figure 8.8 Analysis of *event 2*: Further depressurisation and vaporisation.

Event 3: Overflowing of Storage Tank and Flooding of Pump Room

Forty-one minutes after the initiating event, the local control pump room started to flood with boiling hot water and steam, following an overflow in the storage tank.

The detailed analysis of *event 3* is shown in Figure 8.9 and may be summarised as follows:

1. *Latent System Failures:*
   - The discharge pipe of the storage tank had a smaller diameter than the relief pipe. Consequently, the water over flow coming from the storage tank could not be completely drained out.
     (*Mistake – design error*)

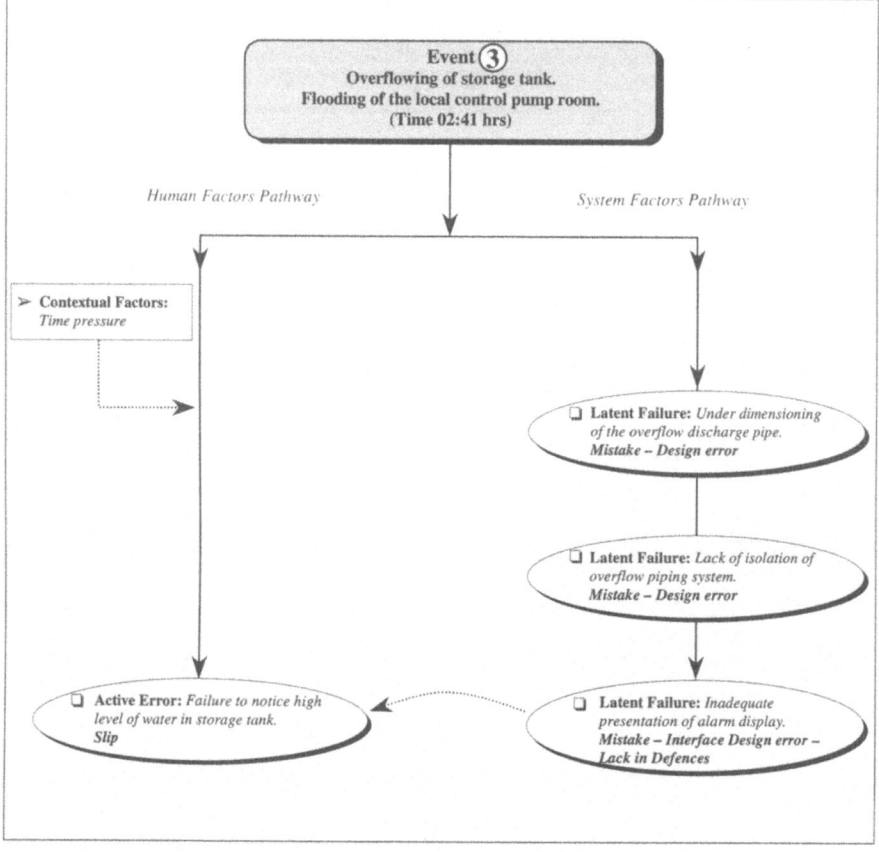

Figure 8.9 Analysis of *event 3*: Overflowing of storage tank and flooding of the local control pump room.

- As the pipe system was not properly isolated, the boiling hot water (80°C) that could not be discharged entered the local control pump room. (*Mistake – design error*)
- Event thought the alarm of *high water level* in the storage tank appeared on the control panel, it was barely visible, cluttered amongst other alarms of minor importance. The synoptic interface representation was poor. (*Mistake – interface design error = > lack in defences*)

2. *Contextual Factor:*
   - The operators were under pressure as a consequence of the emergency.

3. *Local Working Conditions and Lack in Defences:*
   - The displays of the alarm system were not clear and there was no hierarchy of importance or adequate support to distinguish amongst alarms.

4. *Active Errors:*
   - The alarm of high water level in the storage tank was not noticed. (*Slip*)

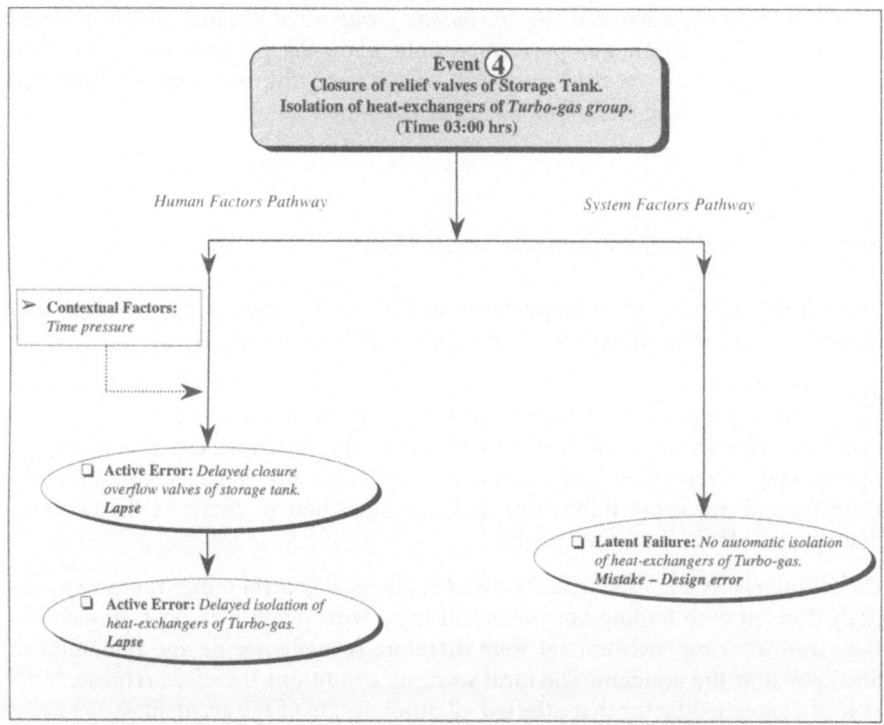

Figure 8.10 Analysis of *event 4*: Closure of relief valves of storage tank and isolation of heat-exchanges of *turbo-gas group*.

Event 4: Isolation of Storage Tank and Heat-exchanges of Turbo-gas Group

Event 4 is particular, as it represents a positive occurrence in the accident sequence. However, the two actions considered in this event, although very relevant in the development of the transient depressurisation, were performed very late in the accident evolution. Therefore, they can be considered errors of delay of necessary safety and protection actions.

The detailed analysis of *event 4* is summarised as follows (Figure 8.10):

1. *Latent System Failure:*
   - The by-pass of the heat exchanger of the *turbo-gas group* was not automatically started by the alarm signals of *low pressure* or *low water flow*. This action had to be performed both manually and locally.
     (*Mistake – design error*)
2. *Contextual Factor:*
   - One hour into the accident, the operators could still be considered under pressure as consequence of the emergency.
3. *Active Errors:*
   - The relief valves of the storage tank were closed with severe delay. This favoured fast depressurisation and overflowing of the storage tank.
     (*Lapse*)

- The heat exchangers of the *turbo-gas group* were isolated with a relevant delay. This action was performed only when the pressure inside the heat exchanger had reached very low values, and the exchange of heat had favoured boiling and vaporisation of the water.
  (*Lapse*)

Event 5: Beginning of Accident Management from Local Sites

Event 5 was of paramount importance in the development of the accident and human actions, even though it was not particularly complex per se.

The "station blackout" due to the intervention of the fire protection system caused the temporary loss of the electrical supply power to the entire plant. The operators were able to re-establish electrical power supply. However, the centralised control system remained "out-of-service." This made any management operation from the control room impossible and operators had to continue the accident management from local control sites.

Unfortunately, as a consequence of *event 3*, the local control pump room was partially flooded with boiling hot water and filled with steam. The operating conditions and working environment were therefore severely hostile and remained so for the rest of the accident. The local working conditions therefore represented a crucial contextual factor that affected all other events of the accident.

In order to describe the sequence of steps, *event 5* has been subdivided in two sub-events connected by backwards logical links:

- out of service of the remote control system (*event 5_2*);
- intervention of fire protection system (*event 5_1*).

The root cause analysis of the entire *event 5* does not contemplate any active error. However, the *human factors* pathway was developed taking into consideration the latent design failure of the absence of any information display from remote control locations to support the operators. A number of organisational and management issues were identified, as summarised in the following ISAAC tree (Figure 8.11):

1. *Latent System Failure and Lack in Defences* (*event 5_2*):
   - The beginning of accident management from local sites, due to the unavailability of the centralised control system, was a direct consequence of the "out-of-service" of the centralised control system. The operating system was equipped with a continuity group, as a back-up system in the case of a blackout. During the transfer from external to back-up electrical power, the centralised supervisory system failed and could not be reinstated. This event was quite common and was due to an error in the design of the system.
     (*Mistake – design error = > lack in defences*)

2. *Local Working Conditions:*
   - From this moment, all control operations had to be carried out manually and locally, without a comprehensive overview and understanding of the plant conditions. Some of the "*local working conditions*" were very unfavourable

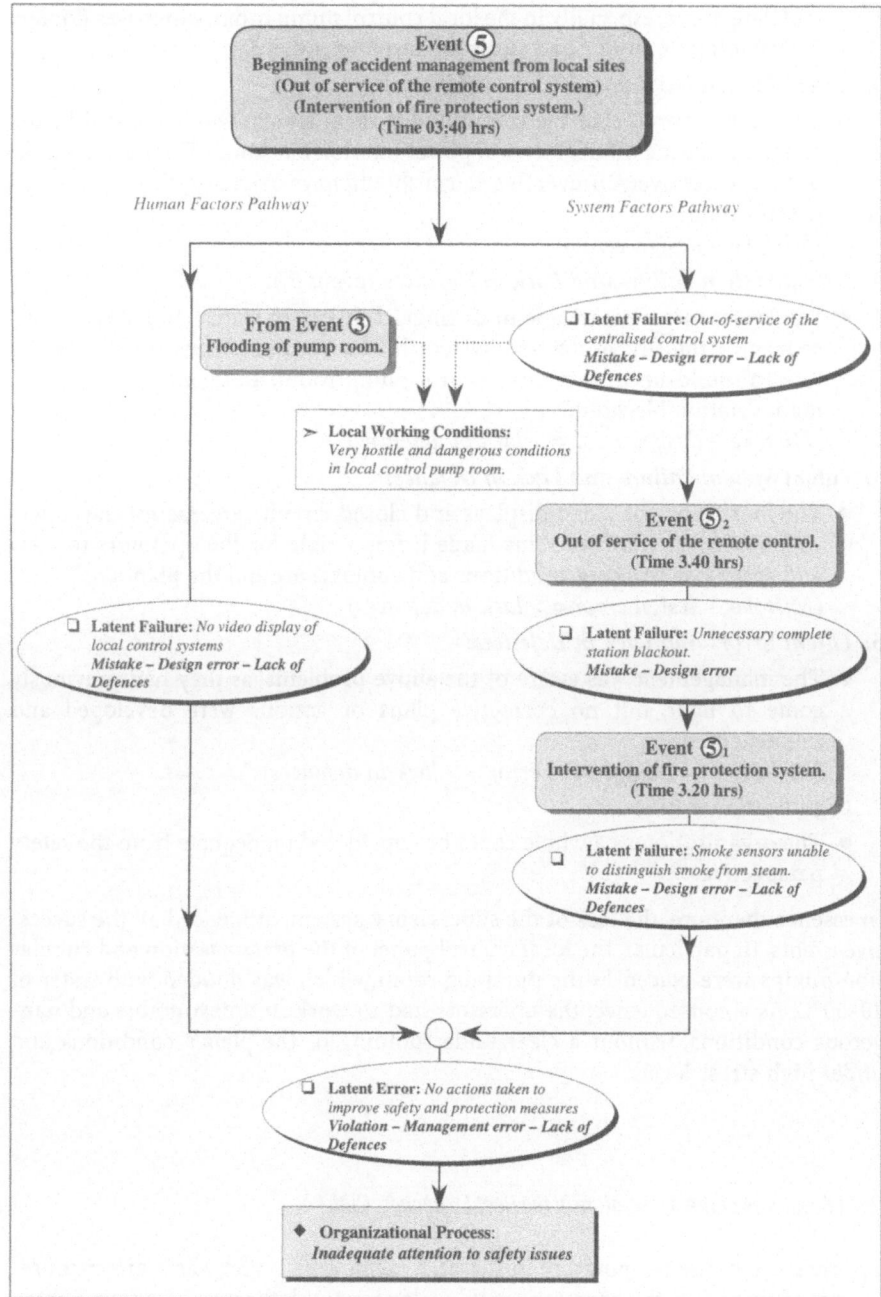

Figure 8.11 Analysis of *event 5*: Beginning of accident management from local sites; out of service of remote control system; and intervention of fire protection system.

and dangerous, especially in the local control pump room, which was flooded with boiling hot water and steam, as result of *event 3*.

3. *Latent System Failure and Lack in Defences:*
   - The "out-of-service" of the centralised control system was generated by the complete cutting off of electrical power (station blackout). The station black-out was excessive. A preventive alarm should have been installed in the main control room.
     (Mistake – design error)

4. *Latent System Failure and Lack in Defences (event 5_1):*
   - Smoke sensors were unable to distinguish between smoke and steam. Consequently, the fire protection system intervened, as a consequence of the "steam" build-up in the local control pump room, and generated an automatic "station blackout."
     (Mistake – design error = > lack in defences)

5. *Latent System Failure and Lack in Defences:*
   - The inexistence of video displays and closed circuit cameras for the supervision of local control rooms made it impossible for the operators to view and supervise working conditions and contexts around the plan.
     (Mistake – design error = > lack in defences)

6. *Latent Errors and Lack in Defences:*
   - The management was aware of the above problems, as they had previously come to light, but no corrective plans or actions were developed and implemented.
     (Violation – management error = > lack in defences)

7. *Organisational Process:*
   - The organisation as a whole could be considered inadequate from the safety perspective.

In essence therefore, the loss of the supervisory system, influenced all the successive events. In particular, the local control panel of the pressurisation and circulation pumps were placed in the pumping room, which was flooded with water of 40–50°C. As a consequence, the operators had to work in unfavourable and dangerous conditions, without a clear understanding of the plant's conditions and under high stress levels.

Event 6: Increase of the Temperature in the Heat Exchangers Outlet

The increase of the temperature in the heat exchangers outlet was a direct consequence of event 5, as the operators could no longer check the overall system behaviour, having to manage the system from local positions only. The whole protection system turned out to be inadequate to manage this type of emergency.

The root cause analysis of *event 6* covers management issues as well as individual aspects, and may be summarised as follows (Figure 8.12):

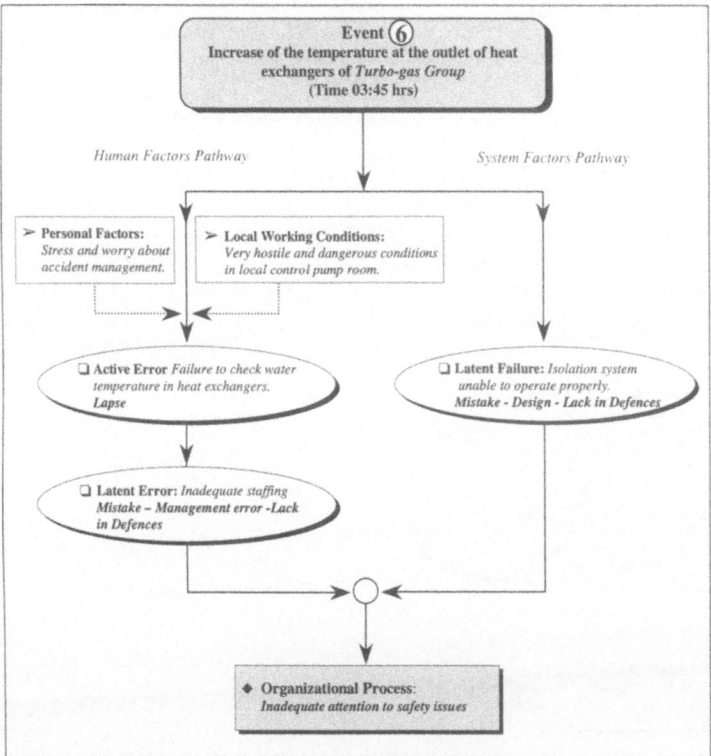

Figure 8.12 Analysis of *event 6*: Increase of the temperature in the heat exchangers outlet.

1. *Latent System Failure and Lack in Defences:*
   - The isolation system of the *turbo-gas group* was unable to fulfil its requirements.
   (*Mistake – design error = > lack in defences*)

2. *Contextual and Personal Factors:*
   - The operators were located in a very unfavourable and dangerous environment, in the local control pump room.
   - The operators were stressed and worried about the management of the heat generators. The time was 1 hour and 45 minutes into the emergency.

3. *Active Error:*
   - The operators did not keep the temperature of heat exchangers in the *turbo-gas group* under control after isolation.
   (*Lapse*)

4. *Latent Error and Lack in Defences:*
   - The number of operators involved in managing the emergency was no longer sufficient, given the working conditions and the state of the system.
   (*Mistake – management error = > lack in defences*)

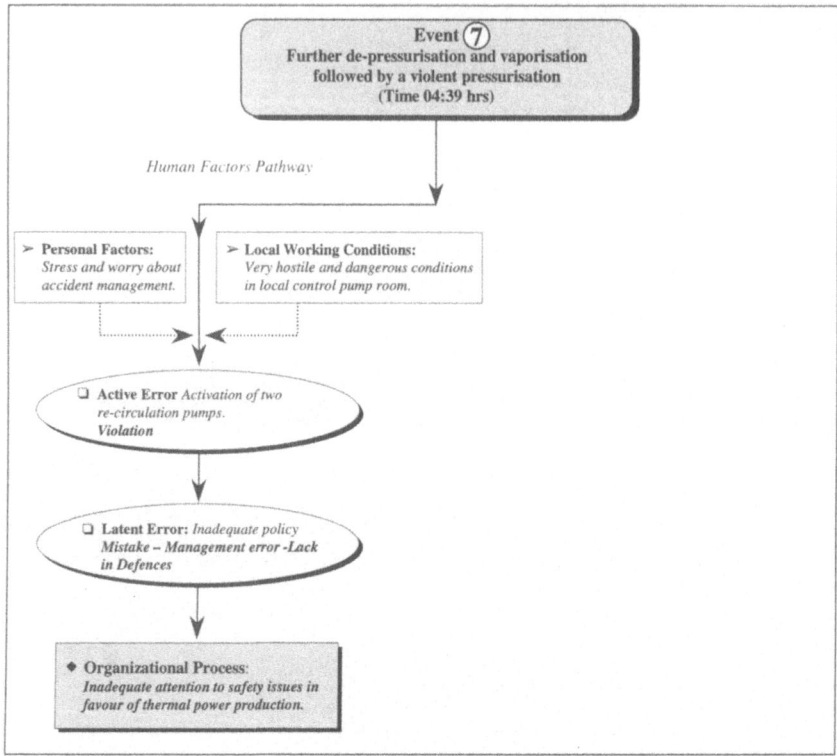

Figure 8.13 Analysis of *event 7*: Further depressurisation and vaporisation, followed by a violent pressurisation.

5. *Organisational Process:*
   - The organisation as a whole could be considered inadequate from the safety perspective.

Event 7: Further Depressurisation, Vaporisation Followed by a Violent Pressurisation

Event 7 was the result of a typical management directive that favoured the production of thermal power versus safety.

The lack of clear situation awareness and the difficulties due to the hostile working environment favoured another active error leading to a violent increase of pressure in the system. This event was totally due to human actions and therefore no *system factor pathway* was developed in the ISAAC tree. The root cause analysis of *event 7* may be summarised as follows (Figure 8.13):

1. *Contextual and Personal Factors:*
   - The operators were operating in a very unfavourable and dangerous environment, in the local control pump room.

- The operators were stressed and still very worried about the management of the heat generators, 2 hours and 39 minutes into the accident.

2. *Active Error:*

- The operators, after three failures, succeeded in activating two circulation pumps by by-passing the low-pressure protections, causing a further reduction of pressure. This was followed by a violent pressurisation. The operators were under high psychological pressure to restore the thermal power distribution according to management policies and therefore bypassed all protections systems of the recirculation system.
  (*Violation*)

3. *Latent Error and Lack in Defences:*

- The company policy privileged the distribution of thermal energy to users in any circumstances. Therefore, the operators, with the aim of restoring that service, made a number of errors that compromised the plant integrity.
  (*Mistake – management error = > lack in defences*)

4. *Organisational Process:*

- The organisation showed inadequate attention to safety issues in favour of thermal power production.

Event 8: Start-up of Heat Recovery from Turbo-gas Group System

Event 8, as event 7, was the result of the management directive that favoured the production of thermal power.

The lack of clear situation awareness and the working environment difficulties completed the scenario of conditions that favoured the mistake made by the operating team. The root cause analysis of *event 8* may be summarised as follows (Figure 8.14):

1. *Contextual and Personal Factors:*

- The operators were operating in a very hostile and dangerous environment.
- The operators were stressed and focused on the management issue concerning heat generations. Three hours and 15 minutes had already passed from the beginning of the accident.

2. *Active Error:*

- The operators restarted heat recovery and production system from the *turbo-gas group* without ensuring that the appropriate pressure level had been reached in the power plant.
  (*Mistake*)

3. *Latent Error and Lack in Defences:*

- The company policy was not particularly focused on training operators, especially in emergency management. Therefore, the operators did not have the necessary background for facing a serious and long-lasting emergency.
  (*Mistake – management error = > lack in defences*)

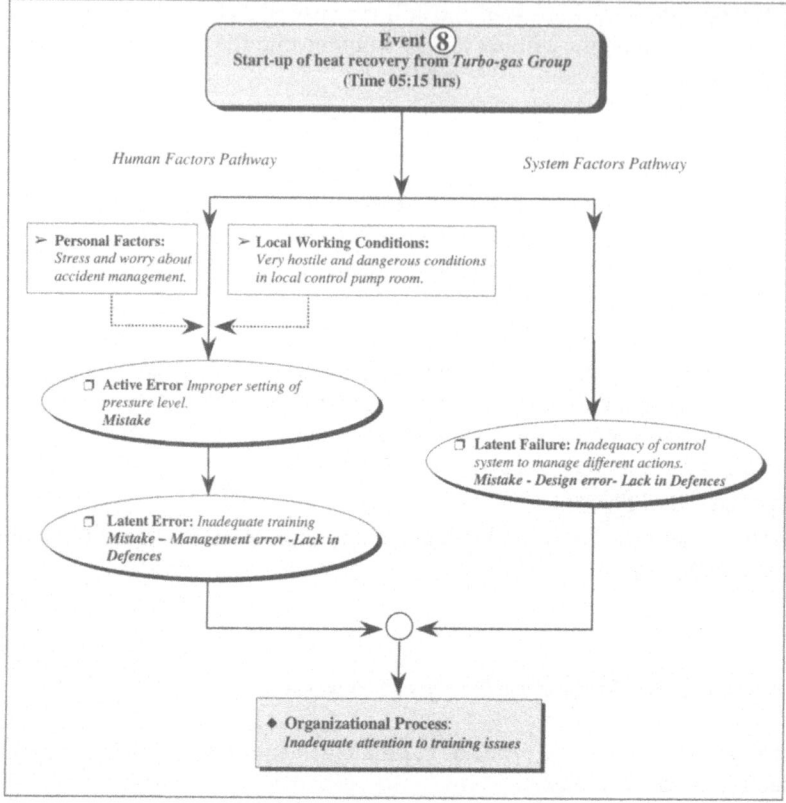

Figure 8.14 Analysis of *event 8*: Start-up of heat recovery from *turbo-gas group* system.

4. *Latent System Failure:*
   • The centralised control system was inadequate for managing the several functions performed by the power plant. Therefore, the actions carried out in different locations of the plant could not be monitored. This made their inhibition impossible in the case of hazard or disagreement with the safety measures.
   (*Mistake – design error = > lack in defences*)
5. *Organisational Process:*
   • The organisation showed inadequate attention to safety issues dedicated to training emergency management.

Event 9: Shock Wave

Event 9 was the conclusive event of the accident that completed the long sequence of inadequate performances dictated by unavoidable situations and inadequate management directives.

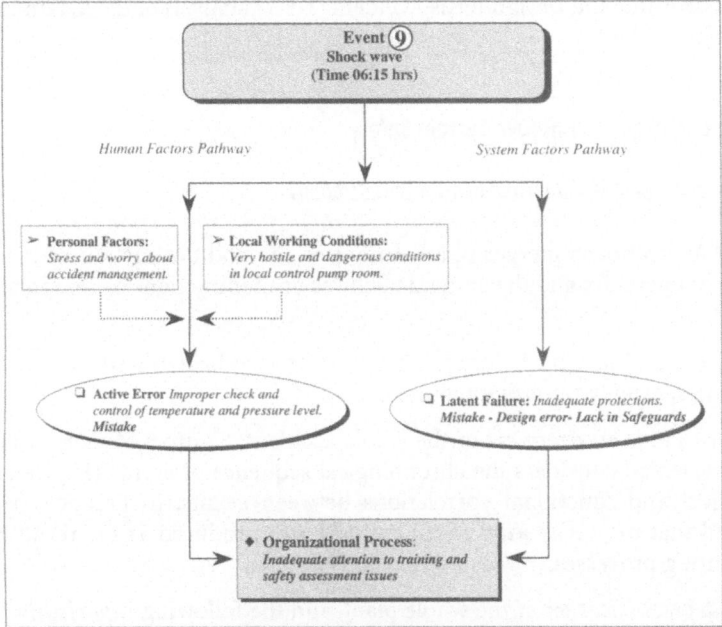

Figure 8.15 Analysis of *event 9*: Shock wave.

The root cause analysis of this final *event 9* may be summarised as follows (Figure 8.15):

1. *Contextual and Personal Factors:*
   - The operators were operating in a very hostile and dangerous environment.
   - The operators were stressed and focused on management issues concerning heat generations. The time was 4 hours and 15 minutes into the accident.

2. *Active error:*
   - Lack of control of pressure evolution in the heat exchangers. Thanks to the action of the recirculation pumps, the vaporisation point was rapidly attained within the heat recovery system, which was affected by the restarted *turbo-gas group*.
   (*Mistake*)

3. *Latent System Failure and Lack in Safeguard:*
   - The shock wave is the most hazardous event to be considered in this type of plant. This event however had not been properly accounted for during the safety assessment of the plant. Consequently, inadequate safeguards and barriers had been designed and implemented that were unable to contain and limit the effects of a shock wave.
   (*Mistake – design error = > lack in safeguards*)

4. *Organisational Process:*
   - The organisation showed inadequate attention to training operators for emergency management and for safety assessment issues, especially

concerning the Design Basis Accident (DBA) analysis of the maximum credible accident.

Recommendations to Improve System Safety

Summary of Accident Evolution and General Considerations

The ISAAC approach merges in a harmonised method the analysis of inadequate actions of operators and the systemic failures or malfunctions of components and subsystems.

In this case, each specific failure and human error has been traced back to its latent root causes, usually of design nature.

In order to give an overview of the whole accident, Figure 8.16 merges the event time line, which considers the chronological sequence of events (Figure 8.5), and the logical and functional correlations between events. In practice, it can be observed that the *shock wave* (final *event 9*) was produced as the result of three contributing processes:

- The *depressurisation* of the whole plant, and the following *vaporisation*, due to the initiating event (*event 1*) and to the loss of electrical power (station blackout) (*event 5*).
- The *vaporisation* derived from the heat exchange in the *turbo-gas group*, due to the delayed isolation of the heat exchangers (*event 4*), the poor isolation of the piping system (*event 6*), and the restarted heat recovery and production system (*event 8*).
- The violent *pressurisation* that followed the activation of the pump system in the attempt to reinstate thermal power production (*event 7*).

In Figure 8.16, it is clearly apparent that certain *Events* played a crucial role in the accident sequence. In particular, a major difference resulted from the change over of accident management from the main control room to the local control pump room (*event 5*). The other critical event was the delay in isolating the storage tank and heat-exchanges of *turbo-gas group* (*event 4*), which affected the processes of depressurisation and vaporisation during the first part of the accident. The analysis of the accident has led to the following principal considerations:

1. The identification of a number of major factors, mostly of latent nature, related to the initiating event.
2. The recognition of a number of *active errors* made by the operators in the management of the accident.
3. The identification of an important design and management *latent errors*, of an organisational nature, that strongly affected the "accident evolution."

These three types of findings will now be reviewed and the safety recommendations that can be drawn from them will be discussed in detail.

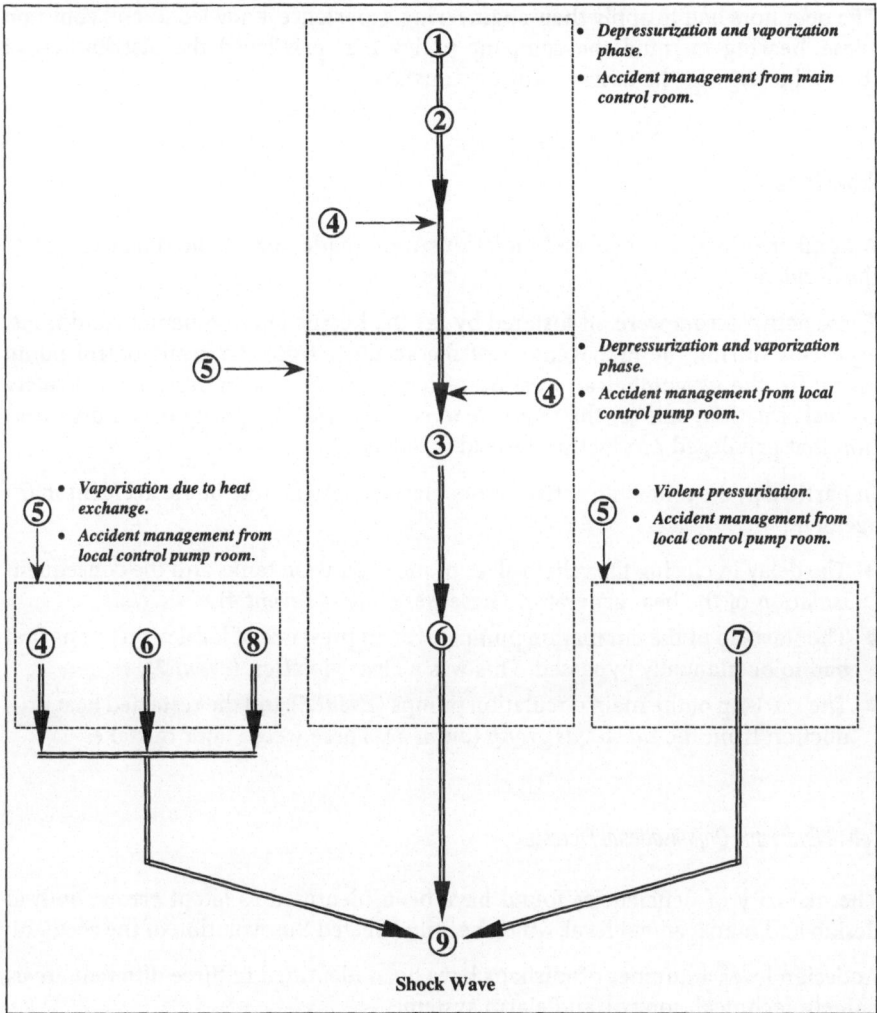

Figure 8.16 Logical and functional structuring of accident sequence.

Initiating Event

The initiating event (*event* 1) has shown the inadequacy of the safety measures.

On the day of the accident the adverse weather conditions enhanced the inadequate positioning of the pressure transducer that was located in the open air with no protection from weather conditions. In addition, the electrical circuits and monitoring system were poorly designed and the supervisory support system led to an unclear understanding of the situation.

The operators were not sufficiently trained, and were not supported by adequate procedures to deal with emergency conditions.

The operators had to apply their engineering experience, knowledge, and common sense, bearing in mind the company policy that privileged the distribution of thermal energy to the users in any circumstance.

Active Errors

A number of active errors and violations were made during the management of the accident.

These active errors were all fostered by (a) the hostile environmental conditions, especially during the management of the accident from the local control pump room; (b) the psychological pressure of having to manage an unexpected safety critical situation; and (c) the requisite to comply with the policy of the organisation that privileged production instead of safety.

In particular, the following active errors played a crucial role in the accident management:

- The delay in closing the relief valves of the expansion tanks and the consequent isolation of the heat generator. These were lapses (*event 4*)
- The start-up of the circulation pumps, even in presence of local protections that had to be manually bypassed. This was a clear violation (*event 7*);
- The start-up of the main circulation pumps (*Event 2*) and the restarted heat production from the *turbo-gas group* (*event 8*). These were major mistakes.

Latent Errors and Organisational Processes

The majority of deficiencies found have been identified as latent errors, both at design and management level, which heavily affected the evolution of the accident.

At design level, a number of mishaps have been identified in three different areas, namely, technical, control, and alarm system:

- At technical system design level, the following errors were found:
  1. Inappropriate dimensioning of the discharge lines in the case of excess water release from the storage tanks (*event 3*).
  2. Un-isolated piping system (*event 3*).
  3. Fire sensors unable to distinguish between steam and smoke (*event 5*).
  4. Isolation of the control rooms during the management of emergency conditions (*events 5–8*).
- At control system design level, the main fault was:
  1. Complete cut-off of all electrical power, and consequent blockage of the pumping and pressurisation power (*events 5–8*).
- At alarm system design level, the following error was found:
  1. Confusing set-up and presentation on the VDU of alarms and indicators for emergency management (*event 1*).

Amongst all these latent errors, the sequence started by the improper intervention of the smoke sensors is a symptomatic indication of the inadequate safety policy existing in the plant at the time of the accident.

During the accident, following the shutdown caused by the fire protection system, the automatic supervisory system for the thermal power generation remained out of service, even after the restart of the electrical system. This type of malfunction, as well as the inappropriate design of the smoke detectors, was known to the plant management. The inappropriate activation of the fire protection system with the consequent power cut-off, and the failed reactivation of the centralised automatic control system are therefore latent errors due both to design and management deficiencies.

These two events generated a very difficult accident management environment, as all control actions had to be carried out at a local level with poor coordination and very little overall supervision. In practice, the operators lost coordination and situation awareness, as each team started to carry out its own tasks and functions with no information concerning the system's overall situation. In particular, there was a critical lack of coordination and interference between the team of operators located in the local control pump room and those dedicated to the control of the electricity and thermal energy production. Moreover, the operators acting in the local control pump room had to perform their tasks in extremely unfavourable conditions, i.e., in presence of very high temperatures and steam in the environment, and with boiling hot water on the floor of the room.

Safety Recommendations

The positive attitude of the management towards the investigation carried out independently by the HF Team on specific human-related and organisational aspects, after the formal enquiry, is certainly the most relevant factor that contributed to the success of the study.

The management was aware that the HF Team had already identified relevant mishaps and faults at the organisational level during the ethnographic studies. However, rather that raising barriers and hamper the continuation of the investigation, with the excuse to protect "industrial secrets" or "company-specific policies," the management recognized that the outcome of the study could bring serious and beneficial improvements to the whole plant and company as a whole.

The attitude of the management and the corresponding frankness of the operators in discussing the problems encountered during the accident led to the development of the recommendations that will be discussed hereafter.

Some of the recommendations have already been implemented and have certainly contributed to improving the safety level of the plant. Other recommendations are being put into practice at a slower pace, as they require longer period of development and preparation.

In general, the recommendations focus on organisational and management issues, as most active failures have been connected and rooted to higher-level management or latent factors.

The recommendations are presented in sequence of logical development and implementation, and not as they were identified during the investigation:

1. *Design Issues*
   The plant control system needed a complete design revision and updating, especially the safety devices, logic interconnections, and human machine interfaces. In particular, it was identified that:
   - New safety devices should be installed, such as closed circuit TV cameras to monitor remote control stations;
   - A new central control system should be designed and developed according to modern criteria of human-centred design approaches and high safety principles.

2. *Standard and Emergency Procedures*
   The need to develop written standard and emergency procedures, combined with training issues (see next recommendation) was considered essential in improving safety and effectiveness of plant control. These would offer a valid element of support to the experience and know-how of operators. The existence of written procedures could have offered a valuable barrier to the development of the shock wave accident.

3. *Training Issues*
   A complete training programme was deemed absolutely necessary, in order to ensure adequate and standardised dissemination of company policies and procedures, as well as the recovery and preservation of experience, know-how and practices of most expert operators. The implementation of this recommendation was recognised as important as the one relative to design issues.

4. *Risk Assessment*
   The need to carry out a new and complete safety assessment study for the plant, following the implementation of new safety devices, represents a major endeavour for the management of the plant. The performance of both deterministic and probabilistic safety assessment studies (design basis accident and quantitative risk assessment) would ensure that a vast variety of initiating events and accident configurations could be safely handled by the defences, barriers and safeguards. Most of the critical mishaps that occurred during the shock wave accident would have been recognised and examined by either a deterministic or a probabilistic study, offering solutions and countermeasures.

5. *Definition of Safety Audit Procedure*
   It was suggested that the company should adopt and implement a formal safety audit method and procedure for regularly checking and auditing its safety levels and the maintenance of required standards.

6. *New Management Policy*
   Finally, the development and implementation of new company policies, written in practical procedures and taught during the training, was recognised essen-

tial for the future of the company. In particular, it was suggested to put "safety" in the first position of the Company goals and objectives.

The implementation of the above recommendations would ensure a series of very efficient and effective defences, barriers, and safeguards for preventing and managing the occurrence of accidents, and minimising their possible consequences for society and environment, as well as for the company itself.

8.3 Summary: Application of HERMES for the Accident Investigation of a Power Plant

8.3.1 Application of the HERMES Methodology

This chapter has presented a practical application of the HERMES methodology in the area of accident investigation for the analysis of a real accident that occurred in a complex plant for combined electrical power and thermal energy production.

The five standpoints identified as the basic methodological stands to perform a sound HF approach (Chapter 2) have been respected. In particular:

1. The performance of an accident investigation and root cause analysis study was identified as the specific area of application of the methodology (standpoint 4).

2. The specific goals and objectives of the investigation were set from the viewpoint of human error and accident management principles, i.e., identify all human-related root causes that contributed to the accident (standpoint 1).

3. The assessment of the working contexts and the evaluation of the social climate existing within the organisation, in conjunction with all engineering practices and existing procedures for plant management in normal and emergency operations were thoroughly examined. This represents the basis for understanding the working contexts and in selecting adequate paradigms and models for the analysis (standpoint 3).

4. A cognitive model and associated taxonomy of reference for classifying "human errors" and "system failures" were selected and applied throughout the investigation (standpoint 2).

5. Accident investigation is a typical retrospective type analysis, which identifies improvements and safety aspects that can be of impact on the safety level for the whole system. Therefore, the recommendations that were developed represent a reference and roadmap for future safety audits on the plant (standpoint 5).

The application of HERMES to this case study showed that the methodology is applicable in practice even for complicated investigations and offers modelling architectures using simple and easy structuring.

The methodology was applied in two phases (Figure 8.17). Phase 1 of the work focused on the evaluation of the socio-technical context of work and the famil-

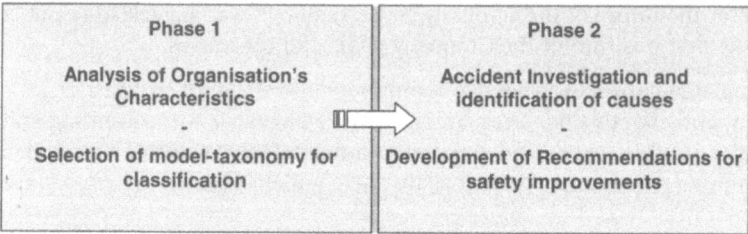

Figure 8.17 Phases of development of an accident investigation.

iarisation with the practices, habits and tasks existing within the company. These ethnographic studies led to the definition of a model and approach for structuring the accident sequence and human behaviour, as well as for developing safety recommendations at more strategic level.

Phase 2 focused on the detailed evaluation of the causes of the accident, mainly from the human factors and management viewpoints, and on the development of recommendations for improving safety and for setting the basis for future audits and assessments of the plant and organisation as a whole.

8.3.2 Conclusions

The typical process management approach DSME (Define, Select and Implement, Monitor, and Evaluate) for ascertaining the implementation of a correct methodological process and for evaluating the effectiveness of the outcome was applied.

The goals of the study were clearly set by the organisation and further elaborated in collaboration with the HF Team. It was apparent that the organisation was facing a relevant human factors problem. Moreover, the possibility that some critical aspects could involve top management and even design errors were accepted and understood by the organisation, showing a very open safety culture and attitude.

The HF Team, after selecting the ISAAC approach to study the accident, ensured that the procedure could be implemented not only by external specialists in Human Factors, but also by safety engineers and internal personnel of the organisation. This is also one of the reasons that favoured the selection of a simple but very efficient approach. In this way it could be ensured that future assessments of the safety level of the plant/organisation could be performed in a rapid and straightforward manner.

The implementation of the ISAAC phases were actually carried out with the collaboration of the organisation at all levels, from top management to shift supervisors and line operators. This is an aspect of paramount importance for the success of any accident investigation. At the same time the transfer of know-how from the HF Team to members of the organisation could be easily attained.

Finally, the effectiveness of the study could be ascertained by evaluating the improvement of safety reached thanks to the implementation of new measures. Some simple modifications defined in the recommendations could in practice be implemented in a short time. Other modifications required multiple interventions on the plant design and quite substantial engineering changes.

All in all, a substantial increase of the safety level within the organisation could be achieved.

Bibliography

AAIB. Air Accidents Investigation Branch (1990). Report on the accident to Boeing 737-400 G-OBME near Kegworth, Leicestershire. Aircraft Accident Report 4/90, The Department of Transport, UK.

ACAS (2002). Automotive Collision Avoidance Systems Development Program. (*http://www.nhtsa.dot.gov/people/injury/research/pub/ACAS/Acknow.htm*)

Amalberti, R. (1996). *La coduite de systèmes à risque*. Presse Universitaire de France, Paris.

Amat, A.-L., Bacchi, M., Cacciabue, P. C., Kjaer-Hansen, J., Larsen, M. N., Lie, A. M., and Ralli, M. (1997). The Ispra Crew Resource Management Course (CRM Course): *Facilitator Preparation Manual*. EUR 17691 EN

Andreone, L., Eschler, J., Kempf, D., Widlroither, H., Amditis, A., and Cacciabue, P. C. (2002). An application of the human-centred design approach to the interaction between the driver and the vehicle: the EUCLIDE project. *9th World Congress on Intelligent Transport Systems* (ITS-2002). Chicago, Illinois, USA, 14–17, October, 2002.

Agar, M. H. (1980). *The Professional Stranger: An Informal Introduction to Ethnography*. Academic Press. London.

Apostolakis, G. (Ed.) (1998). Risk Perception and Risk Management. Special Issue of *Reliability Engineering and System Safety, RE&SS*, 59, (1), Elsevier, Amsterdam.

Apostolakis, G. E., Mancini, G., and van Otterloo, R. W. (Eds.) (1990). Human Reliability Analysis, Special volume, *Reliability Engineering and System Safety, RE&SS*, 29 No 3.

Ashby, W. R. (1956). *An Introduction to Cybernetics*. Chapman and Hall, London.

AWARE (2002). Anti-Collision Warning and Avoidance Radar Equipment, Esprit Project EP-22966. (*http://www.tech.volvo.se/aware/aware.htm*)

Bagnara, S., Di Martino, C., Lisanti, B., Mancini, G., and Rizzo, A. (1989). *A Human Error Taxonomy Based on Cognitive Engineering and on Social Occupational Psychology*. EUR 12624 EN. CEC-JRC, Ispra.

Bainbridge, L. (1983). The ironies of automation. *Automatica* 19, 6, 775–780.

Baldwin, L. P., Paul, R. J., and Williams, D. S. (1999). Managing soft knowledge in a hard workplace. *International Journal of Cognition Technology and Work, IJ-CTW*, 1 (3). Springer-Verlag, London. pp. 133–141.

Baranzini, D., Bacchi, M., and Cacciabue, P. C. (2001). A tool for evaluation and identification of training needs in aircraft maintenance teams. *Int. Journal of Human Factors and Aerospace Safety (HF&AS)*, 1 (2).

Baron, S., Kleinman, D. L., and Levison, W. H. (1970). An optimal control model of human response. Part I: Theory and applications, *Automatica*, 6, pp. 357–369.

Barriere, M. T., Bley, D. C., Cooper, S. E., Forester, J., Kolaczkowski, A., Luckas, W. J., Parry, G. W., Ramey-Smith, A., Thompson, C., Whitehead, D. W., and Wreathall, J. (1998). *Technical Basis and Implementation Guidelines for a Technique for Human Event Analysis* (ATHEANA). NUREG – 1624, US-NRC, Washington DC.

Barua, A., Chellappa, R., and Whinston, A. B. (1997). Social computing: Computer-supported cooperative work and groupware. In G. Salvendi (Ed.), *Handbook of Human Factors and Ergonomics*. J. Wiley & Sons, New York, pp. 1760–1782.

Bellorini, A., Cacciabue, P. C., Larsen, M. N., and Facinelli, E. (1997). Human factors computer-based training in air traffic control. *9th International Symposium on Aviation Psychology*, Columbus, Ohio, April 27–May 1, 1997.

Berry, J. W. (1980). Introduction to methodology. In H. C. Triandis, and J. W. Berry (Eds.), *Handbook of Cross-Cultural Psychology*, 2 (pp. 1–28). Boston, MA: Allyn & Bacon, Inc.

Bevan, N. (1999). Design for usability. *Proceedings of International Conference on Human Computer Interaction (HCI)*, 22–26 August, Munich, D.

Bevan, N., and Macleod, D. (1994). Usability measurement in context, *Behaviour & Information Technology*, vol. 13, Nos.1 and 2, 132–145.

Bieder, C., Le-Bot, P., Desmares, E., Bonnet, J-L., and Cara, L. (1998). MERMOS: EDF's new advanced HRA method. In A. Mosleh, and R. A. Bari (Eds.), *Proceedings of PSAM-4 Conference*, New York, 13–18 September, Springer-Verlag, New York, 129–134.

Billings, C. E. (1997). *Aviation Automation: The Search for a Human-Centered Approach*. Lawrence Erlbaum Associates, Mahwah, NJ.

Bilton, T. (1997). The evaluation of Virgin Atlantic Airways' crew resource management training program. In R. A Telfer, and P. J. Moore (Eds.), *Aviation Training: Learners, Instruction and Organization*. Ashgate, Aldershot, UK.

Bonar, J., Collins, J., Curran, K., Eastman, R., Gitomer, D., Glaser, R., Greenberg, L., Lajoie, S., Lesgold, A., Logan, D., Magone, M., Shalin, V., Weiner, A., Wolf, R., and Yengo, L. (1986). *Guide to Cognitive Task Analysis*. Learning Research and Development Center (LRDC), University of Pittsburgh, Pittsburgh, OH.

Brooking, A. (1999). *Corporate Memory*. International Thomson Business Press, London

Brown, J. S., and Duguid, P. (2000). *The Social Life of Information*. Harvard Business School Publishing, Cambridge, MA.

Brunswik, E. (1952). The conceptual framework of psychology. In *International Encyclopaedia of Unified Science*, (Vol. 1, N. 10), University of Chicago Press, Chicago, IL.

Brunswik, E. (1956). *Perception and the Representative Design of Psychological Experiments*. University of California Press, Berkeley, CA.

Bucciarelli, L. L. (1988). An ethnographic perspective on engineering design. *Design Studies*, 9 (3), pp. 159–168.

Button, G., and Sharrock, W. (1997). The production of order and the order of production. *Proceedings of 5th European Conference on Computer-Supported Cooperative Work*. Kluwer, Dordrecht, pp. 1–16.

Byrom, N. T. (1994). The assessment of safety management systems using an auditing approach. In P. C. Cacciabue, I. Gerbaulet, and N. Mitchison (Eds.), *Safety Management Systems in the Process Industry*. EUR 15743 EN, pp. 150–156.

Cacciabue, P. C. (1997). A methodology for human factors analysis for system engineering: theory and applications. *IEEE-System Man and Cybernetics, IEEE-SMC*, 27 (3), pp. 325–339.

Cacciabue, P. C. (1998). *Modelling and Simulation of Human Behaviour in System Control*. Springer-Verlag, London.

Cacciabue, P. C. (1999). Modelling and simulation of human behaviour in process control: needs, perspectives, and applications. In D. Harris (Ed.), *Proceedings of 2nd Engineering Psychology and Cognitive Ergonomics Conference*. Oxford, UK, October 28–30, 1998. ISBN 1-84014-545-5. Ashgate, Aldershot, UK. pp. 3–20.

Cacciabue, P. C. (2000a). Principi di gestione dell'errore umano (in Italian). In G. Mantovani (Ed.), *Ergonomia: lavoro, sicurezza e nuove tecnologie*. pp. 81–122, Il Mulino, Bologna.

Cacciabue, P. C. (2000b). Human factors insight and data from accident reports: the case of ADREP-2000 for aviation safety assessment. *3rd International Conference on Engineering Psychology & Cognitive Ergonomics*, EP&CE-III, Edinburgh, October 25–27, 2000.

Cacciabue, P. C., Amditis, A., Bekiaris, E., Andreone, L., and Tango, F. (2001). The importance of user needs analysis on HMI design: the EUCLIDE example. *Proceedings of Panhellenic Conference on Human–Computer Interaction*, PC-HCI-2001, 7–9 December 2001 Patras-Greece, pp. 39–44.

Cacciabue, P. C., Carpignano, A., and Piccini, M. (1998). Human factors assessment for highly automated systems: a case study of a power plant. In A. Mosleh, and R. A. Bari (Eds.), *Proceedings of PSAM 4 – International Conference on Probabilistic Safety Assessment and Management*. New York, 13–18 September, 1998. ISBN 3-540-76262-0. Springer-Verlag, London, pp. 123–128.

Cacciabue, P. C., Cojazzi, G., Hollnagel, E., and Mancini, S. (1992). Analysis and modelling of pilot-airplane interaction by an integrated simulation approach. *Proceedings of 5th IFAC/IFIP/IFORS/IEA Symposium on Analysis, Design and Evaluation of Man–Machine Systems*. The Hague, The Netherlands, 9–11 June. Pergamon Press, Oxford, UK, pp. 227–234.

Cacciabue, P. C., Donato, E., and Rossano, S. (2002). *Designing Human–Machine Systems for Automotive Safety: A Cognitive Ergonomics Perspective*. European Commission, EUR 20378 EN.

Cacciabue, P. C., Fujita, Y., Furuta, K., and Hollnagel, E. (Eds.) (2000). Human factor analysis of JCO criticality accident cognition, Special Issue. *Int. Journal of Cognition Technology and Work, IJ-CTW*, **2** (4). Springer-Verlag, London.

Cacciabue, P. C., Gerbaulet, I., and Mitchison, N. (Eds.) (1994). *Safety Management Systems in the Process Industry*. European Commission, **EUR 15743 EN**.

Cacciabue, P. C., and Hollnagel, E. (1995). Simulation of cognition: applications. In J-M. Hoc, P. C. Cacciabue, and E. Hollnagel (Eds.), *Expertise and Technology: Cognition & Human–Computer Cooperation*. Lawrence Erlbaum, Hillsdale, NJ, pp. 55–73.

Cacciabue, P. C., Martinetto, M., Montagna, S. M., and Re, A. (2003). Scenario development for testing safety devices in automotive environments. *Proceedings of International Conference on Human Computer Interaction (HCI)*, 22–27 June, Crete, Greece.

Cacciabue, P. C., Martinetto, M., and Re, A. (2002b). The effect of car drivers' attitudes and national cultures on design of an anticollision system. *11th European Conference Cognitive Ergonomics* (ECCE-11). Catania, Italy, 8–11, September, 2002.

Cacciabue, P. C., Mauri, C., and Owen, D. (2003). Development of a model and simulation of aviation maintenance technician task performance. To appear in *Int. Journal of Cognition Technology and Work, IJ-CTW*. Springer-Verlag, London.

Cacciabue, P. C., and Pedrali, M. (1997). Human factors in proactive and reactive safety studies. *Symposium on High-Consequence Operations Safety*, SANDIA Nat. Lab., Albuquerque, NM, July 29–31, 1997.

Cacciabue, P. C., Ponzio, P., and Bacchi, M. (1999). Intranational cultural differences resulting from human factors analysis of complex socio-technical systems. *7th European Conference on Cognitive Science Approaches to Process Control*. Villeneuve d'Ascq, F, 21–24 Sept. 1999.

Carpignano, A., and Piccini, M. (1999). Cognitive theories and engineering approaches for safety assessment and design of automated systems: a case study of a power plant. *International Journal of Cognition Technology and Work, IJ-CTW*, **1** (1). Springer-Verlag, London. pp. 47–61.

Carpignano, A., Piccini, M., and Cacciabue, P. C. (1998). Human reliability for the safety assessment of a thermoelectric power plant. *European Safety and Reliability Conference 1998*. In S. Lydersen, G. K. Hansen, and H. A. Sandtorv (Eds.), *Proceedings of ESREL 98*, Trodheim, 16–19 June, 1998, A. A. Balkema, Rotterdam. pp. 785–792.

Carroll, J. M. (2000). *Scenario-Based Design of Human–Computer Interactions*. The MIT Press, Cambridge, MA.

Carsten, O. (1999). New evaluation methods: Progress or blind alley? *Transportation Human Factors*, **1**, 2, 177–186.

Carsten, O., and Nilsson, L. (2001). Safety assessment of driver assistance systems. *European Journal of Transport and Infrastructure Research, EJTIR*, 1, 3, 225–243.

COMUNICAR. (2002). Project: Communication Multimedia Unit Inside Car. (*http://www.comunicar-eu.org*).

CNSC, Canadian Nuclear Safety Commission (2003). Human factors verification and validation plans. Regulatory Guide, G-278.

Davenport, T. H., and Prusak, L. (1998). *Working Knowledge: How Organizations Manage What They Know*. Harvard Business School Publishing, Cambridge, MA.

Davis, J. H., and Stanton, M. F. (1988). Small group performance. Past and present research trends. *Advances in group processes*, 5, pp. 245–277.

Davies, J. B., Wright, L., Cortney, E., and Reid, H. (2000). Confidential incident reporting on the UK Railways: The "CIRAS" system. *International Journal of Cognition Technology and Work, IJ-CTW*, 2 (3). Springer-Verlag, London. pp. 117–125.

Degani, A., and Wiener, E. L. (1994a). On the design of flight-deck procedures. NASA Contractor Report 177642. NASA, Ames Research Center, Moffett Field, CA.

Degani, A., and Wiener, E. L. (1994b). Philosophy, policies, procedures and practice: The four "P"s of flight deck operations. In N. Johnston, N. McDonald, and R. Fuller (Eds.), *Aviation Psychology in Practice*, Avebury Technical, Aldershot, UK. pp. 68–87.

Diaper, D., and Stanton, N. (Eds.) (2003). *The handbook of Task Analysis for Human-Computer Interaction*. Lawrence Erlbaum Associates, Mahwah, NJ.

DoD. Department of Defense (1984). System safety program requirements. MIL-STD-882B. DoD, Washington, DC.

Dougherty, E. M., and Fragola, J. R. (1988). *Human Reliability Analysis. A System Engineering Approach with Nuclear Power Plant Applications*. J. Wiley & Sons, New York.

Dreyfus, H. L., and Dreyfus, S. E. (1986). *Mind over Machines*, Blackwell Ltd., New York.

EC. European Commission (1998). Commission Mandate to the European Standard Bodies for standardization in the field of information and communication technologies (IT) for disabled and elderly people. EC, 1998, M/273 – EN.

EC. European Commission (1999). Commission Recommendation of 21 December 1999, "on safe and efficient in-vehicle information and communication systems: A European statement of principles on human machine interface (notified under document number C(1999)4786), (2000/53/EC).

Edwards, E. (1972). Man and machine: Systems for safety. In *Proceedings of British Airline Pilots Association Technical Symposium*. British Airline Pilots Association, London. pp. 21–36.

Edwards, E. (1988). Introductory overview. In E. L. Wiener, and D. C. Nagel (Eds.), *Human Factors in Aviation*, Academic Press, San Diego, CA. pp. 3–25.

Embrey, D. E., Humphreys, P. C., Rosa, E. A., Kirwan, B., and Rea, K. (1984). *SLIM-MAUD: An Approach to Assessing Human Error Probabilities Using Structured Expert Judgement*. NUREG/CR-3518, USNRC, Washington, US.

Endlsey, M. R. (1994). Indvidual differences in pilot situation awareness. *The International Journal of Aviation Psychology*, 4, 3, pp. 241–264.

Ericsson, K. A., and Simon, H. A. (1984). *Protocol Analysis: Verbal Reports as Data*. The MIT Press, Cambridge, MA.

EUCLIDE (2002). Project: Enhanced human machine interface for vehicle-integrated driving support system. (*http://www.euclide-eu.org*).

FAA. Federal Aviation Administration (1978). Line-Oriented Flight Training. Advisory Circular AC 120-35A. Washington, DC.

FAA. Federal Aviation Administration (1990). Special FAA Regulation 58. Advanced Qualification Program. Federal Register, Vol. 55, No 91, Rules and Regulations pp. 40262–40278). FAA, Washington, DC.

FAA. Federal Aviation Administration (1996a). Line-oriented simulators (LOS). Advisory Circular, 120-35. Federal Aviation Administration, FAA, Washington, DC.

FAA. Federal Aviation Administration (1996b). The interfaces between flightcrews and modern flight deck systems. Federal aviation administration human factors team report, June 1996. FAA, Washington, DC.

FAA. Federal Aviation Administration (1997). *Aviation Safety Aviation Programs*. Advisory Circular, 120-66. FAA, Washington, DC.

FHWA. Federal Highway Administration (1999). *Development of Human Factors Guidelines for Advanced Traveller Information System (ATIS) and Commercial Vehicle Operations (CVO): Driver Memory for In-vehicle Visual and Auditory Messages*, Publication N. FHWA-RD-96-148.

Flach, J., Hancock, P., Caird, J., and Vicente, K. (Eds.) (1995). *Global Perspective on the Ecology of Human–Machine Systems*. Lawrence Erlbaum Associates, Hillsdale, NJ.

Foushee, H. C., and Helmreich, R. L. (1988). Group interaction and flight crew performance. In E. L. Wiener, and D. C. Nagel (Eds.) *Human Factors in Aviation*, Academic Press, San Diego, CA.

Garfinkel, H. (1967). *Studies in Ethnomethodology*. Prentice-Hall, Inc. Englewood Cliffs, NJ.

Gibson, J. J. (1966). *The Senses Considered as Perceptual Systems*. Houghton-Mifflin, Boston.

Gibson, J. J. (1979). *The Ecological Approach to Visual Perception*. Houghton-Mifflin, Boston.

Goeters, K-M. (Ed.) (1998). *Aviation Psychology: A Science and a Profession*. Avebury, Aldershot, UK.

Gould, J. D. (1988). How to Design Usable Systems. In Helander, M. (Ed.), *Handbook of Human–Computer Interaction*, New York: North-Holland, 1988. pp. 757–789.

Gow, H. B., and Otway, H. (Eds.) (1990). *Communicating with the Public About Major Accident Hazards*. Elsevier Applied Science, London, UK.

Gras, A., Moricot, C., Poirot-Delpech, S. L., and Scardigli, V. (1994). *Face à l'automate: le pilote, le contrôleur et l'ingénieur*. Publications de la Sorbonne, Paris.

Hancock, P., Flach, J., Caird, J., and Vicente, K. (Eds.) (1995). *Local Applications of the Ecological Approach to Human–Machine Systems*. Lawrence Erlbaum Associates, Hillsdale, NJ.

Hancock, P., Parasuraman, R., and Byrne, E. A. (1996). Driver-centered issues in advanced automation motor vehicles. In R. Parasuraman, and M. Mouloua (Eds.), *Automation and Human Performance: Theory and Applications*, Laurence Erlbaum Associates, Mahwah, NJ.

Hannaman, G. W., and Spurgin, A. J. (1984). *Systematic Human Action Reliability Procedure* (SHARP). EPRI NP-3583, Project 2170-3, Interim Report, NUS Corporation, San Diego, CA, US.

Hannaman, G. W., Spurgin, A. J., and Lukic, Y. D. (1984). *Human Cognitive Reliability Model for PRA Analysis*. NUS-4531, NUS Corporation, San Diego, CA.

Hansen, J. P. (1995). Representation of systems invariants by optical invariants inconfigural displays for process control. In P. Hancock, J. Flach, J. Caird, and K. Vicente (Eds.), *Local Applications of the Ecological Approach to Human–Machine Systems*. Lawrence Erlbaum Associates, Hillsdale, NJ, pp. 209–228.

Hawkins, F. H. (1987). *Human Factors in Flight*. Gower, Aldershot, UK.

Heath, C., and Luff, P. (1991). Collaborative activities and technological design: task co-ordination in London underground control room. *Proceedings of 2th European Conference on Computer-Supported Cooperative Work*. Kluwer, Dordrecht, pp. 65–80.

Helander, M. (Ed.) (1988). *Handbook of Human–Computer Interaction*. Elsevier, Amsterdam.

Helmreich, R. L. (1984). Cockpit management attitudes. *Human Factors*, 26, pp. 63–72.

Helmreich, R. L., and Merritt, A. C. (1998). *Culture at Work in Aviation and Medicine: National Organisational, and Professional Influences*. Ashgate, Aldershot, UK.

Helmreich, R. L., Merritt, A. C., Sherman, P., Gregorich, S., and Wiener, E. (1993). The flight management attitude questionnaire. Aerospace Crew Research Project Technical Report 93–4.

Helmreich, R. L., and Wilhelm, J. A. (1991). Outcome of crew resource management training. *International Journal of Aviation Psychology*, 1, 4, pp. 287–300.

Hofstede, G. (1980). *Culture's Consequences: International Differences in Work-Related Values*. Sage, Beverly Hills, CA.

Hollnagel, E. (1991). The phenotype of erroneous actions: implications for HCI design. In G. R. S. Weir, and J. L. Alty (Eds.), *Human–Computer Interaction and the Complex Systems*, Academic Press, pp. 73–121.

Hollnagel, E. (1993). *Human Reliability Analysis: Context and Control*. Academic Press, London.

Hollnagel, E. (1998). *Cognitive Reliability and Error Analysis Method*. Elsevier, London.

Hollnagel, E. (1999). Accident and barriers. Proceedings of 7th European Conference on Cognitive Science Approaches to Process Control, Villeneuve d'Ascq, France, pp. 175–180.

Hollnagel, E. (Ed.) (2003). *Handbook of Cognitive Task Design*. Lawrence Erlbaum Associates, Mahwah, NJ.

Hollnagel, E., and Cacciabue, P. C. (1991). Cognitive Modelling in System Simulation. Proceedings of Third European Conference on Cognitive Science Approaches to Process Control, Cardiff, UK, September 2–6.

Hollnagel, E., and Woods, D. D. (1983). Cognitive Systems Engineering: New Wine in New Bottles. *International Journal of Man–Machine Studies*, 18, 583–606.

Hopkin, V. D. (1995). *Human Factors in Air Traffic Control*. Taylor & Francis, Bristol, PA.

Howard, R., and Matheson, J. G. (1980). Influence Diagrams. SRI International, Menlo Park, CA, US.

Hudson, P., Reason, J., Wagenaar, W., Bentley, P., Primrose, M., and Visser, J. (1994). Tipod-Delta: proactive approach to enhance safety. *Journal of Petroleum Technology*, 40, 1994, pp. 58–62.

Hunns, D., and Daniels, B. K. (1980). The methods of paired comparison and the results of the paired comparisons consensus exercise. *Proceedings of the 6th Advances in Reliability Technology Symposium*, NCSR R23, Culcheth, UK.

Hunt, G. J. F. (Ed.) (1997). *Designing Instructions for Human Factors Training in Aviation*. Avebury, Aldershot, UK.

ICAO. International Civil Aviation Organisation (1978). KLM B-747, PH-BUF and Pan Am, B-747, N736, collision at Tenerife Airport, Spain. ICAO Circular 153-AN/56.

ICAO. International Civil Aviation Organisation (1984). *Accident Prevention Manual*, 1st Edition. International Civil Aviation Organisation, Montreal, Canada.

ICAO. International Civil Aviation Organisation (1986). *Manual of Aircraft Accident Investigation*, 4th Edition. International Civil Aviation Organisation, Montreal, Canada.

ICAO. International Civil Aviation Organisation (1987). Accident/incident reporting manual-second edition, 1987 DOC 9156-AN/900. International Civil Aviation Organisation, Montreal, Canada.

ICAO. International Civil Aviation Organisation (1988). International Standards and Recommended Practices on Aircraft Accident Investigation, Annex 13 of the Convention on International Civil Aviation, 7th Edition, International Civil Aviation Organisation, Montreal, Canada.

ICAO. International Civil Aviation Organisation (1991). *Human Factors Digest, No. 3: Training Operational Personnel in Human Factors*. ICAO, Circular 227-AN/136. International Civil Aviation Organisation, Montreal, Canada.

ICAO. International Civil Aviation Organisation (1993). *Human Factors Digest, No. 7: Investigation of Human Factors in Accidents and Incidents*. ICAO, Circular 240-AN/144. International Civil Aviation Organisation, Montreal, Canada.

ICAO. International Civil Aviation Organisation (1997). Accident/Incident Reporting Manual-ADREP 2000 draft (1997). ICAO, International Civil Aviation Organisation, Montreal, Canada.

ISO. International Standard Organisation (1993). *Guidance on Usability*. ISO/DIS 9241-11.

ISO. International Standard Organisation (1999). *Human-centred design processes for interactive systems*. ISO/FDIS 13407(E).

Jensen, R. S. (1995). *Pilot Judgment and Crew Resource Management*. Avebury, Aldershot, UK.

Johnson, C. W., and Botting, R. M. (1999). Using Reason's model of organisational accidents in formalising accident reports. *International Journal of Cognition Technology and Work, IJ-CTW*, 1 (2). Springer-Verlag, London. pp. 107–118.

Johnston, A. N., McDonald, N., and Fuller, R. (Eds.) (1995). *Aviation Psychology: Training and Selection*. Avebury, Aldershot, UK.

Kantowitz, B. H., and Sorkin, R. D. (1983). *Human Factors: Understanding People–System relationship*. J. Wiley & Sons, New York.

Kirlik, A. (1995). Requirements for psychological models to support design: Toward ecological task analysis. In J. Flach, P. Hancock, J. Caird, and K. Vicente (Eds.), *Global Perspective on the Ecology of Human–Machine systems*. Lawrence Erlbaum Associates, Hillsdale, NJ, pp. 68–120.

Kirwan, B. (1994). *A Guide to Practical Human Reliability Assessment*. Taylor and Francis, London.

Kirwan, B., and Ainsworth, L. K. (1992). *Guide to Task Analysis*. Taylor and Francis, London.

Kirwan, B., Kennedy, R., Taylor-Adams, S., and Lambert, B. (1997). The validation study of three human reliability assessment techniques: THERP, HEART, and JHEDI: Part 1 – Results of Validation Exercise. *Applied Ergonomics*. **28**, 17–25.

Lauber, J. K. (1984). Resource Management in the Cockpit. *Air Line Pilot*, **52**, 20–23.

Luczak, H. (1997). Task Analysis. In G. Salvendi (Ed.), *Handbook of Human Factors and Ergonomics*. J. Wiley & Sons, New York, pp. 340–416.

Marsden, P. (1987). *The actual frequency of encounter of American Presidents*. Department of Psychology, University of Manchester.

Masson, M., and Koning, Y. (2001). How to manage human error in aviation maintenance? The example of a JAR 66-HF education and training programme. *International Journal of Cognition Technology and Work, IJ-CTW*, 3 (4). Springer-Verlag, London. pp. 189–204.

Maurino, D. E., Reason, J., Johnston, N., and, Lee, R. B. (1995). *Beyond Aviation Human Factors*, Avebury Aviation. Aldershot, UK.

McRuer, D. T, Graham, D., Kredel, E., and Reisener, W. (1965). Human pilot dynamics in compensatory systems – theory, models and experiments with controlled elements and forcing function variations. Report AFFDL-TR65-15, Wright Patterson AFB, Ohio.

Merritt, A. C., Helmreich, R. L., Wilhelm, J. A., and Sherman, P. J. (1996). Flight Management Attitudes Questionnaire 2.0 (International) and 2.1 (USA/Anglo). Technical Report 96-04.

Minsky, M. (1975). A framework for representing knowledge. In P. Winston (Ed.), *The Psychology of Computer Vision*. McGraw-Hill, New York.

Mitchell, C. M. (1987). GT-MSOCC: A Domain for Research on Human–Computer Interaction and Decision Aiding in Supervisory Control Systems. *IEEE Transaction on Systems, Man, and Cybernetics, IEEE-SMC*, **17** (4), pp. 553–572.

Mitchell, C. M. (1999). Model-based design of human interaction with complex systems. In A. P. Sage, and W. B. Rouse (Eds.), *Handbook of Systems Engineering and Management* J. Wiley & Sons, New York, USA. pp. 745–810.

Mitroff, I. I., Pauchant, T. Finney, M., and Pearson, C. (1990). Do some organisations cause their own crises? The cultural profile of crisis-prone vs. crisis-prepared organisations. *Industrial Crisis Quarterly*, 3, 269–283.

Moray, N. (1997). Human Factors in process control. In G. Salvendi (Ed.), *Handbook of Human Factors and Ergonomics*. J. Wiley & Sons, New York. pp. 1945–1971.

Moray, N., Ferrell, W. R., and Rouse, W. B. (Eds.) (1990). *Robotics, Control and Society*. Taylor and Francis, Bristol, PA.

Moricot, C. (1996). Pilots' representations in response to the evolution of technology and as for their role in safety. 3rd ICAO Global Flight Safety and Human Factors Symposium. April 9–12, 1996, Auckland, New Zealand. ICAO Human Factors Digest 13. CIRC 266-AN/158, pp. 113–120.

Nagel, D. C. (1988). Human error in aviation operations. In E. L. Wiener, and D. C. Nagel (Eds.), *Human Factors in Aviation*. Academic Press, San Diego, CA. pp. 263–303.

Naikar, N., and Saunders, A. (2003). Crossing the Boundaries of Safe Operation: An Approach for Training Technical Skills in Error Management. *International Journal of Cognition Technology and Work, IJ-CTW*, 5 (3), Springer-Verlag, London.

Nardi, B. A., and O'Day, V. L. (1999). *Information Ecologies: Using Technology with Heart*. MIT Press, Cambridge, MA.

Nardi, B. A., and O'Day, V. L. (1999). *Information Ecologies: Using Technology with Heart*. MIT Press.

Neisser, U. (1967). *Cognitive Psychology*. Appleton-Century-Crofts, New York.

Newell, A., and Simon, H. A. (1972). *Human Problem Solving*. Prentice-Hall, Englewood Cliffs, NY.

NHTSA. National Highway Traffic Safety Administration (1993). *Preliminary Human Factors Guidelines for Crash Avoidance Warning Devices*, NHTSA Project No. DTNH22-91-C-07004, USDOT, Draft Document, COMSIS Co.

NHTSA. National Highway Traffic Safety Administration (1996). *Preliminary Human Factors Guidelines for Crash Avoidance Warning Devices*, Project No. DTNH22-91-C-07004.

Norman, D. A. (1981). Categorization of action slips, *Psychological review*, 88, pp. 1-15.

Norman, D. A., and Draper, S. W. (1986). *User-Centered System Engineering*. Lawrence Erlbaum Associates, Hillsdale, NJ.

Orasanu, J. M. (1993), Decision making in the cockpit. In E. L. Wiener, B. G. Kanki, and R. L. Helmreich (Eds.), *Cockpit Resource Management*, Academic Press, San Diego, CA.

Orasanu, J., and Salas, E. (1993), Team decision making in complex environments. In G. Klein, J. Orasanu, R. Calderwood, and C. Zsambok (Eds.), *Decision Making in Action: Models and methods*, Norwood, NJ, Ablex.

Orlady, H. W., and Orlady, L. M. (1999). *Human Factors in Multi-Crew Flight Operations*. Avebury, Aldershot, UK.

Parry, G. W. (1994). Critique of Current Practice in the Treatment of Human Interactions in Probabilistic Safety Assessment. In T. Aldemir, N. Siu, A. Mosleh, P. C. Cacciabue, and B. G. Göktepe (Eds.), *Reliability and Safety Assessment of Dynamic Process Systems*, Springer-Verlag, Hidelberg, Germany. pp. 156-165.

Paz Barroso, M., and Wilson, J. R. (2000). Human error and disturbance occurrence in manufacturing systems (HEDOMS): A framework and toolkit for practical analysis. *International Journal of Cognition Technology and Work, IJ-CTW*, 2 (2). Springer-Verlag, London. pp. 51-61.

Payne, S. J., and Green, T. R. G. (1986). Task-action grammars: A model of the mental representation of task languages. *Human Computer Interaction*, 2, 93-133.

Perry M. J., Fruchter, R., and Rosenberg, D. (1999). Co-ordinating distributed knowledge: Study into the use of an organisational memory. *International Journal of Cognition Technology and Work, IJ-CTW*, 1 (3). Springer-Verlag, London. pp. 142-152.

Pew, R. W., and Baron, S. (1983). Perspectives on Performance Modelling. *Automatica*, 19, No. 6, 663-676.

Pew, R. W., Baron, S., Feehrer, C. E., and Miler, D. C. (1977). Critical Review and Analysis of Performance Models Applicable to Man–Machine Systems Evaluation. Tech. Rep. No. 3446, Cambridge, MA, Bolt, Beranek and Newman.

Piccini, M. (1998). Affidabilità umana in sistemi altamente automatizzati: il caso studio di un impianto termoelettrico in cogenerazione e teleriscaldamento. (in Italian) Master Graduation Thesis in Nuclear Engineering. Politecnico di Torino.

Piccini, M. (2002). I Fattori Umani nel progetto di sistemi di controllo ed interfacce uomo-macchina in sistemi complessi altamente automatizzati. (in Italian) Doctorate Thesis in Nuclear Engineering. Politecnico di Torino.

Polet, P., Vanderhaegen, F., and Wieringa, P. A. (2002). Theory and safety-related violations of systems barriers. *International Journal of Cognition, Technology and Work, IJ-CTW*, 4 (3). Springer-Verlag, London. pp. 171-179.

Polychronopoulos, A., Kempf, D., Martinetto, M., Amditis, A., Cacciabue, P. C., and Andreone, L. (2003). Warning Strategies Adaptation in a collision avoidance/vision enhancement system. *Proceedings of International Conference on Human Computer Interaction (HCI)*, 22-27 June, Crete, Greece.

Prince, C., Prince, A., Salas, E., and Brannick, M. (2001). Improving LOS crew resource management debriefs: what do we need to know? In D. Harris (Ed.), *Engineering Psychology and Cognitive Ergonomics (Vol. 5): Aerospace and Transportation Systems*. Avebury Technical, Aldershot, UK. pp. 261-266.

Ralli, M. (1993). *Fattore Umano* (in Italian). Libreria All'Orologio. Roma, Italy.

Rankin, W., and Krichbaum, L. (1998). Human Factors in Aircraft Maintenance. "Integration of Recent HRA Developments with Applications to Maintenance in Aircraft and Nuclear Settings." June 8-10, 1998, Seattle, Washington, USA.

Rasmussen, J. (1983). Skills, rules and knowledge : signals, signs and symbols; and other distinctions in human performance model. *IEEE Transactions on Systems, Man, and Cybernetics, IEEE-SMC*, 13, 3, pp. 257–267.

Rasmussen, J. (1986). *Information Processes and Human–Machine Interaction. An Approach to Cognitive Engineering.* North Holland. Oxford.

Rasmussen, J., Leplat, J., and Duncan, K. (Eds.) (1987). *New Technology and Human Error*, J. Wiley, London.

Rasmussen J., Pedersen, O. M., Carnino, C., Griffon, M., Mancini, G., and Cagnolet, P. (1981). Classification System for Reporting Events Involving Human Malfunction. (Risø-M-2240, EUR-7444EN). Risø National Laboratory, Roskilde, Denmark.

Rasmussen, J., Pejtersen, A. M., and Schmidt, K. (1990). Taxonomy for cognitive work analysis. In J. Rasmussen, B. Brehmer, M. de Montmollin, and J. Leplat (Eds.) Proceedings of the 1st MOHAWC Workshop, Liege, May 15–16. Vol. 1 ESPRIT Basic Research Project 3105, European Commission, Brussels, Belgium. pp. 1–153.

Rasmussen, J., Pejtersen, A. M., and Goodstein, L. P. (1994). *Cognitive System Engineering.* J. Wiley, New York.

Rasmussen, J., and Vicente, K. (1989). Coping with human errors through system design: implications for ecological interface design. *Int. Journal of Human Machine Studies, IJ-HMS*, 31, pp. 517–534.

Rasmussen, J., and Vicente, K. (1990). Ecological Interfaces: A technological imperative of high tech systems? *Int. Journal of Human Computer Interaction, IJ-HCI*, 2, pp. 93–111.

Reason, J. (1986). Recurrent errors in process environments: some implications for the design of intelligent decision support systems. In E. Hollnagel, G. Mancini, and D. D. Woods (Eds.), *Intelligent Decision Support in Process Environment.* NATO ASI Series, Springer-Verlag, Berlin, FRG, 1986. pp. 255–270.

Reason, J. (1987). Generic error modelling system (GEMS): A cognitive framework for locating common human error forms. In J. Rasmussen, K. Duncan, and J. Leplat (Eds.), *New Technology and Human Error.* J. Wiley, London, UK. pp. 63–83.

Reason, J. (1990). *Human Error.* Cambridge University Press, Cambridge, UK.

Reason, J. (1994). Comprehensive error management in aircraft engineering. *Proceedings of Aerotech 94.* London, Institute of Mechanical Engineers.

Reason, J. (1997). *Managing the Risks of Organisational Accidents.* Ashgate, Aldershot, UK.

Reason, J., and Mycielska, K. (1982). *Absent-Minded? The Psychology of Mental Lapses and Everyday Errors.* Prentice-Hall, Englewood Cliffs, NJ.

Redding, R. E., and Seamster, T. L. (1994). Cognitive Task Analysis in air traffic controller and aviation crew training. In N. Johnston, N. McDonald, and R. Fuller (Eds.), *Aviation Psychology in Practice*, Avebury Technical, Aldershot, UK. pp. 190–222.

Rouse, W. B. (1980). *Systems Engineering Models of Human–Machine Interaction.* North Holland, Oxford.

Rouse, W. B. (1991). *Design for Success: A Human-Centered Approach to Designing Successful Products and Systems.* J. Wiley & Sons, New York.

Rouse, W. B. (1990). Human resource issues in system design. In N. Moray, W. R. Ferrell, and W. B. Rouse (Eds.), *Robotics, Control and Society.* Taylor and Francis, London, UK. pp. 177–186.

Rouse, W. B., and Rouse, S. H. (1983). Analysis and classification of human error. *IEEE Transaction on System, Man, and Cybernetics*, 13 (4), pp. 539–549.

Rouse, W. B., and Sage, A. P. (1999). Information technology and knowledge management. In A. P. Sage, and W. B. Rouse (Eds.), *Handbook of Systems Engineering and Management.* J. Wiley & Sons, New York, USA. pp. 1175–1208.

Sage, A. P., and Rouse, W. B. (Eds.) (1999). *Handbook of Systems Engineering and Management.* J. Wiley & Sons, New York.

Salvendi, G. (Ed.) (1997). *Handbook of Human Factors and Ergonomics.* J. Wiley & Sons, New York.

Sanders, M. S., and McCormick, E. J. (1976). *Human Factors in Engineering and Design.* McGrow-Hill Inc. Singapore.

Sarter, N. B., and Woods, D. D. (1994). Pilot Interaction with Cockpit Automation II: An Experimental Study of Pilots' Model and Awareness of the Flight Management System (FMS). *The International Journal of Aviation Psychology*, 4 (1), 1–28.

SAVE (2002). System for Effective Assessment of Driver State and Vehicle Control in Emergency Situations, TR1047 D3.1 (*ftp://ftp.cordis.lu/pub/telematics/docs/tap_transport/save_d3.1.pdf*).

Scapin, D. L., and Pierret-Golbreich, C. (1990). Toward a method for task description: MAD. In L. Berlinguet, and D. Berthelette (Eds.) Work with display units 89, Elsevier, Amsterdam, The Netherlands. pp. 371–380.

Schank, R. C., and Abelson, R. P. (1977). *Scripts, Plans, Goals, and Understanding: An Inquiry into Human Knowledge Structures*. Erlbaum Associates, Hillsdale, NJ.

Schraagen, J. M., Chipman, S. F., and Shalin, V. J. (Eds.) (2000). *Cognitive Task Analysis*. Erlbaum Associates, Mahwah, NJ.

Seamster, T. L., Redding, R. E., and Kaempf, G. L. (1997). *Applied Cognitive Task Analysis in Aviation*. Avebury, Aldershot, UK.

Seaver, D. A., and Stillwell, W. G. (1982). *Procedures for using Expert Judgement to Estimate Human Error Probabilities in Nuclear Power Plant Operations*. NUREG/CR-2743, USNRC.

Sebillotte, S. (1995). Methodology guide to task analysis with the goal of extracting relevant characteristics for human–computer interfaces. *International Journal of Human–Computer Interaction, IJHCI*, 7 (4), 341–363.

Sebillotte, S., and Scapin, D. L. (1994). From user's task knowledge to high-level interface specification. *International Journal of Human–Computer Interaction, IJHCI*, 6 (1), 1–15.

Senders, J. W., and Moray, N. P. (1991). *Human Error: Cause, Prediction, and Reduction*. Lawrence Erlbaum Associates, Hillsdale, New Jersey.

Sheridan, T. B. (1985). Forty-five years of man–machine systems: history and trends. Keynote Address. Proceedings of 2nd IFAC/IFIP/IFORS/IEA Symposium on Analysis, Design and Evaluation of Man–Machine, Varese, Italy, 10–12 September. pp. 5–13.

Sheridan, T. B. (1992). *Telerobotics, Automation and Human Supervisory Control*. The MIT Press, Cambridge, MA.

Sheridan, T. B. (1999). Human supervisory control. In: *Handbook of Systems Engineering and Management* (A. P. Sage and W. B. Rouse Eds.), pp. 591–628. J. Wiley & Sons, New York.

Sheridan, T. B., and Ferrell, W. R. (1974). *Man–Machine Systems: Information, Control and Decision Models of Human Performance*. MIT Press, Cambridge, MA.

Skinner, B. F. (1957). *Verbal Behaviour*. Appleton-Century-Crofts. New York.

Smallwood, T. W., and Fraser, M. (1995). *The Airline Training Pilot*. Ashgate, Aldershot, UK.

Stassen, H. G., Johannsen, G., and Moray, N. (1990). Internal representation, internal model, human performance model, and mental workload. *Automatica*, 26 (4), 811–820.

Stokes, A. F., and Wickens, C. D. (1988). Aviation Displays. In E. L. Wiener, and D. C. Nagel (Eds.), *Human Factors in Aviation*, Academic Press, San Diego, CA. pp. 387–431.

Swain, A. D., and Guttmann, H. E. (1983). Handbook on Human Reliability Analysis with Emphasis on Nuclear Power Plant Application. NUREG/CR-1278. SAND 80-0200 RX, AN. Final Report.

Taggart, W. R. (1987). CRM a different approach to human factors training. *ICAO Bulletin*, 42 (5).

Taggart, W. R. (1994). Crew Resource Management: Achieving enhanced flight operations. In A. N. Johnston, N. McDonald, and R. Fuller (Eds.), *Aviation Psychology in Practice*. Avebury, pp. 309–339.

Telfer, R. A., and Moore, P. J. (1997). *Aviation Training: Learners, Instructions and Organization*. Avebury, Aldershot, UK.

Tennery, Y. J., Rogers, W. H., and Pew, R. W. (1995). Pilot Opinions on High Level Flight Deck Automation Issues: Toward the Development of a Design Philosophy. NASA Contractor Report 4669, NASA-Langley Research Center, Hampton, Virginia.

Turner, R. (1974). *Ethnomethodology*. Harmondsworth, Penguin.

Vicente, K. (1999). *Cognitive Work Analysis. Toward Safe, Productive, and Healthy Computer-Based Work*. Laurence Erlbaum Associuates, Mahwah, NJ.

Vicente, K., and Rasmussen, J. (1990). The ecology of human–machine systems II: Mediating "direct perception" in complex work domains. *Ecological Psychology*, 2 (3), pp. 207–249.

Vicente, K., and Rasmussen, J. (1992). Ecological interface design: Theoretical foundations. *IEEE Transactions on Systems, Man and Cybernetics, IEEE-SMC*, 22, pp. 589–606.

Van Maanen, J. (1979). The fact of fiction in organisational ethnography. *Administrative Science Quarterly*, 24, 539–550.

Whalley, S. P. (1987). *PHECA computer code*. Lihou Loss Associates Ltd. AND Whalley, S. P. (1988). *Minimizing the Cause of Human Error*. In Proceedings of Advances in Reliability Technology Symposium, Libberton G. P. (Ed.). London: Elsevier.

Wickens, C. D. (1984). *Engineering Psychology and Human Performance*. Charles Merrill, Columbus, OH.

Wickens, C. D., and Flach, J. M. (1988). Information Processing. In E. L. Wiener, and D. C. Nagel (Eds.), *Human Factors in Aviation*. Academic Press, San Diego, CA, pp. 111–155.

Wiener, E. L. (1981). Complacency: is the term useful for air safety? Proceedings of the 26th Corporate Aviation Safety Seminar, Denver, pp. 116–125.

Wiener, E. L., Kanki, B. G., and Helmreich, R. L. (Eds.) (1993). *Cockpit Resource Management*. Academic Press, San Diego, CA.

Wiener, E. L., Kanki, B., and Helmreich, R. L. (Eds.) (1993). *Cockpit Resource Management*. Academic Press, Inc. San Diego.

Wiener, N. (1948). *Cybernetics*. MIT Press, Cambridge, MA.

Wilde, G. J. S. (1994). *Target risk*. Toronto.

Williams, J. C. (1993). *A User Manual for the HEART Human Reliability Assessment Method*. Nuclear Electric Ltd., Barnett Way, Barnwood, Gloucester, GL4 7RS. C2547-1.001.

Wilson, J. R., Cordiner, L., Nichols, S., Norton, L., Bristol, N., Clarke, T., and Roberts, S. (2001). On the right track: systematic implementation of ergonomics in railway network control. *International Journal of Cognition Technology and Work, IJ-CTW*, 3 (4). Springer-Verlag, London. pp. 238–252.

Wilson, J. R., and Haines, H. M. (1997). Participatory ergonomics. In G. Salvendi (Ed.), *Handbook of Human Factors and Ergonomics*. J. Wiley & Sons, New York. pp. 490–513.

Wixon, D., and Ramey, J. (Eds.) (1996). *Field Methods Casebook for Software Design*. John Wyley & Son, New York.

Wood, L. E. (1996). The ethnographic interview in user-centered work/task analysis. In D. Wixon, and J. Ramey (Eds.), *Field Methods Casebook for Software Design*. J. Wiley, New York, 1996, pp. 35–56.

Wreathall, J. W. (1982). Operator Action Tree, An Approach to Quantifying Operator Error Probability During Accident Sequences. NUS Report 4159, NUS Corporation, Gaithersberg, Maryland.

Wreathall, J., Luckas, W. J., and Thompson, C. M. (1996). Use of multidisciplinary framework in the analysis of human errors. In P. C. Cacciabue, and I. A. Papazoglou (Eds.), Proceedings of ESREL'96 – PSAM III International Conference on Probabilistic Safety Assessment and Management. Crete, Greece, 24–28 June, 1996, Springer-Verlag London, UK. pp. 782–787.

Yoshida, H., Takeda, M., Hayashi, Y., and Hollnagel, E. (1997). Cognitive control behaviour of the operator in the emergency. Proceedings of the 13th Triennial Congress of the International Ergonomics Association, June 29–July 4. Tampere, Finland. Vol. 3 pp. 82–84.

Zaff, B. S. (1995). Designing with affordances in mind. In J. Flach, P. Hancock, J. Caird, and K. Vicente (Eds.), *Global Perspective on the Ecology of Human–Machine Systems*. Lawrence Erlbaum Associates, Hillsdale, NJ, pp. 238–272.

Index